Dog Behaviour, Evolution, and Cognition

Dog Behaviour, Evolution, and Cognition

Ádám Miklósi

Department of Ethology, Eötvös Loránd University, Budapest

OXFORD

UNIVERSITY PRESS

OXFORD
UNIVERSITY PRESS

Great Clarendon Street, Oxford OX2 6DP

Oxford University Press is a department of the University of Oxford.
It furthers the University's objective of excellence in research, scholarship,
and education by publishing worldwide in

Oxford New York

Auckland Cape Town Dar es Salaam Hong Kong Karachi
Kuala Lumpur Madrid Melbourne Mexico City Nairobi
New Delhi Shanghai Taipei Toronto

With offices in

Argentina Austria Brazil Chile Czech Republic France Greece
Guatemala Hungary Italy Japan Poland Portugal Singapore
South Korea Switzerland Thailand Turkey Ukraine Vietnam

Oxford is a registered trade mark of Oxford University Press
in the UK and in certain other countries

Published in the United States
by Oxford University Press Inc., New York

First published 2007
First published new in paperback 2009
Reprinted 2009 (twice), 2010, 2011 (twice)

British Library Cataloguing in Publication Data

Data available

Library of Congress Cataloging in Publication Data

Data available

Typeset by Newgen Imaging Systems (P) Ltd., Chennai, India
Printed and bound by
CPI Group (UK) Ltd,
Croydon, CR0 4YY

ISBN 978–0–19–929585–2 (Hbk)
ISBN 978–0–19–954566–7 (Pbk)

To my mother and father who have always believed that
I can do it, and to Zsuzsanka, Betty, and Gergö
who made doing it possible.

Prologue: *comparare necesse est**

In 1994, after some discussion, we decided to clear our laboratories of the aquaria that had been in use for many years in a research programme on the ethology of learning in the paradise fish (Csányi 1993). To be honest, the exact reason for this move at that time was not exactly clear to me, but I had no great regret for the research topic because we were the only laboratory studying learning processes associated with antipredator behaviour in this little East Asian labyrinth fish.

However, the idea of approaching dog–human social interactions from an ethological perspective did not seem to be much of an improvement in that respect, because literature on the subject was simply non-existent. Thus József Topál, my colleague and friend, and I were a bit uncertain about the future when Professor Vilmos Csányi, the head of the department at that time, began to argue enthusiastically that the study of dog behaviour in the human social context could be very important in understanding cognitive evolution, with many parallels to human behaviour (Csányi 2000). We were told hundreds of casual observations of dog–human interaction (many people would call these anecdotes), and it seemed that the task would be to provide an observational and experimental background to these ideas. Csányi pointed out that in order to be successful in the human social world dogs had to achieve some sort of social understanding, and very likely this came about in course of their evolution. Accordingly, the social skills of dogs can be set in parallel with corresponding social skills in early humans. I do not know what exactly József thought about all this, but at least he owned a dog.

After some thinking about what to do and how to do it, we saw some light at the end of the tunnel when Karin Grossman, a famous German child psychologist, introduced us to Ainsworth's Strange Situation Test, which is used to describe the pattern of attachment in children. Watching the videos on how the children behaved when a stranger entered the observation room or when their mother left, made us each realize independently that dogs would behave in just the same way!

It took us another two years to publish our first study on the behavioural analysis of dog–human relationships based on the Strange Situation Test in the *Journal of Comparative Psychology*, but from that time on we had a quite clear idea of our research programme, which was focused on looking for behavioural parallels between dogs and humans.

Actually, the idea of behavioural similarity between humans and dogs was not novel at all. Scott and Fuller (1965) devoted a considerable part of their work to human and dog parallels. For example, in the first paragraph of their last chapter they write: 'These facts suggest a hypothesis: the genetic consequences of civilized living should be intensified in the dog, and therefore the dog should give us some idea of the genetic future of mankind....' In retrospect it is interesting that although the achievements of this research group have always been recognized at the highest level, these conclusions were neither debated nor praised (or, more importantly, followed up in research). However, one point is important: although Scott and Fuller realized the special social status of dogs in human groups in their behavioural work, they emphasized parallels between the dog puppy and the human child. In contrast, our aim was to provide an evolutionary framework that hypothesizes behavioural convergence between the two species.

*Comparison is essential; analogous to the Latin motto *navigare necesse est*, which can be translated as 'trade is essential'.

Accordingly, we argued that evolutionary selective pressures for dogs might have moulded their behaviour in such a way that it became compatible with human behaviour.

Since then, 12 years have passed and during that time many research groups have started to study dog behaviour. Although we have continued to work according to our research programme, we have realized that the field begs for integration. In recent years many books on dogs have been published by researchers working in various fields, as well as by experts with different backgrounds. The goal of most of these books was to explain dog behaviour from an author's particular point of view, often based on an assorted array of arguments where scientific facts were often treated at the same level as anecdotes, stories, or second-hand information. In this book I want to break this mould by presenting only what we know about dog behaviour and suggesting possible directions for future research. The main aim is to provide a common platform for scientific thinking for researchers coming from the diverse fields of archeozoology, anthrozoology, genetics, ethology, psychology, and zoology.

The increased amount of contemporary research has made it impossible to refer extensively to older work, much of which is, however, available in other textbooks. For similar reasons I have omitted to mention research that is not published in refereed journals, or the many folk beliefs about dogs. In addition, there is no attempt to 'bridge' gaps in our knowledge by 'facts that everyone knows', in the absence of published evidence. Some readers may see this as a serious fault which makes the presentation of the topic uneven, but I have preferred to use these opportunities to indicate directions in which research should be pursued.

Perhaps this is not the first book on dog ethology, but it has been written with the intent to place this species (once again) in the front line of ethology, which is the science of studying animal (and human) behaviour in nature. From the start we believed that the whole project makes sense only if dogs are studied in their natural environment where they share their life with humans in small or large groups. But we soon felt that such an endeavour can only be insightful if it is put in a comparative perspective. This gave us the idea of socializing some wolves (and also some dog pups) in order to obtain comparative data. This research not only opened our eyes to the very different world of 'wild' canids but also taught us to be very cautious about coming to hasty conclusions about behavioural differences between dog and wolf. Naturally, observations on these two species suggested many differences; however, the real trick was to find the ways in which these differences could come to light under the conditions of a scientific experiment. Later this comparative work was broadened to include cats and horses, but first of all human children. We believe strongly that dog behaviour can be understood only if it is studied in a comparative framework that takes into account evolutionary and ecological factors and rests on a solid methodological basis.

Today, research inspired by ethology or behavioural ecology is characterized by a functional perspective. Researchers focus their interest on those aspects of behaviour that contribute to the survival of the species. In the present case the focus is on a species, dogs, and on how collaboration among different scientific disciplines can lead to a more complete understanding of their evolution and present state. For many years scientists have looked with suspicion at dogs and denied them the status of 'real' animals. Thus the main goal of this book is to provide evidence that dogs can be studied just as well as other animals (including humans) and even that they have the potential to become one of the most well-researched species in the near future. In this regard dog ethology could play a role in providing raw material for disciplines that are studying genetic and physiological aspects of behaviour, and also for those who are interested in applied aspects such as dog training, problem behaviour, dog–human interaction, or the use of dogs in therapeutic intervention.

I am very lucky to be a member of a wonderful research team with colleagues who have always been supportive. I am grateful to Vilmos Csányi who gave us all the opportunity to embark on this research programme. Over the years József Topál became the best colleague and friend that one could wish for in collaborative work, without whom I would never have had the chance to get this project started. I owe a lot to Márta Gácsi who has gently helped me in coming to understand the 'world of dogs' over the years. I will never forget our first (and only) visit to Crufts. Antal Dóka, who has been an indispensable colleague without whom the research group could not have functioned so smoothly. Over the years we were lucky to have Enikő Kubinyi, Zsófia Virányi, and Péter Pongrácz join our group, all of whom have made important contributions in particular fields of dog social behaviour and cognition.

Over the years our research was supported by the Eötvös Lóránd University, the Hungarian Scientific Research Fund (OTKA), the Hungarian Academy of Sciences, the European Union, the Ministry of Health, and the Dogs for Humans Foundation.

Our research group owes much to those enthusiastic dog owners and their dogs, who contributed by offering their time for our research. In addition we would like to express our thanks to Zoltán Horkai and to the keen students (Bea Belényi, Enikő Kubinyi, Anita Kurys, Dorottya Ujfalussy, Dorottya Újvári, Zsófia Virányi) who participated in the Family Wolf Project and persisted in doing this job under difficult conditions.

I am very grateful to Antal Dóka for drawing and redrawing many figures and graphics for the book. Being untalented at producing pictures, I am thankful for the photos that were shot by Márta Gácsi (if not indicated otherwise). She and Enikő Kubinyi also made great efforts to help reading the proof.

I would also like to thank to Richard Andrew, Colin Allen, László Bartosiewitcz, Vilmos Csányi, Dorit Feddersen-Petersen, Simon Gabois, Márta Gácsi, Borbála Győri, Enikő Kubinyi, Daniel Mills, Eugenia Natali, Justine Philips, Peter Slater, József Topál, Judit Vas and Deborah Wells for reading and commenting on single chapters or the whole manuscript. Although these colleagues did everything in their power to point out my weaknesses, I shall take the responsibility for any mistakes left in the book.

I am also grateful to Oxford University Press and in particular to Ian Sherman for taking on this project without much hesitation, and also helping to polish my raw Hungarian version of English.

Finally, a note to the critical reader. Please do not hesitate to point out the weaknesses of this book. Not only to make the next version even better, but also to urge others to provide facts in the form of well-designed experiments that will separate scientific knowledge from beliefs and stories. If researchers and many others interested in dogs are provoked to do better research then the book and I have achieved our goal.

Budapest, 2 February 2007
Ádám Miklósi

Contents

Dogs in historical perspective, and conceptual issues of the study of their behaviour

1.1 Introduction

This book is about the biological study of dog behaviour, based on the programme summarized so clearly by Tinbergen in 1963. He, Lorenz and others have always pointed out that the main contribution of ethology is the biological analysis of animal behaviour based on observations in nature. Unfortunately, however, only a handful of mainstream ethologists have applied these concepts to dog behaviour. In contrast to sticklebacks, honeybees or chimpanzees, not to mention a few tens of other species, dogs received relatively little attention from ethologists or comparative psychologists. It seems that these creatures ('man's best friends') have somehow become outcasts from mainstream science, for reasons that are not obviously clear but which may be guessed.

Dogs are often referred to as 'artificial animals', probably because their history of being 'domesticated'. Here the image is that of a 'savage' stealing a wolf cub from its mother (e.g. Lorenz 1954), which then 'became' dog after many years and generations in the hands of humans. Today most researchers disagree with this simplistic view of dog domestication (e.g. Herre and Röhrs 1990), and it is much less clear on what grounds the evolution of such 'real' and 'artificial' animals can be differentiated. The kind of goal-directed selective breeding implied by the category of 'artificial animal' probably started much later than has been assumed. Logically, an 'artificial animal' cannot have a natural environment, so in order to allow the dog into the club of 'real' animals we would have to find a natural environment for it (Chapter 3, p. 47).

The study of dogs did not fit well with the increasing influence of behavioural ecology, which was partially initiated by the call for a more functional approach to behaviour by Tinbergen (1963). Obviously, dogs are not the best candidates for studying survival in nature, mainly because most present-day dogs live with humans and have access to vets, and we do our best to save our companions from the challenges of nature. In this sense dogs can be regarded as being special (but not necessarily 'artificial').

More surprisingly, interest in the study of dogs did not emerge with the cognitive revolution in ethology. Griffin (1984), one of the initiators of this movement, seems to have carefully avoided reference to dogs in most of his works on this subject. We are introduced to miraculous behaviour of ants, starlings or dolphins, which we look at with admiration, but similar behaviour in dogs is often regarded as suspicious. To some extent this attitude is understandable, as early workers were often tricked by so called 'dog artists' who showed remarkable skills for 'talking' or 'counting' (e.g. Pfungst 1912, Grzimek 1940–41) (Figure 1.1). After it was found out that such apparently clever behaviour could be explained by the dog responding to minute bodily cues produced either consciously or unconsciously by the owner or trainer (the Clever Hans effect, see Pfungst 1907 and Chapter 2.5, p. 37), dogs were banished from laboratories for being unreliable subjects.

However, it seems that dogs are showing signs of making a real comeback. Ethologists, comparative psychologists, and many others are now working hard to find a place for dogs in the biological study

Figure 1.1 (a) Stuppke, a counting dog artist, was observed by Bernhard Grzimek, a German zoologist. Stuppke barked the number shown to him. The remarkable talent of the dog was based on recognition of a 'start' and a 'stop' signal given by his master, Mr Pilz. (b) No wonder that Stuppke could also read numbers with his eyes covered (photos taken from Grzimek 1940–41). (c) Oskar Pfungst (1912) reported on Don, the talking dog (photo from Candland 1993, Oxford University Press).

The Dog at the Convent Door.

Figure 1.2 The 'cultural transmission' of dog anecdotes. Menault (1869) reports the story of a dog that, after observing beggars ringing the bell at the door of the convent and receiving some soup, went to the door and pulled the string. The ability to learn by observation of humans has only recently been demonstrated experimentally (Chapter 8, e.g. Kubinyi *et al.* 2001; Box 8.6).

of behaviour. This is difficult, but the steep increase in research papers over the last 10 years already shows the fruit of this work. Thus there is every chance that dog ethology will revive.

1.2 From behaviourism to cognitive ethology

Early researchers, including Darwin (1872), regarded the dog as a special animal that is comparable to humans. Many people shared this anthropomorphic attitude and it is not surprising that dogs ended up at the top of the ladder representing intelligence and emotional behaviour in animals (Romanes 1882a, b). It did not take long for the situation to change, and dogs could not avoid their fate when under the increasing influence of behaviourism they were then treated as a sort of stimulus–response automaton. The interest in wolves and social behaviour in general has helped dogs regain a foothold in the behavioural sciences, and this has led to an ethologically oriented understanding of dog behaviour. The history of the study of dogs reflects the changes in our views of animals, and although much time has passed and a lot of knowledge has been gained, the basic questions of present-day research are more or less the same as they were 100 years ago.

1.2.1 Dog heroes visit the laboratory

Dogs have long been the favourite heroes of animal stories. Sharing our daily life with these animals has offered endless opportunities to observe or witness the varieties of dog–human interactions. One famous collector of such stories was George Romanes (1982a). His descriptions of dogs provided evidence for often very intelligent behaviour which prompted him to argue that such performances should be explained by human-like thinking mechanisms (Candland 1993).

Interestingly, Lloyd Morgan (1903), who was a strong critic of the methods used by Romanes, did not refrain from telling such stories when he wanted to illustrate a particular behavioural phenomenon. At one point he describes how his fox terrier Tony grappled with the problem of how to carry a stick with unequal weights at its ends. After describing the dog's behaviour Morgan concludes that he has seen little evidence for assuming that the dog 'understood the problem'. Instead, during repeated attempts to carry the stick the dog learned the solution by trial and error. Thus 'intelligent' behaviour on the dog's part could often be based on relative simple learning processes. For Morgan, stories provided opportunities for formulating hypotheses and did not serve as explanations for mental abilities. Nevertheless he did not deny that dogs could have a mental representation for an object, such as a bone.

Thorndike (1911) was among the first to develop a method to objectively measure learning in animals. He put hungry cats and dogs into a box which could be opened from inside by manipulating a simple latch. Observing the animals repeatedly in this situation, he found that it took them less and less time to get out. In agreement with Morgan, he also thought that the final 'intelligent' behavioural solution was the result of a step-by-step process of 'trial and error' learning. Thus the systematic observations of both Morgan and Thorndike seemed to contradict the conclusions of Romanes, who argued that, for example, cats and dogs have some idea about the properties of locks. Interestingly, Thorndike noted a difference between dogs and cats, because, despite being starved for some time,

dogs were much inferior in escaping. From his descriptions it seems that, in comparison with the cats, dogs were less inclined to get out, and they were also very cautious in interacting with the latch, which probably indicates a different social relation between people and these dogs. Thus it is less surprising that in the textbooks the fame of representing Thorndike's concept of trial-and-error learning was left to the cats. From further experiments Thorndike did not find support for the long-held view that dogs learn by imitation (see Chapter 8.6, p. 191) because animals did not escape any earlier from the box if they were shown how to open the lock.

In 1904 Pavlov received the Nobel Prize for Medicine for the physiological study of the digestive system, for which dogs had served as subjects. By this time he had noted that not just the presence of food in the mouth but also other external stimuli (the sound of the food put in the bowl or the approaching experimenter providing the food) have the potential to elicit salivation. For many years after that dogs remained one of the most preferred subjects in the research that led to the development of the conditioned reflex principle (Pavlov 1927), which was extended by Pavlov's pupils. Pavlov was not only a good experimenter, however, but also a good observer. Thus he noted early on that there are marked individual differences among the dogs, which could be also observed in their response to the training (Teplov 1964). Dogs were categorized as belonging to one of the classic temperament types described by Hippocrates (sanguine, choleric, phlegmatic, melancholic) (see also Box 10.1). Even at that time Pavlov pointed out that observed behavioural traits are the outcome of complex processes having both genetic and environmental components, and he was probably the first to suggest separating these two effects by raising dogs in different environments before subjecting them to training. The generality of Pavlov's work on the conditioning reflexes provided the basis for comparative work on dogs and humans. Based on this experimental approach, dogs can be regarded as the first animal models of human personality (Chapter 10, p. 221). This makes it less surprising that in contrast to some other laboratories Pavlov's researchers respected the individuality of the animal. Most dogs were given names, and the observation of their spontaneous behaviour in the laboratory or outside was used as additional information for understanding their reaction in the training situations. Importantly, in contrast to recent research on personalities, Pavlov and his colleagues based their investigations on single dogs and then generalized the results to other individuals belonging to the same personality type.

1.2.2 Dogs in the comparative psychology laboratory

One cannot avoid being emotionally touched on reading many of the papers published on dog behaviour in laboratories working on a Pavlovian model of learning. Professional scientists, often having a good 'personal' relationship with these dogs, often do not seem to realize what they are doing. There is no way that anyone today could or would do many experiments like these. The purpose of reviewing these experiments is to show how the lack of ethological thought can misdirect scientific efforts.

A subjective survey of the literature shows that by the 1920s rats and pigeons had become the main subjects of research. Thus we might wonder why some research programmes seemed to prefer dogs. Having adopted a clearly anthropocentric programme in looking for appropriate animal models of human behaviour, we could reason that for some features of human behaviour dogs seemed to offer a more appropriate model. By doing this, these researchers have implicitly acknowledged that dogs are more similar to humans than are other species. Indeed, in discussing dog behaviour they often relied on comparison with humans (children), assuming similar underlying mental mechanisms (e.g. Solomon *et al.* 1968, see Box 1.4). Interestingly, this argument was not extended to subjective states. Thus the dogs' suffering in many of these experimental procedures was never really a concern.

Another important aspect of these experiments was that the experimental context had very little, if any, relevance to the natural behaviour of the dog, and there was very little correspondence between the experimentally manipulated variables and the variables that may relate to a natural situation. The presence of humans was also confusing for the dog, because the good/positive social relationship before and after the experiment was often contradicted by the role of humans in the training trials.

One aim of this research was to provide a behavioural model for neurosis, or traumatic experience (Lichtenstein 1950, Solomon and Wynne 1953). For example, dogs were shut into an experimental chamber and exposed to electric shock ('helplessness': Seligman *et al.* 1965). After this experience they were tested in a task in which they were given the possibility of avoiding similar shocks by escaping from the dangerous place. Many experiments found that after such an experience the dogs did not learn. They showed low responsiveness and seemed 'to give up and passively accept' the shock (Seligman *et al.* 1965). We might question the ethological basis of this behaviour. Is there a natural situation when dogs experience such pain? The most likely, if not only, situation is when a dominant conspecific inflicts a physically dangerous attack finished off by a persistent bite. In such a case the attacked animal's only chance is to show all possible signs of submission with as little movement as possible ('freezing'). Some of the dogs might have associated painful experience with their interactions with humans, which certainly contributed to the dog's 'neurosis' apart from the effect of their lack of control over the situation (Seligman *et al.* 1965).

A better aspect of this period is that many early studies provided a detailed description of the dogs' behaviour, and it became obvious that their reactions to the treatments were very variable. This suggests that despite being 'laboratory dogs' animals differed in their previous experience, including their relationship with the humans inside or even outside the laboratory. A further important lesson from these studies is that training methods using painful punishments can have unforeseeable (and mostly negative) consequences on the behaviour of dogs, either because of their genetic endowment or their earlier experience with humans (socialization).

These traditions of comparative psychology were left behind when more ethologically inspired questions dominated laboratory research (Figure 1.3.).

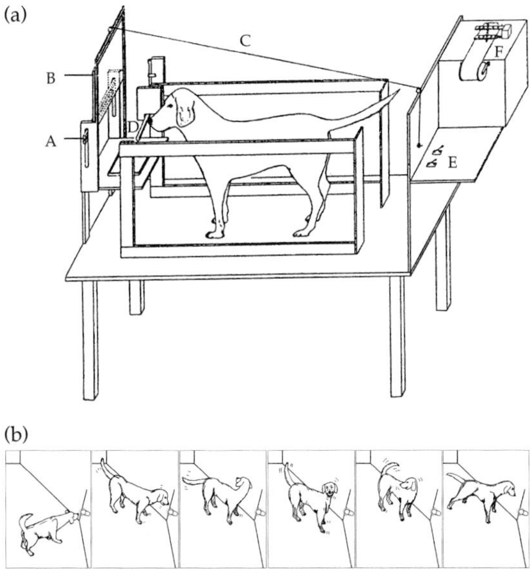

(a)

(b)

Figure 1.3 Dogs under study. (a) A dog in a Pavlovian stand as illustrated in Woodbury (1943). The dog is trained to recognize differences in acoustic sound patterns. (b) An illustration from Jenkins *et al.* (1978) showing 'Dog 7' which after being conditioned to the light stimulus (at the front) signalling food, displays a range of social behaviours towards the light stimulus and the food tray (behind the dog, not shown on the illustration).

In 1978 Jenkins and co-workers contrasted the Pavlovian stimulus substitution theory (Pavlov 1934) with the ethological analysis of the dog 'begging' for food (Lorenz 1969). Pavlov's theory assumed that the (conditioned) stimulus (e.g. light or bell) signalling the food will actually replace the original (unconditioned) stimulus (e.g. food); that is, when it sees the light come on the dog displays preparatory acts which reflect consummatory actions towards the conditioned stimulus (e.g. licking, snapping at the light). In contrast Lorenz argued that the conditioned stimulus acts as a releaser for appetitive behaviours. Thus the dog searches for the food or displays 'begging', as when pups solicit food from older conspecifics. Jenkins *et al.* (1978) trained dogs to approach a lamp which signalled the presence of a food reward. In the course of the training dogs showed very variable behaviour, but nevertheless many social behaviour patterns emerged, such as play signals, tail wagging, barking, nosing. Thus we could argue that dogs interpreted the experimental situation in a

social context with which they were familiar. For them the conditional stimulus (light) was not just signalling the arrival of food but it was also a social stimulus. In this more natural context, 'request' for food (from humans) is usually preceded by some sort of signalling (e.g. tail wagging, barking) and behaviour actions (e.g. nosing, pawing). These motor patterns are derived from the species-specific behavioural repertoire of the dog, which is later modified during the period of socialization. The social experience and habitual behaviour of the individual dogs markedly influences the behaviour during these observations. The important conclusion is that 'one must examine how dogs react to natural signals of food outside the laboratory setting' (Jenkins *et al.* 1978)—one of the first signs of a need for collaboration between comparative psychologists and ethologists. Such an approach opens up a new way of combining methods that rely on controlled laboratory settings with those that emphasize observations on natural behaviour, including knowledge of the individual's previous experience.

1.2.3 Naturalistic experiments

Especially during the first half of the last century, dogs were popular subjects for investigators who rejected arbitrary laboratory observations. This work, which culminated just before the Second World War, was mostly carried out in Germany and the Netherlands. These researchers continued the tradition of Morgan and others recognizing the importance of (more or less) controlled experiments, but they wanted to rely, to a greater extent, on the natural behaviour of dogs. Many of them were pupils or followers of Köhler (1917/1925), who emphasized the role of 'insight' in solving new problems, and Uexküll (1909), who stressed the importance of recognizing the features of the natural environmental (*Umwelt*) of the animal under study. Importantly, both Köhler and Uexküll had a marked influence on early ethological thought (Lorenz 1981), thus to some extent Buytendijk and Fischel (1936), Sarris (1937), Fischel (1941), Grzimek (1942) and others can be regarded as forerunners of present-day dog ethologists. Although most of their experiments were performed in the laboratory or in an enclosed yard they always stressed that dogs should be observed and tested in tasks that correspond to challenges of their natural environment. Most of these investigators also emphasized the need for comparative work with children that could also help in developing theories for explanations of dog behaviour, but there was a disagreement over the extent to which the experimenter should put himself in the dog's place (Chapter 1.5). For example, Fischel (1941) found that both dogs and children solve a simple problem with similar amounts of training, but children are much superior when they are presented with the reversed version of the problem. These results were interpreted as evidence that children are able to rely on 'insight', in contrast to dogs. Nevertheless, observations have also shown that even such cases of insightful behaviour (which have also been described for the dog, e.g. Sarris 1937) depend on previous experience with similar situations, and any success is preceded by earlier partial solutions in analogous problems.

Given the variability in the dogs used for these observations, including their experience, relationship with the investigator, and the procedures used it is not surprising that many investigations provided contradictory results. For example, Sarris (1937) found evidence for *means–end understanding* in one dog. After repeated experience of pulling ropes with meat attached to the other end or not, the dog learned not to pull if there was no physical connection between the meat and the rope (but see Osthaus *et al.* 2005; Chapter 7.6.1, p. 161). Apparently his dogs did not rely on the human pointing gesture, in contrast to what we know today about this ability (Miklósi and Soproni 2006; Chapter 8.4.1, p. 181).

Most of these investigators rejected the then-prevalent reductionist view that behaviour is based on a chain of Pavlovian reflexes. One counter-argument was based on the processes controlling behaviour during search. Buytendijk and Fischel (1936) stressed that such behaviour would be impossible without some sort of 'mental image' in the brain, which emerges step by step after repeated experiences of the object. In contrast, Fischel (1941) thought that the behaviour of dogs is driven by 'action schemas' which develop after repeated experience with a positive or negative outcome of the action. Fischel denied the existence of mental

images because he often saw dogs acting in a habitual manner, without taking into account that the situation had changed. For example, a dog would try to retrieve an object even if there were no more objects left, and Fischel explained this by assuming that human commands release actions schemas and do not activate mental images of the objects. The predatory nature of dogs could have facilitated the organization of behaviour around actions and not objects.

By this time others had shown that dogs are able to differentiate among objects on the basis of different commands. A German shepherd tested by Warden and Warner (1928) showed that he could perform the same action with a different outcome (retrieval of object A or B) depending on a verbal command (Chapter 8.4.2, p. 189). These results seem to contradict Fischel's theory that dog behaviour is purely action driven.

A strong proponent of the mental image concept was Beritashvili (1965), who worked in Georgia in parallel with Pavlov's school but became unsatisfied with the explanatory value of the Pavlovian model of behaviour. It was again the search task that led him to doubt the purely reflexive or action-driven behaviour of the dog. In his laboratory dogs had to search for a piece of hidden food. Beritashvili varied the time elapsed between hiding and the possibility for search, the nature of hidden targets, and the number of hiding locations. In one such experiment dogs observed that the assistant hid a piece of bread close by, but a piece of meat at a greater distance. When permitted to search, the dogs went invariably for the preferred meat. Beritashvili argued that this preferential choice can be only explained by assuming that the image of the meat 'took over' the control of behaviour. This and many similar observations prompted Beritashvili to argue that at the beginning of the learning process the behaviour is controlled by an image which develops as a result of attention to the situation. However, after repeated exposure to the same situation ('conditioning') the dog learns a conditioned behaviour over which the mental image has less control. By causing brain damage to certain animals, Beritashvili (1965) found further evidence for his theory. These dogs were still able to remember the places where they saw food being hidden but they did not show a preference for going for the meat first, which was taken as evidence that these experimental dogs had lost the ability to construct a mental image.

These natural observations gave also other clues to the understanding of dog behaviour, many of which have been forgotten until recently. For example, Sarris (1937) noted the importance of looking at individual differences, especially with regard to behavioural skills reflecting variability in 'intelligence' (see below). Buytendijk and Fischel (1936) noted that the attachment of the dog to its owners is fundamental in understanding its behaviour. Many investigators also emphasized the importance of these scientific investigations in improving methods of dog training.

1.2.4 Time for comparisons

Along with the development of ethology into an independent field of scientific inquiry there was an increased interest in gathering data about wolf behaviour. This began an ongoing tradition of studying the surviving wolf population in the USA (e.g. Murie 1944, Mech 1970) and to a lesser extent in Europe (e.g. Okarma 1995). In parallel, many observations were carried out on captive populations in which the main focus was on the comparative

aspects of social behaviour (e.g. Fox 1975, Schotté and Ginsburg 1987, Zimen 2000). Lorenz's idea of ethology providing important insights into evolutionary processes by comparative analysis of behaviour probably influenced this research strongly. In particular, Fox (1975) aimed to present a broad view of the social behaviour in Canidae (but see also e.g. Bekoff 1977, Fentress and Gadbois 2001), whereas others aimed mostly to compare only wolves and dogs (e.g. Schotté and Ginsburg 1987, Frank and Frank 1982). The comparative study of dogs and wolves also gained a foothold in laboratories, although the many methodological problems hindered these projects (Chapter 2.3, p. 30). Moreover increasing concern about the ecological aspect of behavioural research turned researchers' interest to species living in the 'wild', and the dog was regarded as an 'artificial' animal without any ecological validity (see Chapter 4) (Box 1.1).

In parallel to this, John Paul Scott and John Fuller (1965) utilized extreme variation in dog social behaviour for comparative studies investigating genetic effects on social behaviour. Many results of this project still have a strong influence on our understanding of dog behaviour in spite of the fact that the circumstances and the research questions were often relatively arbitrary (Chapter 9, p. 201).

Box 1.1 A framework for behavioural comparisons

Timberlake (1994) categorized comparative behavioural investigations along two independent dimensions, providing four different possibilities. This framework is useful for conceptualizing comparative work in dogs with reference to *Canis* or humans. Behavioural convergence facilitates interspecies comparisons with high ecological relevance, for example, in the case of social behaviour, but it is not based on genetic relatedness. Within-species comparisons have both high ecological relevance and genetic relatedness and could be important in finding out the nature of adaptation to the species' actual environment. Phylogenetic comparisons can look for divergent evolution in the case of homologous relationship when the ecological relevance is relatively low. Finally, comparisons

lacking ecological relevance and genetic relatedness are mainly of categorical interest.

		Concern with genetic relatedness	
		Low	High
		(Convergence)	(Microevolution)
Ecological relevance	High	Dog vs human (e.g. communicative behaviour)	among subspecies of wolf or wolf vs coyote and jackal
	Low	(Classification) Dog vs human (e.g. manipulating ability)	(Homology) Wolf vs dog (e.g. territorial behaviour)

1.2.5 The cognitive revolution hits dogs

The renewed interest in thought processes in animals initiated by both psychologists (e.g. Hulse *et al.* 1978, Roitblat *et al.* 1984) and ethologists (Griffin 1976, Ristau 1991) contributed to the "rediscovery" of the dog (Devenport and Devenport 1990). The Information Processing Project at the University of Michigan directed by Frank (1980) was the first to apply the concepts of this cognitive approach to behavioural research in Canidae, and later Bekoff (1996) followed this path. In their arguments for studying cognitive processes in animals, the behavioural observations on dogs play an important role. In a critical reinterpretation of the work of many early investigators, Bekoff and Jamieson (1991) argue that dogs kept in the laboratory are unable to show their natural capacities and therefore they should be observed in nature. They advise that 'good ethologists think themselves into the minds of the animals' (see p. 15) but at the same time they dismiss simulation theory in the case of human–animal relationship because it is not possible to simulate the mental state of the other by using a mental structure which evolved for a different purpose and gained its experience in a different environment. Although they call for an experimental approach and regard anecdotes only as pilot observations, they seem to be less worried about using a rich cognitive vocabulary and referring to complex mental states on the basis of behavioural observations.

Ethologically oriented research, which also relies heavily on cognitive concepts, is currently experiencing a golden age. The breakthrough probably took place in 1998, when two research groups independently embarked on the same project aimed at understanding dog–human communication (Miklósi *et al.* 1998, Hare *et al.* 1998, Box 1.2; see also Chapter 8). Since then the number of publications has risen sharply, and at present it seems that dogs could become one of the major subjects for understanding behavioural evolution including underlying mental mechanisms.

1.3 Tinbergen's legacy: four questions plus one

Ten years before receiving the Nobel Prize, Tinbergen (1963) summarized the main goals of the biological study of behaviour. Since then 'Tinbergen's four questions' have became the basic theses of ethology, and feature in the introductory pages of most textbooks. Thus it seems useful to set the ethological study of dogs in the framework provided by Tinbergen. Although Tinbergen raised four issues which need to be addressed by ethology, he also pointed out that this endeavour should be rooted in the description of natural behaviour. Thus we will also start with this, mostly forgotten, aspect.

1.3.1 Description of behaviour

An ethologist begins any investigation by observing the species in its natural environment. Although many scientists doubt that ethologists sitting in the branches of trees or lying in the grass looking through binoculars are actually 'doing science', detailed knowledge of behaviour is important for at least two reasons. First, the observable behaviour is the phenotype under investigation, and for any scientific study there is a need to make behaviour 'measurable' (Martin and Bateson 1986). Thus the first task is to decompose the behaviour into units with the goal of producing a species-specific behaviour catalogue for the species (an *ethogram*). Second, observation of animals in their natural environment prompts a 'good' ethologist to ask questions such as 'Why does this animal behave as it does?' (Tinbergen 1963). Thus, observing animals in nature is the best way of finding questions which demand scientific explanations.

Although dog ethograms are available (based on behavioural descriptions of the wolf, see Chapter 2.6, p. 38), these have rarely been employed in describing the spontaneous behaviour of dogs in their natural environment (but see Bradshaw and Nott 1995, Bekoff 1995a). Comparative investigations are also lacking, most notably in the case of breeds. Nevertheless there are some steps in this direction (e.g. Goodwin *et al.* 1997, Fentress and Gadbois 2001, Feddersen-Petersen 2001a,b). Such descriptive work is especially important for acknowledging the difference between spontaneous behaviour in the 'wild' and under laboratory conditions. Knowledge about dog behaviour helps enormously in planning experiments under more controlled conditions.

Box 1.2 Do dogs utilize the human pointing gesture?

Pointing is one of the most widely used human non-verbal gestures for indicating objects, and even superficial observation reveals that humans also use this form of gesturing when interacting with dogs. Pointing is not only observed during spontaneous interactions or training but is also used during work, e.g. indicating to a sheepdog the direction for approaching the herd. Assuming that the ability to work with humans was an important factor at some point in dog evolution, the utilization of human gestures (including pointing) could have been advantageous for dogs.

Based on the work of Anderson *et al.* (1995) we have used a standardized method (*two-way choice task*) testing for this ability in dogs. In this task the experimenter points with extended hand to one of two containers, one of which hides a small piece of food. Briefly, the procedure is as follows. After

the dog is allowed to take food from the containers (flowerpots) a few times, in order to get used to the situation, it faces the experimenter who is standing 2.5–3 m away from the dog and equidistantly from the pots which are placed 2.5 m apart. Now she (1) calls the dog's name, (2) waits until the dog looks at her, (3) moves her arm towards the baited pot and keeps it in this position for 1–2 s, (4) pulls her hand back to her chest, and (5) only then is the dog allowed to choose. This form of the pointing gesture was termed *momentary distal pointing*, because the pointing is shown only for a short duration, the dog cannot see the outstretched arm when making a choice, and the tip of the pointing finger is about 60–70 cm from the pot. (In other parts of the book we will come back to this method of testing dog–human communication to show other aspects of the dog's performance).

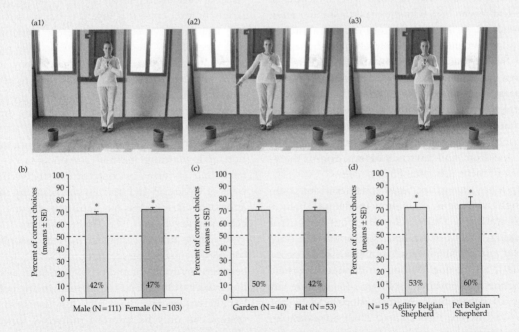

Figure to Box 1.2 (a 1–3) In the first phase of the test the experimenter draws the dog's attention to her (a1). If the dog looks at her face she enacts a short (1–2 s) pointing gesture (a2). Only once the hand is returned to the starting position is the dog released to make a choice (a3). The performance of dogs in the pointing tests does not depend on gender (b), is not influenced whether the dog lives in the garden or shares a flat with humans (c), and is not affected by previous training experience in agility trials (d) (Belgian shepherd dogs). Dotted line, chance level; *, significantly above chance performance. The percentages in the column refer to the ratio of dogs that choose significantly over chance (binomial test, p < 0.03, at least 15 correct out of 20 trials). Gácsi *et al.* 2008.

1.3.2 The first question: function

Defined simply, the functional approach is interested in finding out how any behaviour pattern contributes to the survival of the species. Any such investigation is successful only if the ethologist knows the actual environment of the animal well. Thus this question cannot be answered unless we provide a description of the environment in which the dog lives. Along with others, we think the natural environment of the dog is that ecological niche which has been created by humans (e.g. Herre and Röhrs 1990, Serpell 1995; see also Chapter 3, p. 47). The dog as a species emerged as a result of evolutionary processes which affected some canid species a few tens of thousands of years ago. This means that we can look for those behavioural traits that enhanced the survival of dogs in human-dominated environments. In some cases these environments differ enormously, as reflected in high levels of phenotypic variability in dogs. This fact puts to the test researchers who are used to smaller environmental variation in the case of natural niches. A village where it can roam freely at night or during the day, a fifth-floor flat, and the streets and parks can all be (often physically discontinuous) places where the dog is at home. In some cases (feral) dogs live in environments where humans are rarely present, but this situation is probably secondary. Nevertheless it represents one end of the spectrum, and the study of feral dogs is therefore not futile (Chapter 4.3.5, p. 86).

In many cases functional considerations come to light when some dogs show inadequate behaviour patterns. Object chewing, out-of-control barking, or out-of context aggression not only upset and frighten owners but can also be problematic for the dog. Without understanding their functional importance, solutions for getting rid of such behavioural malformations will be not easy to find (Fox 1965, Overall 2000). For example, recent investigations indicate that contrary to previous assumptions, barking may have some function in dogs as a means for communicating with humans (e.g. Yin 2002, Pongrácz *et al.* 2005; Chapter 8.4.2, p. 185).

1.3.3 The second question: mechanism

Although for many scientists 'behavioural mechanisms' meant looking for the genetic or neurobiological underpinnings, when ethologists talk about this aspect of behaviour they mean either the identification and experimental investigation of those environmental or inner events which contribute to the occurrence of the behaviour, or how behaviour is organized (e.g. Baerends 1976). Typically ethologists practice a top-down approach (p. 118), being interested in higher organising principles of behaviour that assume that the animal of interest is in the position to display the richness of its natural skills because it has the experience of its natural environment. Thus laboratory investigations on laboratory animals that have little relevance to natural behaviour are to be avoided, unless their usefulness can be clearly stated.

In the case of wolves and dogs the study of behavioural mechanisms includes problems such as the effect of various signals on the behaviour of others in the context of play (e.g. Bekoff 1995a), mate choice (e.g. Dunbar 1977), or aggression (e.g. Harrington and Mech 1978). The training of dogs also raises many important questions with regard to how dogs learn about natural and artificial aspects of the environment (Lindsay 2001). Especially in the latter case it provides a battlefield for contrasting different models of the underlying mental processes which control behaviour. Although there is a tradition of explaining learned components of dog behaviour in terms of complex associative processes of Pavlovian and operant conditioning, other approaches stress a less mechanistic explanation of behaviour (e.g. Csányi 1988, Timberlake 1994, Toates 1998). They aim to construct models describing complex mental processes that provide an interface between environment and behaviour. Such modelling is very difficult because there are many potential alternatives and the actual components of the system can only be inferred indirectly through observation of behaviour. There is some hope that cognitive ethology can provide a general framework for this field of research by emphasizing the evolutionary and comparative study of animal mental processes (Kamil 1998).

1.3.4 The third question: development

The study of behavioural development has usually been the battlefield for arguments aimed at separating behaviour into 'innate' and 'acquired' components. Describing development as a series of complex interactions between the unfolding genetic information and the actual environment (*epigenesis*) has calmed down the debate but has not solved the actual problems.

In the case of the dog we are lucky that the work done by Scott and his associates and others (e.g. Fox 1970, Fentress 1993) provided some important starting points, although continuing work seems to be necessary (Chapter 9, p. 201). Some of those early experimental methods (such as long-term deprivation) are no longer available, so we need to look for other ways of finding out how (or whether) early environmental events influence later behaviour, especially given the large variation in dogs as a species and in their living environments. Systematic variation in this respect, which includes both genetic and environmental components, provides the foundation for the concept of personality which has recently become the focus of research (Chapter 10, p. 221).

1.3.5 The fourth question: evolution

The evolutionary study of behaviour is a truly comparative endeavour (Lorenz 1950, Burghardt and Gittleman 1990) and also has a long tradition in behavioural research on canids (e.g. Fox 1975, 1978). The emphasis on the evolutionary study of dogs could be fruitful if we assume that in order to share our niche they have been subject to some sort of selection. Accordingly, there is a need for comparative ethological research in canids in order to see how divergent evolution has changed species-specific behaviour patterns in these species (Chapter 4, p. 67). So far most attention has been paid to the wolf, but a much broader approach is needed, including coyotes and jackals (at least). One reason for this is that *Canis* and some other closely related species show very flexible patterns in the course of adaptation. Various behavioural traits emerge, disappear and reappear in different evolutionary clades; for example, the adaptation to drier and warmer climates occurred in parallel in the coyote, the wolf, the jackal, and somewhat later the dingo.

The living species of Canidae might present different behaviour mosaics which are the most successful in their present environments. Thus comparison of dogs with the present-day wolf, their closest genetic relative, might be too restrictive because since the species split modern wolves may have adapted to a different environment, and the ancestor wolves could have represented a different mosaic pattern of behavioural traits. Lorenz (1954) might have been wrong about the actual ancestors of dogs but he could still have had a good eye for picking out those features of dog behaviour that are not present in the recent wolf but in other species of *Canis*.

The comparison of dog and human behaviour represents the other side of the coin. In this case we can look for answers to questions about behavioural adaptations. Dogs and humans do not share close common relatives, but they seem to share some functionally similar behaviours (Chapter 8, p. 165). This concordance raises questions about the selective nature of the human environment. From the dogs' point of view one could argue that such similarities are actually the results of a selection process, but this argument could be also applied in the other direction by saying that corresponding human behaviours are also likely to have been due to positive selection. Thus the evolutionary study of the dog can not only reveal the path leading to this species but also give some hints about our own past (Box 1.3).

1.4 Evolutionary considerations

Given the perception that dogs seem to be well suited to their actual environment, many cannot resist telling 'adaptive stories' as explanations. Unfortunately, these stories do not distinguish between different kinds of causal factors and also use the concept of adaptation very loosely.

In developing hypotheses of dog domestication we must be careful not to mix ultimate and proximate causes. By *ultimate causes* we usually understand evolutionary or ecological factors which have the potential to explain why some changes took place in the course of evolutionary time. Such ultimate causes

---

Box 1.3 The dog as a convergent behavioural model

Despite large morphological differences between humans and dogs, the notion of some sort of 'spiritual' similarity has been always around. Darwin (1872) also often refers to behavioural or mental parallels between dogs and humans, but it seems to depend from case to case whether the comparison is made on the basis of homology or convergence. Scott and Fuller's (1965) model of development of social behaviour in dogs was intended clearly as a homologous model for humans (Chapter 9), similarly to behavioural models that are based on a general learning mechanisms.

Other approaches recognize the fact that dogs are very successful in living in human social groups. They argue that similarities in the social environment could have resulted in behavioural traits with similar functions, thus representing a case for convergence.

Ideas introduced by Hare *et al.* (2002) suggest that dogs could have gained advantages in communicating with humans. Being able to read specific human communicative cues could be regarded as a case for convergent evolution (see also p. 14). Miklósi *et al.* (1998, 2004) and Topál *et al.* (forthcoming) developed a more general concept of behavioural convergence in dogs, assuming that behavioural changes affected a range of components of dog social behaviour.

Although the degree of these changes might be debated, the authors argue that the affected behavioural traits are responsible for the dog being able to develop, among other things, an attachment relationship with humans showing complex communication and cooperation skills (Chapter 8). These changes formed a species that has achieved surprisingly high levels of human-like social competence (Chapter 8.9.)

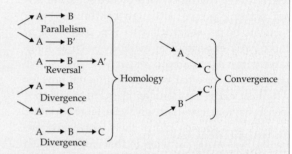

Figure to Box 1.3 Possible evolutionary relationships between phenotypic traits (A–C) based on Fitch (2000). Similarities in phenotypic traits between jackal and coyote might represent a case for parallelism, and the re-emergence of some wolf-like traits in dingoes (e.g. male parental behaviour) might be regarded as reversal (but see p. 89). Depending on the trait dog–wolf relationship presents a case for divergence, and with regard to some social traits the human–dog evolution provides evidence for convergence.

---

are of importance if one wants to understand the reason why dogs, as a novel form of canids, have emerged. *Proximate causes* explain the mechanisms involved in the production of certain phenotypic traits (e.g. behaviour). To study the proximate causation of dog behaviour in relation to wolf behaviour, we have to look for differences (or similarities) in the genetic, physiological, and cognitive factors which control behavioural traits. For example, the retention of certain juvenile characters into adulthood (*paedomorphism*; see Chapter 5.5.5, p. 126) is often used to explain the difference between dog and wolf. However, this does not explain why dogs were domesticated in the first place. Paedomorphism refers to changes in the temporal relationship between two or more phenotypic

traits, assuming that heritable alterations in the genetic control of developmental processes are responsible. Paedomorphism in dogs is often taken as evidence for active human involvement in dog domestication from the beginning, because humans prefer similar features in their offspring. However, even this reasoning does not take us much further because paedomorphism has also been described in other species which evolved without human intervention. For a plausible argument we need to identify those ultimate selective factors which made humans select for certain phenotypic features in ancient canids.

Evolution is conservative in two respects. First, because it works with complex living structures whose features have been already 'tested' over many millions of years, evolution avoids any big change. Second, most novel 'inventions' (e.g. genetic mutation) are more likely to make such a system worse than better. Some evolutionary biologists stress that the constraints of the already established living structures are more interesting than the evolutionary 'progression' (Gould and Lewontin 1979). Thus large 'jumps' in evolution are rare, and in most cases changes take place at a much smaller scale. In addition, there is no evolutionary museum for organisms of failed design, because these are eliminated very early in the process. Thus when looking at the fossil record or living beings, the achievements of the 'blind watchmaker' (Dawkins 1986) are usually overestimated. Only for a naive outsider is evolution a success story.

Gould and Vrba (1982) draw our attention to a further confusion in evolutionary theory concerning the concept of *adaptation*. With regard to dog evolution, adaptation is usually implied in two different ways. First, many assume that the dog is adapted to the human environment, and second, there are arguments that a wolf-like canid is the most likely candidate for being the ancestor because these animals were preadapted to the human social environment. The problem with these statements is that the first disregards the historical aspect of evolution, while the second relies on a confusing argument.

In an evolutionary perspective adaptation becomes a useful concept only if it refers to some novel feature of the organism which emerges in response to the challenge of the novel environment; that is, it has a special function. Gould and Vrba (1982) argued that all other traits should be described as *exaptations* which might have been co-opted by the descendant from its ancestor without any specific changes, or have been changed and are now used for a novel function. The former case of these two possibilities is usually called *pre-adaptation*, that is, when a former adaptive trait is 're-used' without changes in the descendant. Both adaptations and exaptations contribute to the actual fitness of the organism. Thus traits of a species can emerge *de novo* ('adaptations') in the novel environment, or as exapted traits used in a different context, or as exapted traits that are utilized without any change. Gould and Vrba (1982)

assumed that because of the conservative nature of evolution most traits of the species are exaptive.

Applying this concept to the dog, it is clear that dogs can be said to be 'adapted' to the human environment only if we can show that novel traits have emerged. Similarly, wolves are not preadapted to the human niche but they inherit a set of exaptive traits which contribute to the survival of dogs in the human environment. Thus from the evolutionary point of view research has to separate 'true' adaptations from exaptive traits which have either been modified or not. Actually, the short time since the divergence from wolves (despite the intensive selection in the last few thousand years) makes it unlikely that dogs have evolved a large set of specifically adaptive novel characters (in the strict sense). In the case of exaptive traits, some changes might be traceable. For example, in dogs barking has a very different and much broader communicative function than in wolves (Chapter 8.4.2, p. 185).

Another way of dealing with adaptive changes of phenotypic traits is based on comparing species either on the basis of phylogenetic relatedness or sharing similar environments (see Box 1.3). If two species share a common ancestor at some distant time, the relationship of their traits is described as *homologous*. If at some point in time a split results in two species, any subsequent adaptation will increase the difference between the traits in the two species. However, we usually do not have a full record of speciation events, so the comparison of either fossils or extant species will be often based on inference. Homology of certain traits is a relative concept because it depends on how far we go back in time, since at some point in time all species had a common ancestor. The concept of homology is useful in finding out more about the last common ancestor, and piecing together evolutionary relations among species. For such comparisons ethologists relied on the species-specific behavioural pattern (e.g. courtship behaviour, Lorenz 1950). Thus the comparative study of extant wolves and dogs should shed light on the possible common ancestor of these species. Comparisons based on a homologous relationship focus on the 'resistance' of the complex structure (conservatism, see above) which has been established during earlier stages in evolution.

In both extinct and extant animals there is evidence that unrelated species evolve similar traits that are possibly the result of exposure to the same evolutionary factors in the same or similar environments. Thus the similarity in some phenotypic features is based on the common function of the trait, which is often controlled by a different mechanism (Lorenz 1974). Morphology provides many examples of *convergence* such as the evolution of 'wings' (extremities that enable flight) several times independently in insects, reptiles, birds, and mammals. The verification of convergence is important for the evolutionary argument because it supports the concept of adaptation, that is, species evolve traits as a response to environmental factors. The notion that is sometimes asserted of similarities between social structure of wolves and humans is based on such arguments (Schaller and Lowther 1969, Schleidt and Shalter 2003). More recently arguments have been put forward stressing a similar relationship between social behavioural pattern in humans and dogs (e.g. Miklósi *et al.* 2004, Hare and Tomasello 2005) (see Chapter 5, Chapter 8, Box 1.3, Figure 1.4).

It is useful to distinguish between convergent processes taking place in distantly related taxa, and *parallel* evolutionary changes in more closely related species (Fitch 2000). In the latter case conservative evolution has already determined the direction of possible changes in the ancestor leaving little room for *de novo* adaptation when two descendant species get into a similar environment. Such parallelism probably explains some similar traits in *Canis* species. The genetic heritage from the *Canis* ancestor(s) constrained the direction and magnitude of the possible phenotypic changes in the descendant wolves, jackals, and coyotes. It is likely that many phenotypic similarities between jackals and coyotes are based on such parallelism, despite the fact that their last common ancestor lived many millions of years ago. Thus any member of the genus might respond with similar morphological and behavioural changes to particular ecological circumstances. The phenotypic change in foxes to selection for 'tameness' provides further support for this idea in *Canis* (Belyaev 1979, Chapter 5.6, p. 131) Differentiation of convergence from parallelism is only possible when there are major differences in the starting structure of the organisms;

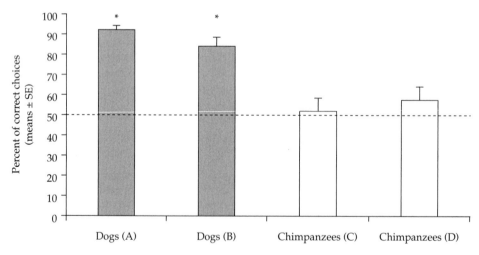

Figure 1.4 The convergent model of communicative skills in dogs gained additional reinforcement by finding that dogs show better comprehension skills in some versions of the two-way choice task (see Box 1.2) than chimpanzees. Here we compare results from studies that used a more or less similar experimental procedure. Note also that the experimenter uses a dynamic sustained proximate pointing gesture; that is, the hand remains in the pointing position during the time the subject makes the choice. (This is why dogs perform better in this case than in the test described in Box 1.2.) However, a recent review showed that it is difficult to make comparative statements because of many experimental and procedural differences in the testing of apes and dogs (Miklósi and Soproni 2006). (A) Miklósi *et al.* 2005, (B) McKinley and Sambrook 2000, (C) Agnetta *et al.* 2000, (D) Itakura *et al.* 1999. Dotted line, chance level; *, significantly above chance performance.

that is, the two species are only distantly related. For example, cooperative hunting in lions and wolves can be considered as a case for evolutionary convergence (independent adaptation) for hunting big game because Canidae and Felidae separated long ago and lions are the only social felid species.

It must be stressed that despite the examples above, it is often very difficult to separate homologous, convergent, and parallel processes. For example, many studies have used skeletal (mostly skull-related) similarities or dissimilarities to argue for (or against) various ancestors of dogs. However, in a large group of closely related species similarity is most often not enough to argue for a homologous relationship which would suggest evolutionary descent and exclude the possibility that the observed similarity is mainly due to convergent or parallel processes simply because of congruent environmental challenges. For example, Olsen and Olsen (1977) noted that some wolves from China have a turned-back apex on the coronoid process of the ascending ramus (Chapter 4, Box 4.8, p. 91) similarly to extant dogs. They assumed that this similarity is based on homology and argued that dogs must have descended from those wolves. However, in passing they also mention that such a turned-back apex is characteristic for animals with an omnivorous diet (e.g. bears). Thus it is as likely that this feature evolves repeatedly in *Canis* species if they adopt an omnivorous diet (parallel evolution), making the character less feasible as a diagnostic signal for relatedness. (However, it seems not to be present in omnivorous jackals.)

1.5 What is it like to be a dog?

Over the years many have toyed with a question, originally put forward by Nagel (1974) in relation to bats. Nagel queried whether natural science could ever offer a method of understanding the subjective conscious state in another creature. Nagel wondered 'What is it like for a bat to be a bat?', but many try to answer a much simpler form of the question 'What is it like for *us* to be a bat?'. Although we have little to offer in answer to the original question, the answers to the second question are usually regarded as demonstrating anthropomorphism when human behaviour and human mental abilities are used as a

reference system to explain the character of an animal or species (see Fox 1990, Mitchell and Hamm 1997).

Recent discussions on anthropomorphism have revealed that whether this method of scientific inquiry is advantageous or disadvantageous depends mostly on the problem at hand (Bekoff 1995b, Burghardt 1985, Fisher 1990; Figure 1.5) Anthropomorphism could be a useful tool in answering questions about the evolution or function of behaviour (Tinbergen's first and fourth questions). For example, animals living in groups might have similar problems to solve (dominance, cooperation, etc.) or similar evolutionary forces have selected them for living in a group in the first place. Thus experiencing that certain behaviours in humans function to reduce anxiety after aggressive interaction ('reconciliation'), one might assume that a similar pattern of behaviour in another species could have the same function (de Waal 1989). Thus in the case of a social mammal like the dog that has evolved some behavioural features making it successful in human communities, we might be entitled to use an anthropomorphic stance in order to look for functional similarities. For example, observing similarities in a behaviour pattern that maintains a close contact between individual group members (e.g. attachment between offspring and parent) could argue for functional similarity between parent–infant and owner–dog relationship (Topál *et al.* 1998; Chapter 8.2, p. 166). Thus such functional anthropomorphism could be a valid way for generating hypotheses on the functional aspects of behaviour because we can assume overlaps in roles played by certain behavioural systems. However, in order to be successful with such a research strategy we must be familiar with the natural behaviour of the species to be compared (Chapter 2.3, p. 30).

The situation is different, however, if on the basis of functional similarity we want to draw a parallel between the mechanisms controlling the behaviour. Such views, which are often referred to as 'arguments by analogy' (e.g. Blumberg and Wasserman 1995), are more difficult to defend, especially if the original functional comparison between the species was based a convergent evolutionary history. Thus the functional similarity in

(a)   (b)

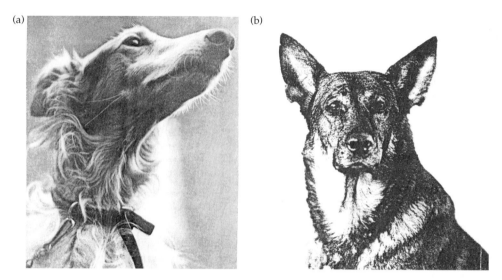

Figure 1.5 (a) Buytendijk's startling image of the dog in his book (1935). The original figure legend indicates an interesting cocktail of baby- and lupomorphism with a flavour of spiritualism. He writes, 'the dog has an attachment to man that is not born out of consciousness and does not become conscious. It is an unreasonable mysterious impulse, strong and imperative, like the primitive forces of Nature'. (b) Fellow, a famous dog from the films of the 1920s. He was able to retrieve objects on commands under strict experimental conditions (Warden and Warner 1928; Chapter 8).

attachment behaviour in dogs and children cannot be used as an argument for similarity in the under-lying behavioural control mechanism. It is more likely that the actual mechanism is different because the ancestors of dogs and humans separated a long time ago and experienced a very different evolu-tionary fate. In the case of the dog, the modifica-tions that took place must have affected the mind of the wolf. Thus looking at the causal (and develop-mental) factors (Tinbergen's second and third ques-tions) it is likely that dogs are actually more 'wolf-like' (Kubinyi, Virányi and Miklósi 2007, Miklósi *et al.* 2007). This seemingly contradictory situation leads to a really interesting question: What kind of changes in the wolf-like behavioural mechanism resulted in human-like functions of behaviour?

1.6 Lupomorphism or babymorphism?

Both researchers and dog experts often refer to one of two extreme behavioural models stressing the importance of either the dog–wolf or the dog–human child similarities. In some respects these views are specific cases of the problems discussed above in relation to anthropomorphism. Models

that stress the homologous relationship between the two *Canis* species use the metaphor of 'wolf in dog's clothing'. These *lupomorph* models (Serpell and Jagoe 1995) assume that domestication changed only the superficial characteristics of wolf behav-iour. For example, this view suggests that the social interactions between humans and dogs should be based on the rules that apply in wolf society. It fol-lows that there is a need for strong hierarchy, which should be established, maintained, and controlled by the human using the behavioural actions and signals on which wolf society is based. Importantly, based on this view we would expect that if dogs inherited the genetic endowment of wolves with-out major differences then equalization of environ-mental differences would result in dog-like behaviour in the ancestors; however, this is not true (Chapter 8, p. 165). Although this behavioural model relies on the well-established notion of a close evolutionary relationship between dog and wolf, it fails to recognize that our understanding of wolf behaviour is very limited (and has changed a lot since the beginning of systematic observational and experimental work; see chapters in Mech and Boitani 2003). Wolf behaviour is also very variable, and there are large differences both over time (ancestors of recent wolves might have had different societies) and in space (different populations of wolves might adopt different patterns of social behaviour). Thus the lupomorph model is often based on 'idealized' wolf behaviour not really supported by current knowledge.

At the other end of the modelling spectrum, experts argue that not only does the adaptation process lead to significant changes in the social behaviour system of dogs, but these individuals actually live in a social world which is in many respects comparable to that of a 1–2 year old human toddler. These analogue models refer to the 'infant in dog's clothing' metaphor, suggesting that the social behaviour of dogs should be understood in terms of the human parental relationships. It is not exceptional that people attribute child-like behaviours to dogs, and say that 'dogs are just like small children'. In one study university students reported only quantitative differences between a typical dog and a school-aged boy on many characteristic anthropomorphic traits like 'moral judgements', 'pleasure', 'imagination', etc. (Rasmussen and Rajecki 1995). Thus these *babymorph* models suggest that dogs are in the social position of a human child with mental abilities corresponding to that of a 1–2 year old. Humans are expected to show parental behaviour towards dogs in terms of affiliative interactions and teaching or education (Meisterfeld and Pecci 2000). However, these models seem to neglect the fact that in human societies dogs often play other social roles than being a child substitute, and human parental behaviour is very variable and probably sensitive to the ecological environment, so the 'Western style' of human–dog interaction may not have been the rule. A further problem is that dogs and infants differ greatly in their experience of the world as well as their cognitive and behavioural capacities.

Actually, both types of extreme model seem to confuse evolutionary arguments and fail to recognize the exceptionally high variability in dog–human relationships, which obviously has several sources (see also Serpell and Jagoe 1995). First, present-day dogs have a wide range of genetically influenced patterns of social behaviour. This means that depending on their selection history and the resulting genetic endowment, dogs will perform differently in different environments. Second, the type of relationship between humans can be very varied. Although some dogs do indeed play the role of a child substitute, others are more of a social companion of equal rank and many dogs live in a working relationship in which their contribution to the family can be measured in financial terms. Third, ecological and cultural traditions have often changed dog–human relationships over time. For example, in some cultures dogs are still part of the human diet, and in other cultures this has ceased only recently.

Thus it seems unlikely that either of the extreme behavioural models can succeed on its own, and it is also not the case that dogs are somewhere between the two extremes. For a comprehensive framework it might be more advantageous to develop behavioural models based on a different approach.

1.7 Modelling of behaviour

Theories developed on the basis of modern biological, psychological, and even technical (computational) knowledge emphasize the possibility of interpreting behaviour in terms of inner states and processes of the mind. Accordingly, Shettleworth (1998) defines cognition as an array of mechanisms by which animals acquire, process, store, and act on information from the environment. The underlying framework for such views is based on the general assumption that the main function of the animal's mind is to provide a representation of the environment. Among others, Gallistel (1990) characterizes such representations as being functionally isomorphic to environmental constructs. It should be pointed out that not everyone agrees with such a view of the mind, and there is an ongoing debate of varying intensity about the nature of mental models.

The so-called *ethocognitive approach* develops metamodels that provide a bridge between models that were developed for conceptualizing behavioural systems (e.g. Baerends 1976, Bateson and Horn 1994, Timberlake 1994) and models that aimed at understanding the central control structure of the mind (e.g. Csányi 1988, Toates 1998), and have the advantage of being particularly useful in a comparative perspective.

Before turning to the description of one possible ethocognitive metamodel of behaviour, it is worth reviewing issues that are associated in general with behaviour modelling.

1.7.1 Top down or bottom up

Sometimes researchers have not much choice in formulating their models. Early cell biologists produced very crude models ('drawings') of the cell, which became more and more detailed as microscopes gained higher powers of resolution. Thus for mainly technical reasons cell biologists had followed a top-down approach to modelling. Meteorologists had (to some extent) the opposite fate. The modelling of wind systems probably started on a smaller scale, but as better technologies allowed for collecting data at high latitudes and in space, global models of wind systems could be established. Here the bottom-up approach was unavoidable. In the case of behavioural sciences both ways of modelling are possible, but unfortunately this situation led to a dichotomy in which researchers campaigned for the advantages of one approach over the other.

Interestingly, the views of researchers on the modelling of mental structures seem to be influenced by the methods used for studying animal behaviour. Proponents of a more naturalistic approach by studying species living in their natural environment often argue in favour of top-down approaches, which means the use of mentalistic descriptions (see below, e.g. Bekoff 1995b, Byrne 1995, de Waal 1989). In contrast, laboratory-based researchers often, but not exclusively, prefer to develop bottom-up models based on utilizing simple mechanistic processes. This does not necessarily reflect the subjective preference of the researcher for a certain view of modelling, but is rather the result of the conditions under which the behaviour is studied.

When animals are observed in their natural or semi-natural environment there is often little chance of controlling the physical and social aspects of the environment or the experience of the animals. Nevertheless, in such situations animals can show their full potential, providing the researcher with a global view of their behaviour. From this perspective it is less surprising that top-down models are more commonly applied to interpret animal behaviour observed or even experimentally probed under diverse conditions.

In contrast, the laboratory offers greater control over external and internal variables, and the (often naive) animal is observed in a simplified environment. Little experience of the subject limits the range of behavioural responses and increases the researcher's chance of predicting behaviour. The close monitoring of input and output offers the possibility of formulating a bottom-up model based on simpler rules which account for a particular local aspect of behaviour without the need to make a connection to the behavioural system as a whole (see also Box 1.4).

Problems arise when researchers try to apply top-down models to account for bottom-up models, or vice versa. In this case bottom-up models are unnecessarily complicated because one has to assume a complex structure consisting of simple rules. In the same vein, top-down models seem to be too vague in accounting for local phenomena, so their validity may be questioned.

1.7.2 Canon of parsimony

Morgan (1903) suggested that behaviour should be explained with reference to mental processes that stand lower on the scale of evolution and development, but he was also careful to add that 'the simplicity of explanation is no necessary criterion of its truth' (Burghardt 1985). Nevertheless, the first part of Morgan's suggestion reinforced approaches that interpret behaviour in terms of simple rules of association because this mechanism seems to be present even in very ancient organisms like the medusa or the flatworm, and also emerges early in ontogeny. Mentalistic interpretations of behaviour were regarded as unnecessarily inflated by assumptions of complex processes.

In terms of the previous discussion, Morgan advocates a bottom-up tactic for the interpretation of behaviour. But even he does not make it obligatory, and allows for top-down modelling if there is independent evidence (Morgan 1903). The main problem with this approach is that the bottom-up modelling is bound to the laboratory where independent variables can be controlled, and many behavioural phenomena are very difficult to observe or elicit under such sterile conditions. The study of 'deceitful' behaviour in primates may be one such example (e.g. Byrne 1995, Whiten and Byrne 1988). Thus researchers describing natural behaviour or abilities, such as navigation, object permanence, or social learning, often use some kind of meta-language for interpretation and altogether avoid reference to simplistic associanism or highly complex cognitivism (see Chapters 7 and 8).

The predictive value of a model is perhaps even more important than the adherence to a certain kind of model (Cenami Spada 1996). If it is indeed the case that the naturalistic and laboratory situations differ in fundamental ways, we should be not surprised that the predictive value of top-down models is low when applied to the laboratory situation (and vice versa). There is an analogous situation when researchers try to reconcile models obtained in *in vitro* or *in vivo* experiments. Biologically active substances which seem to work perfectly in a local system *in vitro* (bottom-up model) often fail as drugs because they do not fit into the whole system *in vivo* (top-down model). Therefore instead of trying to reconcile these two, often fundamentally different, models of behaviour we should look at their predictive value under certain conditions, and rely on the model that offers the better explanation for the underlying mental structures and processes.

1.7.3 Associanism and mentalism

The literature usually distinguishes a mechanistic bottom-up approach which emphasizes that most (if not all) forms of (learned) behaviour can be described as resulting from associative processes which establish a link between an environmental stimulus and a particular response. In this case the mind is described as a flexible associative device which is able to establish causal connections among a wide range of environmental events and behaviour. Some proponents of the view do not deny the emergence of some sort of cognitive structures ('representation of the conditioned stimulus', Holland 1990), but they assume a strong association between the representation and the behaviour and experience which led to its existence. Such models

of behaviour have been variously labelled as being 'low-level' (Povinelli 2000), 'cue-based' (Call 2001), or representing abstract spatiotemporal invariances (Povinelli and Vonk 2003).

Others maintain, however, that without denying the importance of associative processes, the mind also entertains cognitive entities (*representations*) which are not tied directly to behaviour, and are often referred to as *intervening variables*. Such representations can function independently of the direct experience and behaviour which led to their existence; moreover, these representations can also be causal factors for certain behaviours. These models predict more flexible behaviour, especially when the animal experiences a novel situation or problem. Such situation-independent representations are often characterized as 'knowledge' (Call 2001) which allow *mentalizing* (e.g. forming expectations, planning) about possible environmental events and actions, especially in the social environment (see also Box 1.4).

1.7.4 Comparing content and operation

Heyes (2000) suggested that we should distinguish between the content and the operation of the mind. She argued that the content of a mental representation depends on the species because ecological differences will determine 'what and when' is learned. In contrast, operational processes in the animal mind are based mainly on associative processes which do not differ markedly among animal taxa. This view shares many features with the general learning theory (e.g. McPhail and Bolhuis 2001). Accordingly, adaptive changes in behaviour will affect mainly the quantitative aspects of cognitive capacity by affecting only the content without changing the organizational structure of the mind.

Not everyone agrees with such views. Over the years many researchers have put forward experimental evidence for the argument that evolution in certain ecological (or social) environments also resulted in novel rules of operation. Solving complex spatial problems ('cognitive maps') (Dyer 1998), avoiding poisonous food long after eating (Garcia and Koelling 1966), and remembering the type of food cached ('episodic-like memory') (Emery and Clayton 2004) are a few of many such

Box 1.4 Contrasting alternative explanations: how and why dogs learn to avoid eating food

Solomon *et al.* (1968) set out to examine the effect of delay of punishment on withholding some preferred action. The specific question was to find out the effectiveness of punishment if it coincides with the execution of the action. The subjects (beagles) were given a 'taboo training' when the dogs were punished for eating meat but were allowed to eat the same amount of dry laboratory chow. The experimenter punished the dogs by a hard blow on the snout with a tightly rolled-up newspaper. One group of dogs was punished as soon as they touched (with mouth or tongue) the meat (no delay), and dogs in the other group were allowed to eat but were punished after 15 s had elapsed (actually, there were three groups but the one with 5 s delay is ignored here for simplicity.) This procedure was continued until all dogs refrained from eating the meat over a period of 20 days. Before the 'temptation tests' dogs were deprived of food for 2 days. In the test the dogs could chose between 500 g of meat and 20 g of chow without the experimenter being present in the room. Dogs had no additional food during the day, thus they had to live on the food eaten during the tests, which were continued until the dog broke the taboo. Solomon *et al.* also observed the behaviour of the dogs as well as the number of test days elapsed before eating the meat.

1 Dogs in both groups acquired the food taboo in 30–40 days of the training.

2 If the punishment occurred before eating the meat, dogs refrained from eating the meat during 30 days of testing. In contrast, dogs ate the meat within 2 days if they were punished after eating the meat for 15 s.
3 There were marked differences in the behaviour of the dogs both during learning and during the testing phase. Dogs in the 'no delay' group learned to avoid the meat but were a bit hesitant to eat the chow. Later in the training they show 'no obvious signs of fear during the approach to the dry chow and eating it'. Dogs in the '15-s delay group' 'crawled behind the experimenter or to the wall, urinated, defecated … crawled on their bellies to the experimenter' during the training trials.
4 In the 15 s delay dogs 'acted as if the experimenter were still there' but broke the taboo very soon and 'they ate in brief intervals.. appeared to be frightened …' when eating the meat. As soon as 'no delay' dogs dared to eat the meat 'their mood changed abruptly' and 'they wagged the tail' during eating.

Three possible, non-exclusive interpretations (two from the original paper, and the last from us):

1 *Pavlovian:* 'The instrumental behaviour will be shaped by the increases and decreases of fear associated with that behaviour, according to hedonic reinforcement principle.' In the no-delay condition dogs learn to associate fear with

Figure to Box 1.4 A reconstruction of the experimental situation based on the description by Solomon *et al.* (1968). The dog was hit by the newspaper either before (a) or during (b) eating from the bowl containing the meat.

continues

Box 1.4 *continued*

touching the meat, and in parallel eating from the chow will be positively reinforced. Thus in the test the approach to meat arouses fear and delays approach. In the case of using long-delayed punishment dogs have the chance to experience the reinforcing effect of the meat which inhibits the effects of fear on approach behaviour. In the tests these dogs should approach food rapidly.

2 *Cognitive:* 'A theory of conscience' suggests that in both treated groups dogs know 'what they are not supposed to eat.' However, the dogs are uncertain what they should do when the experimenter is not present. Thus in this case of cognitive uncertainty Pavlovian rules take over the control of behaviour.

3 *Ethological:* The experiment replicates a typical social situation when a dominant individual prevents a lower-ranked companion from eating. As in the no-delay group, dominants chase others away from food before they can eat. After extended training the subject learns to avoid the meat, but once in the testing phases it discovers that the meat is freely available and rapidly changes its behaviour. At least in wolves, food already in the mouth is respected by the others (Mech 1970) and is not taken away. The abnormal

stress-related behaviours displayed by the dogs in the delay group and their frequent signalling of submission indicated that these events (punishment after eating) did not correspond with the behavioural rules of dominance. For them the behaviour of the dominant ('experimenter') made 'no sense', thus apart from becoming generally fearful in the presence of the human, they did not learn that the food 'belongs' to the human, and as soon as he was no longer present (in the tests) the dogs grabbed the opportunity and ate the meat.

Conclusion: One might ask which interpretation explains the behaviour better, but indeed they are not exclusive. Interestingly, the authors drew a parallel between the behaviour of dogs and children, and argued that similar mechanisms might operate in both cases. Actually, we think that the best lesson from this experiment is that learning in a social situation depends on whether the subject is in the position to understand the rules of interaction. Finally, it is interesting to note that a very similar protocol was used to find out whether dogs in such situations rely on features of human attention (e.g. Call *et al.* 2003, Chapter 8, p. 179).

cases that have been reported. Thus *adaptionists* emphasize that surviving in different environments may also have selected for differences in the rules about how events are decoded by the mind.

1.7.5 Comparing intelligence

Unfortunately, the term *intelligence* has many different meanings, and is often used in a very superficial way. First, we should not forget that any kind of 'intelligence' reflects only the particular aspect of behaviour which was actually observed and tested under given conditions. Second, intelligence was originally invented as a measure for individual variability in flexible problem-solving abilities (in humans). This means that it is questionable to use the concept of intelligence in a comparative

perspective (Byrne 1995), for example, by looking for breed differences in dogs, or arguing that dogs or wolves are more or less intelligent. The reason for this is simple. Each species has evolved different abilities, and individuals experience a different aspect of the environment in which they grow up. Thus it is particularly difficult to design a task that can pose a problem that is similar to members of different species (Chapter 2.3, p. 30). This is because differential genetic and environmental inputs will also influence the mental potential of the individual to solve the task. Thus it seems to be wiser to retain the use of intelligence in its original meaning, to describe variability among individuals belonging to a genetically well-characterized population, e.g. breed or species. All other use of 'intelligence' should be replaced by reference to differences in cognitive abilities.

1.7.6 Epigenesis, socialization and enculturation

Both bottom-up and top-down models often fail to recognize the complex ways in which genetic endowment can have an influence on mental processes. For example, genetic predisposition might orient the animal to certain aspects of the environment, which will determine what kind of experience is gained. Thus even small genetic differences can result in different kinds of mental representations through complex negative and positive feedback processes. In addition, the full potential of any organism evolves by a continuous interaction between the genetic material and the environment (*epigenesis*) during development that starts right after the fertilization of the egg.

Socialization is an epigenetic process in which a maturing individual is exposed to its social environment and gradually learns about it by interaction with its group members. (The term socialization is often used to describe habituation to the physical environment, which is incorrect.) Obviously, parents have a favoured role in this, but contact with siblings or any other individual facilitates the process which ends when the individual becomes an integral member of its group (or leaves the group). In contrast to other animal species, dogs are exposed to a 'double' socialization process because they are usually exposed to a mixed-species group consisting of both dogs and humans (Chapter 9.3.3, p. 207). A puppy is expected to learn the rules of social life of dogs, as well as many of those of the human community. Often this happens sequentially; that is, dogs are first exposed mainly to conspecifics, and only later join human groups. In some cases researchers distinguish the natural form of socialization to conspecifics from exceptional situations when an animal is exposed only or mainly to the human environment. This later case is often described as *enculturation* (Tomasello and Call 1997), and this term is usually used in reference to apes raised by humans (Savage-Rumbaugh and Lewin 1994). Taking this distinction into account, we could say that dogs are also enculturated when they grow up as members of a human family, even if we recognize that there might be population differences with regard to the profoundness of this experience.

Enculturated apes show many behavioural traits which seem to be absent in their wild companions. Thus researchers have entertained the view that exposure to the complex features of human social environment leads to different kinds of mental abilities which do not emerge in wild individuals. For example, enculturated apes seemed to be better at imitation, understanding attention, etc. (Tomasello and Call 1997). Although the interpretation of the mental capacities of enculturated apes is still debated (see also Bering 2004), the case of dogs is simpler because dogs have been selected in some way to live in human social groups. Thus in the case of dogs enculturation is not a procedural variable but a natural feature of the environment. In other words, social abilities in dogs have been changed in a way that this species 'expects' to be exposed to a human environment. Therefore enculturation should be regarded as a natural process for the dog.

1.8 An ethocognitive mental model for the dog

The model presented here is based on Csányi's *concept model* (1989, 1993) but also includes ideas from both behaviour system and control structure models (see above). The model assumes three different systems that (1) deal directly with environmental input (*perceptual system*), (2) refer to aspects of the environment and inner state (*referential system*), and (3) execute behavioural actions (*action system*). All three systems function in a virtual two-dimensional space defined by a genetic and an environmental component. In the case of each system the interaction of genetic and environmental inputs results in elementary units that are localized somewhere in this space, but importantly their position can change (during a lifetime) according to the actual contribution of the two components. Most often the emerging units are affected strongly by the genetic component, the relative contribution of which might decrease over time because of the interaction of the individual with its environment (Figure 1.6).

In the case of the *perceptual system* the genetic component can be regarded as a default setting for the perception of environmental inputs such as frequency range in hearing or sensitivity for

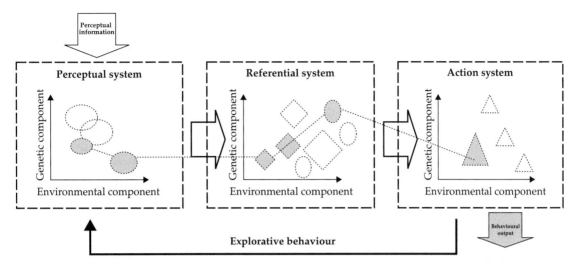

Figure 1.6 A schematic drawing of the ethocognitive model (see also Csányi 1989). The elementary units emerge in a genetic × environmental epigenetic virtual space in the case of all three basic systems. The drawing illustrates how an environmental event activates a 'concept' (grey shapes). The two different geometric shapes in the referential system illustrate separate elementary units for the inner and outer environment. The organism is supposed to continuously update its referential system by exploring and monitoring the environment. The 'concept' emerges through interaction and parallel activation (thin dotted line).

movements (see Chapter 6); however, environmental exposure can also modify the perceptual abilities (Hubel and Wiesel 1998).

The *referential system* consists of two subsystems which represent either the inner environment or the external environment. In the case of the former, different units deal with the actual inner state ('motivation', 'emotions') and other regulatory factors (e.g. 'temperament'). In the case of the latter, elementary units that correspond to certain aspects of the environment are often referred to as *representations*. The nature of such representations can be different; they can refer either to physical entities in the environment, to events or to relations between them. Genetic components of the representational space of the external environment include certain preferences and phobias, the recognition of sign stimuli and tendency for certain behavioural tactics (e.g. win–stay or win–shift).

The main task of the *action system* is to organize the behavioural action by the means of elementary units emerging in the two-dimensional space determined by genetic and environmental interaction (*behavioural schemas*). The interplay between these two components has been the topic of much discussion among ethologists because the early

notion of the *fixed action pattern* seemed not to include the possibility of environmental influence, the recognition of which led to the idea of *modal action patterns* (see also Fentress 1976).

The operational state of the model is described as the emergence of a functional unit (*concept*) which involves the parallel and sequential activation and temporary coupling of a set of elementary units in the perceptual, referential, and action systems. The activation of any concept results not only in an observable behaviour pattern but more importantly, by feedback mechanisms, it also affects ('updates') representations in the referential system ('memory') with regard to both the outer and the inner environment. The operation of the system can be brought about either by environmental stimulation or by internal factors, which is realized by 'exploratory monitoring' behaviour (see Figure 1.6).

In our case we could utilize the power of the ethocognitive model for describing concepts in the mind of the dog, as well as looking for differences between dogs and wolves. For this it is useful to keep in mind that (1) both dogs and wolves have been successful in their respective environments, (2) there is an approximately 0.3% genetic difference between wolves and dogs, (3) exposing wolves to the environment of dogs, including socialization with humans, does not result in dog-like animals, and (4) dogs leaving the anthropogenic environment (stray/feral dogs) do not show wolf characteristics.

The concept model can help us to distinguish two types of questions. First, we should be able to separate genetic and environmental components to some extent, and find out which system (perceptual, referential, action) has been affected by selection and how the genetic compounds have been modified. Such questions can be tackled by wolf–dog comparisons and also by selection experiments (Chapter 5.6, p. 132). Second, one may ask whether genetic changes in parallel with a different environmental input result in an altered structure of concepts and whether as a result novel concepts emerge in dogs. This strategy could involve investigating the relative role of the environment by, for example, raising ('socializing') wolves in a human social setting (see also Chapter 2.3.1, p. 30).

Let's take a few examples. Wolves seem to be keener on meat than dogs: wolf cubs at 6–9 weeks old release a meat bone much later than dogs ('bone competition test with humans': Győri *et al.* 2009). Selection for a wider diet in dogs (especially in the breeds existing today) could have reduced a strong innate preference for meat, but in addition wolves could obtain such a preference *in utero* or during lactation (see Wells and Hepper 2006 for the latter effect). Thus elementary representations of food preference could be affected by both genetic and environmental factors. In addition, as the behavioural manifestation of food preference takes place in a social context the interaction with social behaviours of sharing cannot be excluded.

A further example concerns how the dog's mind might represent humans. There are three (non-exclusive) ways in which such a system could be envisaged. First, dogs utilize basically the same referential system that was originally dedicated to interpreting interaction within the species. Early representations set up by the genetic component will be refined through development by experience and learning through similar channels, as in the case of wolves, adding the peculiarities and features of their human companions to a basically dog-like representation. Such a system would

represent humans as a kind of dog. A second possibility is that domestication largely wrecked the genetic component of the species-specific referential system, and thus in the case of both species the final representations depend crucially on the interaction with the social environment. Therefore the nature and difference between representation of humans and dogs will be basically affected by experience with the social environment. The third version assumes that genetic changes facilitated an early separation of conspecific and human representations, and dogs evolved an ability to set up two separate representational spaces, one for conspecific and another for human companions, both of which have independent genetic and environmental components.

Finally, naturalistic observations agree that feral male dogs (just like their socialized companions) do not participate in raising the young, e.g. they do not take part in feeding the nursing female and the developing pups. Does this indicate a change in the genetic component of the motor schema in males, that is, might they be unable to produce the parental behaviour (e.g. regurgitation)? It might be that they lack proper representations in the referential system for recognizing the behaviours associated with the puppy status of young dogs, or the signals that are emitted by the pups (e.g. eliciting regurgitation by licking the corner of the mouth). Could environmental exposure to pups induce parental behaviour? In addition, male dogs might have altogether lost the ability to recognize the puppy status of young dogs.

The ethocognitive model is not the only way to conceptualize the mind of the dog, and other approaches are also possible (see Frank 1980; Box 1.5). However, its focus on behaviour frees us from the burden of explaining mental processes exclusively in the contentious concepts of associanism or mentalism (see above), both of which could be imported into this model at the level of the referential system if necessary.

1.9 Conclusions for the future

We hope that the dog will find its place (again) among the 'wild' species investigated by ethologists. It seems that the behaviour of the dog can be investigated in the framework provided by Tinbergen and others, including questions that

Box 1.5 Scientific models of behaviour and dog training

Mills (2005) categorizes dog training techniques according to the two main behavioural models used in behavioural sciences. Associative training focuses on establishing a connection between two events, while more cognitive oriented approaches take into account the role of attention and the knowledge of the learner. In a similar vein, Lindsay (2005) assumes mental modules with abilities like 'prediction-control expectancy', 'emotional establishing operation',
and 'goal direction'. From the scientific point of view three points could be thought provoking:

1 Dog training is a means by which the animal is repeatedly exposed to a certain controlled aspect of the environment. Different training methods will provide the dog with a differently structured environment. Importantly, it is to be expected that the referential system of the dog is affected by the method used. Thus, to put it plainly, the 'thinking' of the dog will depend on the method used in the training. A good training method also takes into account the ethology of the species.

2 It is important to consider whether the dog has to be trained because of us or them. There are many dogs out there that enjoy a happy life in the human family without much 'training' in the strict sense. Formal dog training is only one way of interacting with the dog by which skills can be learned. Often our accelerated, city-dwelling lifestyles necessitate our dogs to be formally trained. If provided with a natural environment (just as in the case of our children) many (most?) dogs 'became trained' without much training. Very often dogs are trained formally only when they already show problems in normal social intercourse.

3 Most of the training methods have not been formally validated by scientific research. Thus we do not know whether one method would be superior to others with regard to a given behavioural situation or goal to be achieved, breed, or individual with a particular history or skills of the human owner (see also Taylor and Mills 2006).

(a)   (b)   Imperative/forced

(c) Luring   (d) Clicker training

Figure to Box 1.5 There are several ways of training to a dog to go to a resting place (a). The methods used in training might not only affect actual performance but by setting an environment they also influence the referential system of the dog. (b) imperative/forced; (c) luring; (d) clicker training.

tackle either ultimate or proximate causes of dog behaviour.

Despite the efforts of many scientists behavioural models have still a long way to go, and the present situation is made difficult by the different strategies of model building. Naturalistic observations and a more global view of animal behaviour prompt ethologists to develop top-down models, whereas laboratory-based colleagues having greater control over environmental variables prefer local, bottom-up modelling of behaviour. From this latter perspective top-down models might seem to be unnecessarily complex or even vague, but it is also the case that local models often fail to grasp aspects of 'real' behaviour. Thus it seems that the two approaches should be regarded as complementing and not necessarily replacing each other. This complementary aspect should be also given more attention in dog training, which seems now to rely mostly on bottom-up models.

The ethocognitive model might provide a way to conceptualize the problem of comparing wolves and dogs. This model combines the advantages of behavioural and cognitive models of behaviour. It is not intended to replace traditional models of learning, but it is hypothesized that by pointing out the role and interaction of genetic and environmental components which affect the perceptual, referential, and action systems, more specific observations and experiments can be designed in order to find out the similarities and dissimilarities in the concept structure of wolves and dogs.

Further reading

Lindsay (2001) provides an extensive review of experiments from a learning theory perspective. Shettleworth (1998) and Heyes and Huber (2000) present an overview of the role of evolution in forming animals' cognitive abilities. Johnston (1997) is a useful starter for those who aim at a more holistic (combination of top-down and bottom-up models) view of dog training.

CHAPTER 2

Methodological issues in the behavioural study of the dog

2.1 Introduction

The rediscovery of dogs for behavioural research is probably one of the most exciting developments in recent years. The fact that people with very different scientific training have started to study dogs has led to an increasingly confusing situation where a range of methods is applied, often without a clear understanding of their validity and limitation. Some researchers apply methods only because they seem to be simpler or faster, or because they have been used by others in the past. In some cases one method is clearly preferable to another, but in another situation methods might be complementary. It is not our goal to give an exhaustive review here, partly because there are very good textbooks on the subject in general (e.g. Martin and Bateson 1986, Lehner 1996) and good reviews referring to dogs (Diederich and Giffroy 2006, Taylor and Mills 2006). However, it seems useful to summarize some of the methodological issues from the perspective of dog ethology.

Regardless of the discipline, experimental research must be accounted in terms of validity. *Internal validity* means how well the observed phenomena can be accounted for by the particular experiment in terms of the causal relationship between the manipulated factors and the measured variables. *External validity* refers to the generality of the obtained results, whether the observed effect might be also present in other populations, experimental conditions, at another point in time, etc. (Taylor and Mills 2006).

One reason why dogs have become popular is that they can used in behavioural experiments just as easily as humans. There is no need for an animal house, special animal care staff, a breeding programme, etc.; it is only necessary to get dog owners interested in collaborating with scientists. This means that behavioural observations and experiments can be and are carried out anywhere in the world. In such a situation external validity becomes of great importance because researchers need to be able to replicate each others' results in order to make any progress. This calls for a common agreement and understanding on the methods applied to dogs, and a trend towards standardized testing in at least some special cases (Diederich and Giffroy 2006). In laboratory animals (e.g. rats and mice) researchers speak of *behavioural phenotyping*, which means that a particular genetically homozygous strain will be characterized in a limited number of behavioural tests. Unfortunately, even in such cases the task is very difficult because of the many uncontrolled environmental variables. Thus it seems quite illusory to talk about a similar possibility in dogs. In spite of this it seems worthwhile to identify and describe those genetic and environmental variables which affect behaviour, and which should be taken into account in the planning of behavioural observations and experiments.

2.2 Finding phenomena and collecting data

De Waal (1991) argued that the 'real strength' of ethologists lies in the complementary use of different observational and experimental methods. Although his summary was based on primates it is clear that dogs offer an even better example because there is a wider range of possibilities. First of all, most observations on dogs take place 'in the wild'—that is, in environments which are regularly inhabited by dogs. The environment could be the home of a human family, or even a laboratory which often looks more like a living room than a laboratory. Thus most human environments can be considered as natural for dogs and even a novel place should not present an artificial situation. Even so the methods of observations might differ, so we give here a short summary based on categories used by de Waal (1991).

2.2.1 Qualitative description

People having regular and extensive contact with dogs often witness unique events. Anecdotes or qualitative descriptions of behaviour can be regarded as 'accidental observations' if the events are described in detail, in writing. The popular literature on dogs is filled with such stories, which not only serve as entertainment for the reader but are also presented as a sort of evidence in order to underline assumptions about the complex abilities of dogs. In scientific literature anecdotes are received with mixed feelings. Early investigators such as Romanes (1882a), Lubbock (1888) and many others based most of their arguments on anecdotal evidence observed by them or collected from others. Researchers trained in the scientific method have argued that it is impossible to claim the presence of higher mental abilities in animals on the basis of anecdotal evidence because the observer had no control over the events, and thus might have missed crucial contributing factors and cannot provide a full account of the precedents for the event.

Independent of anyone's personal opinion, anecdotes have always played an important role in generating novel hypotheses for scientists studying animals. In this regard they could be very useful in the case of dogs. However, on the basis of anecdote one cannot argue for any sort of 'understanding' (mental mechanism) of a causal relationship in dogs, because anecdotes describe a 'performance' and are silent with regard to the underlying mental mechanisms. Nevertheless, collecting many similar anecdotes could provide encouragement for initializing an experimental investigation of

alternative hypotheses in order to test for possible mental processes or complex abilities. (Box 2.1).

2.2.2 Quantitative description

Only the systematic collection of quantitative data allows scientific hypotheses to be tested. The explanatory value of such work often depends on the possibility of how well various variables can be controlled in the course of the observations. In the case of so-called *uncontrolled observations*, the main aim is to collect quantitative behavioural data with regard to some specific research question. For example, we might observe dogs sharing their life with inhabitants of a village, and by following the dogs around we note the frequency of interaction between dogs and people or other dogs. Despite often being mainly descriptive, such systematic work can be very important if, for example, it investigates whether the presence of feral dogs has an effect on wild life (e.g. Jhala and Giles 1991).

In *controlled observations* the experimenter waits for a spontaneous occurrence of a behaviour of which the effect needs to be measured. In one study Bekoff (1995a) hypothesized that the play bow serves as a confirmative signal to express willingness to continue playing. Thus he assumed that the play bow should be more frequent before and after actions which cause harm to the partner (e.g. a bite). By comparing the frequency of play bows after harmful and non-harmful interactions he found support for this idea. In other cases researchers collect evidence for certain rare patterns of behaviour under controlled circumstances. This is often the case with unwanted (abnormal) behaviours (e.g. dogs destroy objects in the house when left alone) when owners' accounts need to be validated by trying to reproduce the situation and record the behaviour of the dog.

For *natural experiments* the investigators stage scenarios which closely resemble natural situations but the situations are varied according to some predetermined variables which are in the focus of interest (see also 'trapping' in Heyes 1993). From the dog's aspect the only difference might be that the events follow each other with somewhat higher frequency than they usually do. For example, one

Box 2.1 Do dogs show us what they want? How to utilize anecdotes

Two well-known and experienced scientists and dog experts reported similar stories in their recent book on dogs. Due to space limitations both anecdotes are presented in a condensed form, together with a summary of the interpretations offered by the authors.

- *Csányi (2005, p. 138):* After getting home from a walk in the rain, I had forgotten to dry him. Flip ran after me, got in front of me, stopped, and started to dry his head on the rug. Then he stopped and looked at me questioningly. 'Do you want a towel? I asked. At that he jumped up and ran to the bathroom where his towel hangs.
- *Observer's interpretation:* This is a rare case of miming behaviour in order to make a request. Only on the first occasion can it be regarded as miming, because subsequent similar actions are probably based on learning the contingency between the act and the owner's action.
- *Coren (2005, p. 373):* The game with my little granddaughter involved putting a bath towel over my dog, Darby, covering his head, and asking in a singsong voice 'Where's Darby?' A little pat was the dog's reward for putting up with this indignity. Once, after we had stopped this game Darby caught the towel in his mouth, … looked at me … rolled onto his side … and rolled over … got up … now the towel was hanging mostly over his head and back.
- *Observer's interpretation:* Darby demonstrated a childish attempt to communicate that he wanted to continue playing. If one attributes reasoning, planning, logic, and consciousness to a child performing the same action as Darby in this example, then we should also accept the same abilities in the dog (although in some limited way).

There are intriguing parallels in the stories. First, both dogs' behaviour is interpreted as a request to the owner and second, Flip and Darby spontaneously 'impersonate' the request by seemingly re-enacting a former behaviour. We leave it to the reader to agree or disagree with the interpretations of the observers. However, in general there are two ways of analysing these stories. The sceptics' tactic would be to find separate, alternative explanations for the two cases referring to accidental coincidences and external stimuli driving the behaviour (e.g. wet fur elicits rubbing, etc.). These are actually not difficult to find, so the matter can be put to rest. In contrast, for believers both stories could be convincing enough to make some hypotheses about dog behaviour for subsequent experimental testing. One hypothesis might be concerned with the ability of dogs to recognize the 'attention' of the owner and redirect it to certain parts of the environment. Other assumptions could target the dog's ability to reproduce earlier actions which were learned in a social context and re-enacted under different conditions (see Chapter 8).

Flip

Darby

Figure to Box 2.1 The two 'heroes', Flip (a) and Darby (b) (photos courtesy of Vilmos Csányi and Stanley Coren respectively).

study investigated whether frequency of looking at the location of hidden food depends on the presence of the owner (Miklósi *et al.* 2000; Chapter 8.9, p. 179). Dogs were tested in three different trials, when no food was hidden or when the owner was either present or absent during the observation. Since dogs were accustomed before the experiment to receive food in the living room from places to which they had not had prior access, we assumed the actual tests did not interfere with the everyday life of the dogs.

In some cases it might be necessary to investigate dogs under artificial conditions, but this is not the real strength of working with dogs. Nevertheless, complex procedures including lengthy training cannot be avoided when one tests for perceptual abilities. In these experiments the dog has to learn how to signal by displaying a special behaviour that he has perceived the stimulus or is indicating a choice (Chapter 6.2.2, p. 139). Nevertheless laboratory experiments can play an important role in specific cases when dogs are used as animal models (e.g. looking for animal models of ageing, see Milgram *et al.* 2002, Tapp *et al.* 2003). In general, strictly laboratory work should be the last resort in gaining understanding about dog behaviour. This is because the success of these experiments often relies on populations of laboratory dogs that can hardly regarded as representatives of the species. Even if all their physical needs are fulfilled, they live a very restricted life and have limited social contact with humans or other dogs. Thus instead of designing experiments that are based on captive (and possibly impoverished) dog populations we should seek to find methods which have the potential to test for the same ability under more natural conditions, and which can be applied to dog populations in general (e.g. Range *et al* 2008).

2.3 Making behavioural comparisons

Researchers interested in the evolutionary effect of domestication have often based their arguments on the comparison between dogs and wolves. Although species comparisons seem to be quite a straightforward method for looking at adaptive processes in evolution, in reality nothing can be further from the truth. The main reason for this is that such comparisons often violate the basic condition for any comparative work; that is, that only one independent variable can be changed at a time. Thus in an ideal case if we want to test for species difference we have to ensure that apart from this variable there is no difference in all other variables affecting the behaviour of either of the species. Unfortunately, this condition is hardly ever fulfilled, but this does not distract researchers from claiming species difference, although other factors could also explain the observed difference. Importantly, in any behaviour test we observe the performance of the subjects and not a direct output of their cognitive abilities (Kamil 1988). Performance is the function of many internal and external factors such as motivation and previous experience as well as the particular experimental conditions chosen by the experimenter.

In order to circumvent such problems Bitterman (1965) suggested that species to be compared should be investigated in a series of tests which vary systematically in each potential variable that might influence the performance. However, as pointed out by others (e.g. Kamil 1998), it is difficult to know and control for all such variables and testing for all of them makes any comparative work an unrealistically huge effort. Thus, Kamil (1988) suggested a method of converging operations in which one tests for the same ability by means of different experimental tasks. Although this idea reduces the workload it still allows for the possibility that there might be some independent factors which account for the observed differences. Thus he later extended his advice by suggesting that one should also test the same species in different tasks in which they may not show any difference, or even reverse the order of performance (Kamil 1998).

2.3.1 Wolves and dogs

Unfortunately, dog–wolf studies are not exempt from the problems of comparative research. As a recent example it seems useful to refer to experimental investigations which were aimed at finding out whether dogs are able to rely on human pointing gestures (Box 1.2). In a study designed to find support for the hypothesis that domestication resulted in enhanced communicative skills in dogs (Hare *et al.* 2002), researchers found that dogs were superior to wolves in a test which involved choosing a piece of hidden food on the basis of gestures provided by a human experimenter (*two-way choice task*). The authors concluded that domestication improved the communicative skills in dogs with respect to wolves. Although this interpretation could be correct, the method applied in this study did not exclude alternative explanations. Recently, Packard (2003) listed a few experimental variables which were not controlled for and thus could have influenced the performance of wolves. First, dogs and wolves differ in their level of socialization towards humans, the circumstances of test with wolves were very different, and it was likely that wolves had much less experience with the objects and procedures which were employed in the experiment. Second, because the wolves' performance was uniformly low in any versions of the communicative task, it might be that these animals were not in a position to understand the basic

requirements of the task (Miklósi *et al.* 2004). Wolves that were intensively socialized to humans were later shown to perform better in these pointing tasks (Miklósi *et al.* 2003), probably because they had learned to attend to the human body which gave the signals (Virányi *et al.* 2008, Box 2.2).

Because negative results are difficult to interpret, in addition to Kamil's (1998) suggestions it is important to ensure that subjects of different species have similar prior experience about the environment in general (e.g. socialization to humans) and the requirement of the task in particular (e.g. eating from bowls). Furthermore, it might be useful to test the 'underperforming' species in a simpler version of the test in order to show that the difference is specific to some particular versions of the task.

Important problems might persist even if the advice so far is taken seriously. For example, there might be differences in motivation. Although withholding

---

Box 2.2 Intensive socialization of wolves and effects on performance

In earlier studies wolves were socialized to varying extents (e.g. Fentress 1967, Frank and Frank 1982, Hare *et al.* 2002), which hindered comparative work with dogs. In our research we embarked on an intensive socialization programme for wolves. It was known that successful socialization depends among other things on an early start, when cubs are 4–6 days old (Klinghammer and Goodman 1987, Chapter 9.3, p. 205). The unique feature of this programme was that each cub and puppy had its own human carer, who spent 24 hours a day with the animal for a period of 9–16 weeks. Although the animals had the chance to meet conspecifics regularly (at least weekly), they spent most of their time in close contact with the human carer. The carers often carried the animals on their body in pockets, and they slept together at nights. The animals were first bottle fed and later hand fed. When the subjects' motor skills made it possible, they were trained to walk on leash and execute some basic obedience tasks. The carers carried the cubs and pups to various places either by car or on public

transport. For example, they were regular visitors to the university, participated in dog camps, and frequented dog training schools. From their third week of life, the animals were tested weekly in various behavioural experiments examining social preferences, social and physical neophobia, reaction to dominance, retrieval of objects, communication with humans, and possessivity. After this intensive period, wolves were gradually integrated into a wolf pack at Gödöllő (near Budapest), and the carers visited them once or twice a week (Kubinyi *et al* 2007).

The effect of socialization on wolves was compared in the two-way choice test with momentary pointing gesture. In contrast to earlier findings, intensively socialized wolves developed spontaneous comprehension of a human pointing gesture but at much later age (>1.5 years); younger wolves at 11 months of age had to be trained extensively. Dogs at the age of 2–4 months show reliable performance in this test, but wolves achieve similar level of success only by the age of 2 years.

continues

Box 2.2 *continued*

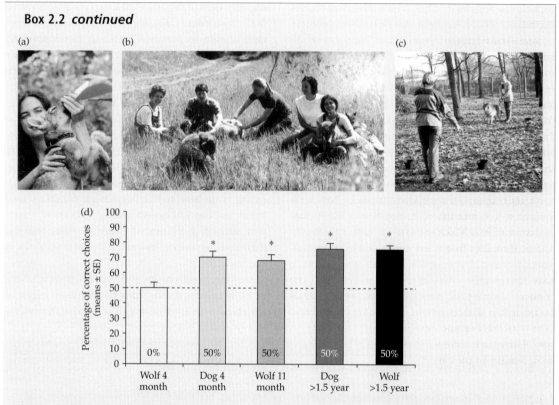

Figure to Box. 2.2 (a, b) Characteristic moments of the wolf socialization programme in Budapest. (c) Two-way choice test with a socialized wolf (photos by Attila Molnár; Ludwig Huber and Enikő Kubinyi). (d) The performance of dogs and intensively socialized wolves with the momentary pointing gesture. Dotted line, chance level; *, significantly above chance performance. The percentages in the columns show the ratio of animals that choose significantly over chance (binomial test, p <0.03, at least 15 correct out of 20 trials) (modified from Virányi *et al* 2008).

food from family dogs seems not to be a practical option, similar duration of fasting might cause different subjective levels of hunger in dogs or wolves partly depending on their current feeding regime. Frank and Frank (1988) noted that social reinforcement (contact with a familiar dog) was a more powerful reinforcement in some learning tasks (barrier test, maze test) in socialized wolves than food reward. Possibly, the eagerness for social rewards is also reflected in the desire to please the human in many trained family dogs. As a consequence, such animals continue "working" in experiments in a kind of "absent-minded" state and show low levels of performance in the test trials. Unfortunately, at present we have little knowledge of how the quality of reward influences the motivation or the performance of dogs.

For many family dogs, favourite play objects (e.g. tennis balls) might be a useful alternative to food reinforcement.

Age is a further complicating factor. On average, dogs mature sexually 1 year earlier than wolves. Although there is little observational evidence in terms of behaviour, most dogs mature only towards the end of their second year (showing adult-like behaviour in general). Thus probably 2-year-old animals would provide the best comparison. However, by this time wolves could be very independent and less willing to cooperate in experiments unless they are intensively socialized and are used to performing in experimental work.

In order to equalize differences in socialization between wolves and dogs, two different solutions are possible. We might 'feralize' ('estrange') dogs in a similar way to wolves; that is, keep both species in semi-wild captive conditions in conspecific groups with reduced human contact. This method was practised to some extent at Kiel (Germany), where the social behaviour of wolf and dog packs was studied by comparative methods (Feddersen-Petersen 2004). Especially when direct contact with the animals is unavoidable during testing, the opposite condition is preferable: both dogs and wolves should be socialized intensively with humans immediately after birth and wolves have to be kept separated from conspecifics for most of their first 4–6 months of life (Klinghammer and Goodman 1987, Miklósi *et al.* 2003; Chapter 9.3, p. 205) (Box 2.2). In a small sample of dogs, which were also intensively socialized like the wolves, we have found some evidence for an enhanced effect of such 'over-socialization' in comparison to dogs raised in the customary way (born in homes and kept with mother and siblings up to 6–8 weeks of age). Thus such intensive socialization might not be necessary in the case of dogs.

2.3.2 The comparison of breeds

Following definitions put forward in the dog literature, existing breeds can be described as intraspecies semi-closed breeding populations that show relatively uniform physical characteristics developed under controlled conditions by human action (e.g. Irion *et al.* 2003). The problem with this definition is that it gives a very static picture of a dog breed. In reality breeds change over time (e.g. Fondon and Garner 2004) because they are subject to both artificial selection by humans, genetic drift, and genetic influx from other dog populations. Dog breeds are certainly more variable than genetically homozygous animal strains kept under laboratory conditions. It is also important to bear in mind that most breeds have been selected for some function. This has resulted in certain patterns of behaviour (and physical characters) which are more pronounced in one type of breed. Thus dogs selected for pulling sledges are expected to be more vigilant. However, in most other respects different dog breeds show a large overlap in behavioural characteristics (Scott and Fuller 1965). Many authors have

also remarked that there is a large inter-individual variation within a breed, which is comparable to the variation found among breeds. This means that breeds tend to differ only in those features for which they have been specially selected, which is only a small percentage of the whole phenotype (Coppinger and Coppinger 2001, Overall and Love 2001). Unfortunately, the physical similarity between individuals of a breed deceives many people who are not experienced in working with dogs. Without providing an exhaustive list, here are a few problems with regard to breed comparisons.

Genetic relation between breeds

As should be clear from the above, dog breeds are artificial categories and are not the results of a genuine evolutionary process. This means that it is not possible to construct an evolutionary tree of breeds. The reason for this is that none of the breeds is derived from a single ancestor population, but is a mixture of different dog populations. In addition, dog breeds have been often recreated over the history using individuals from other breeds. Genetic data show that Pharaoh dogs are a recent 'remix' and only physically resemble the ancient breed depicted on wall paintings (Parker *et al.* 2004; Chapter 5, Box 5.4, p. 107). Thus on the basis of genetic knowledge it is also difficult to claim that one breed is more 'ancient' than another (for details see Chapter 5.3, p. 115).

Behavioural comparisons

It has been fashionable to collect data on behaviour characteristics of breeds by questionnaires (Coren 1994, Hart and Miller 1985, Notari and Goodwin 2006), but this method should not be used to replace ethologically inspired comparative work. The most honest thing to say is that, despite many claims in the literature, breed comparisons do not exist, perhaps with the exception of the Scott and Fuller (1965) study. Here the rules are the same as for the dog–wolf comparisons described above. Given that many breeds have been selected for different types of work with humans, this might have been paralleled by changed behavioural and cognitive capacity. Although at first sight this seems to be an interesting way to look for genetic factors in mental capacities, such comparisons also face the problem that any behaviour observed is a performance which is the result of both genetic and environmental factors. Thus before making any comparison it is necessary to ensure that breeds live in the same environment, have been exposed to the same physical and social stimulation, can be motivated in the same way, and have the same behavioural constitution to solve the task. It is not enough to obtain two breeds of dogs and observe them in a given situation; it is necessary to ensure that the situation has the same 'meaning' for both. In summary, research should be based on well-defined populations living in well-defined environments which are investigated by well-defined (and validated) experimental methods (see also Svartberg 2005). So far this has been achieved only by Scott and Fuller (1965), although whether one agrees with the rearing environment of the breeds or the particular behaviours which were tested for is a different question.

Thus we should be also careful in referring to 'breed difference' (in the sense of genetic difference) upon discovering some difference in behaviour of two or more breeds. Importantly, before such a conclusion can be reached researchers need to exclude environmental differences; for example, many breeds are actually raised in different environments which could also explain the variation. The reason for making this clear is important because often perceived or ill-communicated 'differences' among dog breeds influence people's perception of a breed and could affect legislative issues. Talking about 'intelligent' and 'less intelligent' breeds (Coren 1994; Chapter 1.7.5, p. 21) is probably less harmful, but categorizing a breed as 'aggressive' is a very serious issue (Overall and Love 2001). Any such statements should be made with care and only after researchers have collected convincing evidence. Unfortunately, there is not much knowledge of this kind.

It should be also made clear that breeds can be compared both in breed-specific and non-specific tasks, with very different results. For example, one might expect certain breeds to have better manipulative abilities, thus they should perform better in tasks which involve 'retrieving' or 'pulling by paw', or which are based on certain temperament characteristics like 'playfulness' or 'curiosity' (Svartberg 2005). Unfortunately, ethological descriptions of

breed behaviour are very rare (but see Goodwin et al. 1997). What one expects in a task that could have general relevance is another question. Testing dogs of 10 different breeds (8–10 dogs per breed) Pongrácz et al. (2005) did not find major differences in the solving of a simple detour task. Naturally, the lack of breed specificity does not necessarily mean no genetic difference because complex environmental factors could have a balancing effect.

Comparisons of functional groups
The categorization used by internationally recognized kennel clubs offers the possibility of comparing dogs with regard to their original function. This method is based on the assumption that each given breed was actually selected (and perhaps is still being selected) for that function, and each category used in the comparison contains many breeds as possible. Thus this type of comparison should be based on a few individuals of many breeds, mean values for each breed within a category, or many individuals from a few specially selected breeds. These types of comparisons yield no major differences among breed groups in the case of detour learning (Pongrácz et al. 2005) or with regard to temperament traits (Svartberg 2005) (see Box 2.3).

Geographic and cultural differences
The history of breeds has varied in different countries in recent history. This has occurred because in some cases geographic distance or quarantine laws have limited genetic exchange (some breeds in certain countries were founded by only a few individuals). In addition, cultural differences in the dog–human relationship probably affect the behaviour of dogs, and also the unconscious selection of preferred behavioural traits. It is unfortunate that so far this aspect has been given little attention.

2.3.3 Dogs and children

Interestingly, from the beginning of dog research there have been proposals for comparative work with children. Menzel (1936) and Scott and Fuller (1965) argued for comparative ontogeny in dogs and children; others (Buytendijk and Fischel 1936) emphasized the similarity of the social relationship with humans. In spite of such theoretical discussion very little experimental work has been carried out. Importantly, in primate research such comparative work has long been performed, despite the fact that it is not easy to make the tasks functionally similar for apes and children (but see Savage-Rumbaugh *et al.* 1993). In the case of dogs and children (up to 1.5–2.5 years of age) the comparisons are relatively straightforward because, apart from manual differences, one can assume similar levels of socialization and experience of the environment, as well as using the same observational conditions and experimental apparatus. Recently, abilities relating to object permanence (Watson *et al.* 2001) and reaction to pointing gestures (Lakatos *et al.* 2008) have been tested in dogs and children, using a comparative methodology (Chapter 8, Box 8.4, p. 184).

2.4 Sampling and the problem of single cases (N = 1)

Comparative experimental work often raises the problem whether there are 'typical' dog breeds, or to put it in a different way 'What kind of sample can be said to be representative of dogs?' Unfortunately, there is no simple answer to this question because it would be hard to argue that one or a few breeds are more 'dog-like' than others. This question is also problematic if comparative work includes the wolf. The breeds cannot be ranked along a continuum of difference from the wolf, and it is more likely that dog breeds display a mosaic of traits with regard to wolf behaviour patterns. This would suggest that a mixed sample from many breeds (representing most breed

---

Box 2.3 Are there breed differences in human-directed communicative skills?

Although it is generally assumed that dogs, as a species, have an advantage in communicating with humans, the selective environment might have affected different dog populations ('breeds') in different ways. For example, some dog breeds might have been under stronger human control for developing human-oriented communicative skills (e.g. recent gundogs), whereas other breeds selected for different tasks might not display such abilities. Further, there are some arguments (Hare and Tomasello 2005; Chapter 5.5.3, p. 124) that extant dog breeds represent two stages of evolution. Accordingly, one would expect that breeds that represent earlier stages of evolution might have not evolved such sophisticated communication skills as those breeds that have undergone a selection process for improved working ability. In line with this argument, Hare and Tomasello (2006) report that working dogs (independently of their genetic relationship to the wolf) are better at comprehending a simple human pointing gesture than dog breeds not selected for work. However, the social environment can have an influence on the performance of dogs in this task. In addition,

McKinley and Sambrook (2000) found (on a small sample) that trained working dogs are more skilled in this task than pet working dogs. In addition, the term 'working dog' is often used very loosely because 'terriers', 'sheepdogs', 'protecting dogs', 'sledge dogs' or 'gundogs' are all working breeds but the actual nature of human–animal communication is very different in each case.

In a recent experiment with the two-way choice task we have tested family pet dogs from breeds that are described as cooperative and non-cooperative hunters. Dogs in the former group keep a close contact with the hunter during the hunt (e.g. retrievers) whilst the dogs in the other group work independently either chasing the game (e.g. beagles) or attacking it (e.g. terriers). In the tests cooperative breeds perform better, although all dogs are exposed to a similar human environment (pet dogs). In parallel, we have also found that pedigree dogs show a better performance than mixed-breed dogs, although both are socialized to the same extent in families. At the genetic level this could mean that in mixed breeds selection for such skills has been relaxed.

continues

Box 2.3 *continued*

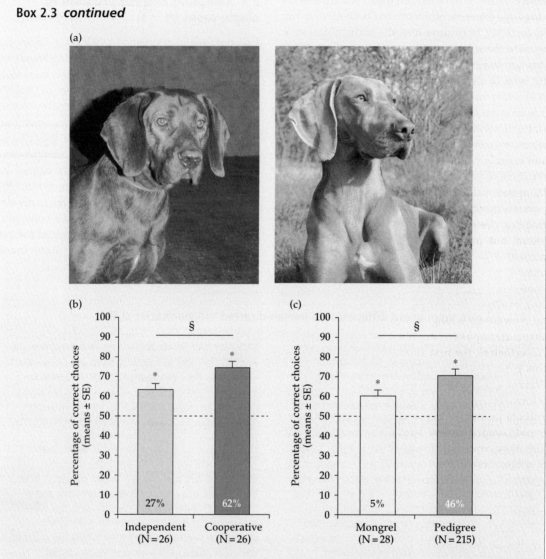

Figure to box 2.3 (a) Two representatives of the non-cooperative (independent hunters) and cooperative hunting breeds. Hanover bloodhound (non-cooperative, left), Weimaraner (cooperative, right). (b) Cooperative hunting breeds are more successful in the two-way choice task than non-cooperative dogs. (c) Pedigree dogs achieve higher level of performance than mixed breed dogs, despite similar socialization history. Dotted line, chance level; *, significantly above chance performance; § significant differences between the two groups. The percentages in the columns show the number of dogs that choose significantly over chance (binomial test, p <0.03, at least 15 correct out of 20 trials) (See Gácsi *et al* (2008).

groups) and perhaps including mixed-breed dogs is the best choice not only in the case of dog–wolf comparison but also when demonstrating 'dog abilities'. However, one must be aware that for physical reasons (e.g. size) certain breeds might be unable to perform the task. Especially in wolf–dog comparison, the use of a single breed should be avoided.

Interestingly, there is a strong bias against research done only on single individuals despite the fact that this approach has been used in psychology, psychiatry, and most fields of medical research. In the animal world apes and dolphins provide the exceptions, the argument being that they are 'rare' species, thus knowledge gained by studying a single individual could be valuable. Why is knowledge gained by studying a single dog not valuable? Actually, we should not complain on behalf of dogs because individuals have often found their place in major scientific journals, although only if they could either 'talk' or 'understand words' (see Johnson 1912, Eckstein 1949, Kaminski *et al.* 2004).

In reality the question is not how much knowledge can be gained from studying a single individual but how this knowledge relates to our present understanding of the phenomenon. In order to show that a biologically important phenomenon exists it is enough to provide convincing evidence in a single individual. People had been aware for a long time that dogs can learn to associate spoken commands with actions. Nevertheless, the first scientific demonstration of this phenomenon was provided by Warden and Warner (1928), who tested a German shepherd which was a show dog used in films. These authors presented convincing evidence for 'command comprehension' because in the systematic tests they controlled for other factors than speech and also provided a statistical evaluation of the results (Chapter 8.4.2, p. 189). Figure 1.5.

This means that if a dog shows indications of a complex skill it is a valid option to carry out carefully designed experiments (see Kazdin 1982 for single-case designs). This happened in the case of a Border collie that was able to retrieve more then 200 objects associated with a particular name (Kaminski *et al.* 2004) and showed rapid learning of object–name associations. However, single-case research is only one way to generate working hypotheses for future studies. Because the history of the subject and its performance is usually not known and there is a limit to the experiments that can be done, such cases are not suitable for detecting mental mechanisms underlying certain complex skills, and for further investigations the number of subjects has to be increased.

2.5 A procedural problem in naturalistic observations: the presence of humans

The ethological study of any animal aims for observations in the natural environment. This means that dogs should be observed under conditions that are 'natural' to them. The most significant compound of the environment for many dogs is the human(s) with whom they maintain a special relationship. Based on this reasoning, we have always observed the dogs in the presence of their owners (e.g. Miklósi *et al.* 2000); in contrast, others avoid the presence of the owner and the dog is managed by a familiar assistant during the experiments (e.g. Call *et al.* 2003).

From a purely methodological point of view, both methods could present problems. If the owner is present the dog will regard the situation as being social and will try to rely on the usual means of interaction. This means that it can be difficult to separate the performance of the dog from the performance of the team (dog plus owner). At the same time, the presence of the owner can make a dog more confident, so that it can maintain the level of performance in a strange environment (just as human infants are tested in the presence of their parent). In contrast, the absence of the owner might generate a fearful state in the dog that interferes with the performance. Thus in this case dogs might need to be habituated to the environment and socialized to the experimenter before the observations.

The presence of the owner can have both direct and indirect effects. Direct effects can surface in problem-solving tasks in which owners might unconsciously give cues that increase the performance of the dogs. This phenomenon, also known as the *Clever Hans effect*, has to be eliminated because it interferes with the goal of the experiment in which the behaviour of the dog should be controlled only by the stimuli provided by the experimenter. For example, it was shown that in search tasks dogs performed better if the owner (handler) knew the location of the hidden item (Becker *et al.* 1962). Although such findings are often interpreted as unintentional cueing by the handler with regard to the location of the hidden item, the presence of the human can be restricted to having only an indirect effect. For example, an informed handler can also influence the dog by behaving in a more 'relaxed' way during the search task, which results in better performance on the part of the dog. In line with this, Topál *et al.* (1997) found that dogs were more active and more successful in getting food by manipulating a lever when the owner encouraged them verbally. Such an indirect effect of the owner's presence might be important when dogs are expected to perform in unfamiliar situations. We should not forget that in most cases these dogs do not 'work for their living' and are not motivated as strongly as other animals tested in a laboratory setting.

The presence of the owner often prompts dogs to communicate if they are put into unfamiliar situations. For example, Scott and Fuller (1965, p. 86) noted that 'in some cases the pups appeared to be trying to figure out what the experimenter wanted them to do'. Such communication seems to be part of the normal interaction, and dogs often do not need to be given any specific signal but only some general assurance that 'everything is OK'.

Thus whether the owner should be present or absent could also depend on the goal of the particular experiment, but probably more emphasis should be placed on having the dog in a naturalistic situation. For example, Scott and Fuller (1965) explicitly reduced and controlled dog–human contact during dog rearing. This might have resulted in dogs which were less disturbed by the absence of particular persons, and were used to the presence of less familiar people. But even in this case one cannot exclude that dogs are influenced by the humans.

In the case of many family dogs it is difficult to exclude the owner, partly because many of them want to know what happens to their pet. In this case it seems to be very important to control the behaviour of the owner and try to prevent them interfering with the experiment in any uncontrolled way. It is also possible to design experiments in such a way that the owner is blind with regard to the experimental question or has restricted perceptual access to the situation (using earplugs or blindfolds). The problem is analogous

to the case of experimental work with 1–2 year old children, where the usual practice is that a parent are also present.

The testing of shelter dogs could present additional problems because of their disturbed social relations with humans. In addition, social interaction with them can rapidly lead to the development of attachment to the experimenter (Gácsi *et al.* 2001). Such procedural problems could become especially important if the goal is to compare the behaviour of shelter and family dogs.

2.6 How to measure dog behaviour?

One key innovation of ethology was to introduce the method of measuring observable categories of natural behaviour which are based on well-described behavioural units defined by their form (Slater 1978, Martin and Bateson 1986). Such catalogues are often organized hierarchically by decomposing functional units of behaviour (e.g. feeding, aggression) into subcategories (flight, fight) and action patterns (bite) (e.g. see Packard 2003). This action catalogue or *ethogram* is then used to record the behaviour in terms of frequency, duration, and the sequence of behavioural units (Lehner 1996). (Unfortunately, the intensity of the behaviour is rarely incorporated in these descriptions, see Fentress and Gadbois 2001.) The application of such a coding system is not easy; observers need to be trained and assessed for reliability. In addition there is no generally useful categorization of behaviour, and often the ethogram has to be redeveloped for particular research questions. Despite all of these hurdles, if applied carefully the ethological method provides the richest description of behaviour. Ethologists advise that at the beginning of behavioural analysis 'splitting' should be preferred to 'lumping' (Slater 1978). Pilot observations can help to reduce the number of observed behavioural variables, or if this is not an option multivariate statistical methods can offer some simplification by introducing secondary variables (see for example, Goddard and Beilharz, 1984, 1985; van den Berg *et al.* 2003). Ethologically derived ethograms for dog behaviour can be found in various studies (e.g. Schenkel 1967, Fox 1970, Feddersen-Pedersen 2001a, Packard 2003; Box 2.4.).

Box 2.4 Behavioural coding in dogs: an example

Various methods have been used to describe the behaviour of dogs. The wide ranging possibilities of describing agonistic behaviour in dogs or wolves are presented in the table below as an example.

Method	Short description	Explanation of the code	Behavioural context used	Main reference
1. Single discontinuous categorical scale	Scaling along a single dimension of aggressiveness	No aggression (1)–threat display (5)	Personality tests	Svartberg (2005)
2. Sum of scores scale	The total score of whether the subject displays an item out of 10 aggressive behaviour elements	Staring = 1 Stiff posture = 1 Bark = 1 . . . Snapping = 1 Total score: XX	Testing for aggression in golden retrievers	van den Berg et al. (2003)
3. Three-way categorization	Each category is characterized by a list of behaviour units	Fight: (chase, face off, holding bite, etc.) Defensive: Bark, crouch, gape, growl, etc.) Flight: (avert-gaze, avoid, crawl, . . ., etc.)	Social interactions in captive wolves	Packard (2003)
4. Independent two-way categorical scaling	A list of 15 behaviour categories is used to classify dominant or submissive state	1. Ears: Erect and forward (aggressive) or flattened and turned down side (fearful/ submissive) 2. Mouth: opened (aggressive) or closed (fearful/submissive) 3. Neck: arched (aggressive) or extended (fearful/submissive) . . . 15 . . .	Not applied	Harrington and Asa (2003)
5. Action centred	The 'position' of head, ear, tail, leg was used to put seven actions (e.g. approach, follow, retreat, etc.) into three categories (low, neutral, high)	E.g. Low posture approach: head low, ears backwards, tail bent low and legs bent.	Social interaction in captive wolves	Van Hooff and Wensing (1987)
6. Pattern coding	The changes at six regions of the face (mouth corner, forehead skin, eye form, etc.) are categorized independently by using region-specific coding categories	E.g. Forehead skin: (A) smooth (B) wrinkled, etc.	Social interaction in captive wolves	Feddersen-Petersen (2004)

continues

Box 2.4 *continued*

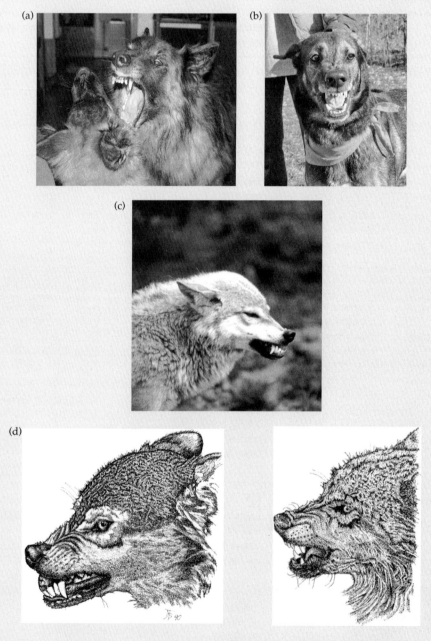

Figure to box 2.4. Characteristic moments of threats in dogs and wolves. (a) Threat displays in Belgian Shepherds. Threatening mixed breed (b) and a socialized wolf (c) (photo by Enikő Kubinyi), and as depicted by an ethologist observer, Feddersen-Petersen (d) (drawings courtesy of Feddersen-Petersen).

In other experimental systems arbitrary categories of behaviour are used mostly because the behaviour of the dog is directed by the experimenter. For example, Scott and Fuller (1965) used five categorical variables with three demerits to describe the behaviour of the puppy during walking on leash (e.g. 'inference with experimenter'). Such behavioural categories are often divided into scores, which could indicate either intensity or presence/absence. The use of such scoring systems often results in adding up scores of different behavioural categories without any real evidence. Thus in this example scores for 'fighting or biting leash', 'vocalization', 'body contact', and so on are added to arrive at a final score of training success (Scott and Fuller 1965, p. 207). The problem is that by doing this we implicitly assume that the different behaviour categories have the same weight in the scoring system. However, how do we know whether, for example, 1 vocalization 'equals' 1 body contact or 1.5 body contact?

When employing a behaviour scoring system, researchers often provide only the range of scores and describe the behaviour only for the extremes (e.g. 1 and 7) and do not give definitions for the categories in the middle range (2–6). A further confusing factor is that in some scoring systems the 'best' score is the median value whereas in others it is the maximum or minimum score.

Other methods, derived mainly from personality research, rely on subjective assessment of dog behaviour. In this case the observer rates the behaviour by means of general descriptors such as 'fearfulness', 'assertiveness', or 'friendliness', which are usually explained by a behavioural definition (Martin and Bateson 1986). Applying this method to dogs, Gosling et al. (2003) found that observers were accurate and consistent in evaluating individual dogs for various behavioural traits. In other experiments and further studies it has also been shown that the judgement of observers predicts future behaviour relatively well and also correlates with objective behavioural measures (Gosling et al. 2003).

This method is based on the well-developed social skills of humans and their ability to process complex behavioural cues rapidly and evaluate individuals on the basis of high-level categories. When used to describe one's own dog this method also offers the advantage that the evaluator can rely on his memory for a very long track record, which is not an option for the observational methods. Similarly, observers can also make a like assessment of dogs in situ by observing an unfamiliar animal for a short period. Thus subjective assessment seems to avoid the difficulty of using the direct observational method to get from the behavioural units to higher levels of behavioural organization. However, unlike observational categories these descriptors are based on a relative scale because scores can depend on the definition provided, on the experience of the rater, and on the relative behavioural difference between the subjects included in the study (Box 2.5).

Box 2.5 Ethological coding and analysis of sequences

There is very little quantitative data about the time pattern of behavioural interactions between dog and human. Such an analysis would need to show that one action by the human is followed in a predictable way by an action of the dog, and vice versa. One reason for the lack of such data is that traditional analyses of such time patterns are very complicated; they can be done only on a large data set and even so the detected pattern is very short.

Recently a novel time structure model (Magnusson 2000) has been developed to detect action patterns in time. This offers a very useful tool for describing dog–human interaction. To do this we staged a simple situation when the owner 'instructs' (by gestural actions and utterances) the dog to help build a tower out of wooden blocks. The human is prevented from getting the blocks; only the dog is in position to carry them to the

continues

Box 2.5 *continued*

human. In this situation, cooperative interactions developed spontaneously, and the dogs' and humans' behaviour was evaluated by a statistical programme (THEME, Magnusson 2000), which looks for significant temporal association among behaviour units (T-pattern) (Kerepesi *et al.* 2005).

In this test 10 dog–owner dyads performed on average 181 interactive T-patterns, which consisted of behaviour units of both the human and the dog. We found also a typical T-pattern in the case of most interacting participants which was the outline of the successfully completed task. The last action of transporting a brick ('dog lets go of the building block') was nearly always part of a statistically validated T-pattern (see below). Thus the dogs were more likely to complete the task if it was preceded by a fine-tuned order of actions, which suggests that this interactive T-pattern did not occur by chance but played a functional role in the task.

(a)

(b)

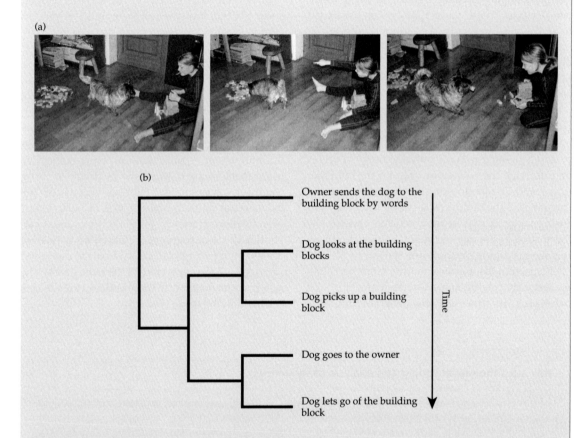

Figure to Box 2.5 (a) By pointing and talk but without verbal command, the child 'requests' the dog to fetch a wood block and carry it over to her. The interaction is repeated until all blocks have been moved over, or after 5 min. (b) Tree-like depiction of one of the interaction sequences (T-pattern) which actually led to successful completion of the owner's request.

In summary, this assessment method is useful when one knows the behaviour of the species, and when an overall characterization of the individual is required (e.g. Sheppard and Mills 2002) but it cannot replace detailed observational analysis of behaviour.

2.7 Asking questions

The fact that many dogs share their lives with humans prompted researchers to look for an alternative (and cheaper) form of data collection by asking the owners. In general, questions target one of four topics:

1 Description and characterization of living conditions (e.g. How often do you walk your dog?).
2 Description of behavioural or personality traits (e.g. Is your dog jealous when you pat another dog?).

3 Description of the perceived relationship with the dog (e.g. Does your dog mind being left alone?).
4 Opinions about certain behavioural traits or abilities (e.g. Could your dog's cognitive skills be equated with those of a 4-year-old child?).

In addition, questions of type 1, 2, and 4 could also be put in a general form asking the owner's opinion on dogs in general or with regard to special breeds (Box 2.6).

Before discussing some problems with this sort of approach, it should be pointed out that asking people about their experience and opinions of their companion animal could be useful for getting ideas. If the possibilities for uncovering problematic issues are limited, such input can be very valuable. However, it should never be assumed without testing that owners, handlers, or other informants provide reliable and valid information (Taylor and

---

Box 2.6 Asking questions about aggression in dogs

Researchers and clinicians have little chance of observing this behaviour directly, and screening for the behaviour in a laboratory setting is also complicated (van den Berg *et al.* 2003). Thus one popular way to collect information on aggressive behaviour in dogs is using questionnaires; however, these differ in the way they obtain information.

There are at least three important dimensions of aggressive behaviour (see also Houpt 2006).

• The competitor could be a conspecific or adult human, or sometimes other less easy categorizable beings such as children or cats.
• The manifestation of aggression depends on whether the dog is at home or away and similarly whether the opponent is familiar (owner, friend) or stranger (male/female)
• The aggressive behaviour has context-dependent properties.

For comparison, we choose a situation when the dog is defending an obtained resource (food or toy) against potential competitors. It is interesting to note that investigators vary in (1)

whether and how they specify the competitor, (2) whether and how 'richly' they describe the aggressive behaviour (compare sections in italics in the list provided below).

Dogs that are not aggressive towards their owners might be so when competing with a stranger. In other cases owners might perceive 'protective or possessive behaviour' as not equivalent to being 'aggressive'. These discrepancies among these questionnaire items could seriously influence data collection, and in addition further distortion could take place if these questions are translated into other languages. Thus in future we cannot avoid some standardization on asking about aggressive behaviour.

Some examples for comparison:

• Line and Voith (1986): Situations in which dogs were *aggressive* (*bared teeth, growling, snapping or biting*) *to owners* (1) took objects and guarded them, (2) food was taken away. (yes/no).
• Podberscek and Serpell (1996): Is the dog *aggressive* at meal times/defending food (yes/no)

continues

Box 2.6 *continued*

- Jagoe and Serpell (1996): *Aggressive* at meal times (in a checklist for behavioural problems) (yes/no)
- Podberscek and Serpell (1997): Was the dog *possessive/protective* of objects? (yes/no); Was the dog *aggressive* when its food was approached? (Score: 1(low) ... 5(high))
- Guy *et al.* (2001a): Does your dog ever *growl* or *snap at anyone* when they try to take away food, toys, or other objects? (yes/no)
- Guy *et al.* (2001b): Does your dog *ever respond to any* of the following situations by *growling, lifting a lip, snapping, lunging,* or *biting*? (1)

touching its food when it is eating; (2) walking past its food when it is eating; (3) adding food to the dish while it is eating; (4) taking away a bone, rawhide, or toy; (5) taking back an object it has stolen (such as a sock) (yes/no).
- Sheppard and Mills (2002): Your dog *becomes aggressive* (i.e. *growl, snap* or *bite*) if *you* try to remove its favourite toy or food. (Score: 1(low) ... 5(high))
- Hsu and Serpell (2003): Dog acts *aggressively* ... when toys, bones, or other objects are taken away *by a member of the household*. (Score: 1(low) ... 5(high).

(a)  (b)

Figure to Box 2.6 There are two ways to maintain control over a possession: (a) The dog threatens the human who tries to take the bone. (b) An alternative tactic is to take away the protected object. Note that the later is effective in avoiding conflicts.

Mills 2006). Information collected by questionnaires can turn out to be very useful for formulating hypotheses, but this indirect method should not used to replace methods relying on direct observational evidence.

- *Problems with the sample*: Questionnaire studies are based on very diverse human populations (readers of a dog magazine, internet users, visitors to vets, university students, any group of dog owners or professionals, e.g. dog handlers, trainers, behaviour counsellors); however, only very rarely is it made clear why the particular sample was chosen as reference. Various biases can distort the results in many ways. For example, readers of a particular dog magazine might have a particular attitude to dogs.
- *Problems with causality:* The findings of many questionnaire studies suggest that some environmental

factor or variable correlates with behaviour. Although researchers are aware that such correlations never refer to a causal relationship, this might mislead someone less knowledgeable. For example, finding that aggression correlates negatively (Podberscek and Serpell 1996) with grooming could either mean that people avoid grooming aggressive dogs, or that dogs are more likely to become aggressive if they are not groomed.
- *Owner biases*: The cooperation of owners might depend on their relationship with the dog. A more 'satisfied' owner is more likely to respond and might also provide a more positive picture of their pet, and the negative aspects of the relationship (e.g. biting) are less likely to be reported honestly. The comparison of two or more populations of dogs also reflects two or more different populations of owners. Thus any difference in the dogs

could be due to differences between the dogs, the owners, or both. For example, based on owners' answer to a questionnaire, Serpell and Hsu (2005) report that 'field' Springer spaniels have a better trainability than 'show' Springer spaniels. This is a quite straightforward interpretation of the results, but it could be also that owners of 'show' Springer spaniels never bothered to train their dog, and/or owners of field dogs are more inclined to report higher levels of trainability just because it is expected from this bloodline.

• *Folk knowledge*: Very often even researchers rely on general folk knowledge of dog behaviour, which can lead to very confusing results. One such misused concept is that of 'intelligence' which was implicated as being different in various breeds (Coren 1994). Careful reading of the original questionnaire shows that by 'intelligence' the author means 'obedient behaviour at dog school'. Even if this was the original intention of the investigators, we may well wonder how easy it would be to train the top-ranking Border collie to pull a sledge for 10 km (Coppinger and Coppinger 2001). Similarly problematic is the comparison of breeds for trainability on the basis of questions that refer to a particular kind of behavioural response. Thus it is not surprising that Siberian huskies and Bassett hounds scored low on a 'trainability' questionnaire which had an item on 'fetching objects' (Serpell and Hsu 2005).

In summary, even if done with care, questionnaire studies can only give an initial hint about the nature of phenomena or problems; they are by no means the solution. Despite recent suggestions these methods have actually very little 'ethological validity' (Notari and Goodwin 2006), and do not have the potential to replace observational and experimental studies.

2.8 Conclusions for the future

This overview of methodological issues indicates that researchers interested in dogs have access to a complex array of tools in order to design experiments that provide answers to Tinbergian questions (Chapter 1). Comparative work, if done carefully, can reveal the function of behaviour, as well as its particular role in the evolution of dogs.

Deliberate manipulation of the actual or developmental environment of the dog could provide a means to study mechanistic questions. In the case of the current environment, repeated systematic observations in problem situations could help in developing a more detailed ethocognitive mental model for the dog (see Chapters 7, 8, and 10). With regard to the developmental environment, investigations on the effect of specific early experience could reveal the influence of the environment on the later expression of behaviour or performance (Chapter 9).

In considering the methodological problems it is important to realize that we know (in terms of scientific validated knowledge) much less about dogs than many of us suppose. There is an urgent need for a much better understanding of methodological problems with the aim of increasing standardization, in the hope that this research field will expand shortly.

Further reading

Lehner (1996) and Martin and Bateson (1986) provide a very good introduction to the ethological method. Kazdin (1982) gives a good introduction into single-case studies which could be helpful in planning such experiments. Cheney and Seyfarth (1990) is a thought-provoking book on how to combine field and laboratory methods for probing into the animal mind, although the subjects in this case are monkeys.

CHAPTER 3

Dogs in anthropogenic environments: society and family

3.1 Introduction

It is only recently that people have begun to think about dogs in terms of populations. Interestingly, the division of the dog into smaller populations often parallels groupings in human society. Very often these subpopulations of dogs are described in isolation, but in reality they are not closed pools. Family dogs, working dogs, or free-ranging dogs are all representatives of the dog as a species, and individuals have the chance to move in this complex network of subpopulations.

Dogs have to follow humans in many aspects of social behaviour. In addition to forming attachment relationships with group members, they have to be able to develop new social relationships rapidly, capitalize on short-term contacts, and be socially tolerant or even ignorant if required. Failure in these forms of social contact reduces the chance of success.

The association between dogs and humans is one of the few cross-cultural features of human societies (Podberscek *et al.* 2000), although some traditions or taboos suppress the public expression of human affection. Even in the most 'dog-loving' societies a considerable part of the human population does not develop individual social relations with dogs although they cannot really avoid regular contact with them. For some dogs the situation is just the opposite. Although in most places dogs are more or less part of human society, there are populations which live outside the boundaries of human-dominated environment. With the increasing burdens of modern society, discussions on how to achieve peaceful dog–human cohabitation intensify. However, any discussion or planning can only

be done on the basis of scientific data, of which there is a huge lack at the moment.

Thus scientists from various disciplines have to work together to develop observation methods and collect comparable data to change this situation. There is a need to collect more data on the population biology and dynamics of both family dogs (Box 3.1) and free-ranging dogs (Beck 1973), and for similar reasons ethologists have a duty to present a behavioural description of dogs in mixed human groups, including working dogs and dogs living in animal shelters. Human environments offer an unexploited source for descriptive observations by 'field ethologists'.

The intense debates on whether people's relationship with their dogs is beneficial or disadvantageous for modern society often obscures the fact that at present dogs provide one of our last contacts with nature. Understanding this species, which has evolved side by side with us, could be important for understanding our broader relationships with the living environment.

3.2 Dogs in human society

Dogs are present in almost every human society around the world. In parallel with the history and present organization of these societies, the role of dogs and their involvement in the economy or culture varies tremendously. Although most people refer to the extreme variation in the appearance of dogs with regard to size, looks, and behaviour, this is only rarely put into the perspective of the manifold relationships that exist between humans and their dogs. The problem is that the role of dogs can be studied from many different aspects, and researchers coming from different disciplines have different goals and use different methods.

Archaeological investigations aim to reconstruct the historical aspects of the relationship (Chapter 5.3.1, p. 101). This work is constrained by the limited amount and uneven distribution of remains found. Thus biases in dog–human relationship either over a period of time or with regard to geographic distribution could be the result of differences in the richness of the archaeological material recovered. Most early

dog fossils come from human burials, which might be indicative of a special relationship, but it may also be that human burials are over-represented in the archaeological record for certain locations and historical periods. Morey (2006) argues for the former case, suggesting that early humans had intimate bonds or mystical/sacral relationships with their four-legged companions. The distribution of dog burials, which are present in most parts of the historical world and originate over an extended time period, could signal that dogs have been 'at least'

Box 3.1 Surveying dog populations: a case from Sweden

In order to provide a background for behavioural studies, as well as supporting the management of dog populations in general, it is important to collect demographic data. Such information can help to resolve the problem of whether a certain population under observation or being examined experimentally is a representative sample of dogs. At present there are only very crude estimates about the nature of the dog population in most countries. Egenvall and co-workers (1999, 2000) published a number of studies reviewing the Swedish dog population from the veterinary perspective, but they also collected data on more general aspects of the dog population which could be also of interest to ethologists. Similar data for

different dog populations could be very useful in estimating the reference population from which dogs are sampled for observations and experiments.

The table below lists the 10 most popular breeds in 3 countries, based on the registrations with the national kennel club in 2005. Although within countries the preferences do not change within a few years, there are considerable differences among countries, except that retrievers, German shepherds and boxers are always on the list. Interestingly the top 10 breeds represent around half of the total registered dogs. Also, the most popular breed has at least double the number of dogs compared to the second most popular breed.

	USA	%	Germany	%	England	%
1	Labrador retriever	15	German shepherd	20	Labrador retriever	17
2	Golden retriever	5	Teckel	8	English cocker spaniel	7
3	German shepherd	5	German Drahthaar	3	English springer spaniel	6
4	Beagle	5	Labrador retriever	3	German shepherd	5
5	Yorkshire terrier	5	Golden retriever	2	Staffordshire bull terrier	5
6	Dachshund	4	Poodle	2	Cavalier King Charles spaniel	4
7	Boxer	4	Boxer	2	Golden retriever	4
8	Poodle	3	Deutsche Dogge	2	West Highland white terrier	4
9	Shih tzu	3	English cocker spaniel	2	Boxer	4
10	Chihuahua	3	Rottweiler	2	Border terrier	3
	Total	52	Total	46	Total	58

(For the year 2005)

continues

Box 3.1 *continued*

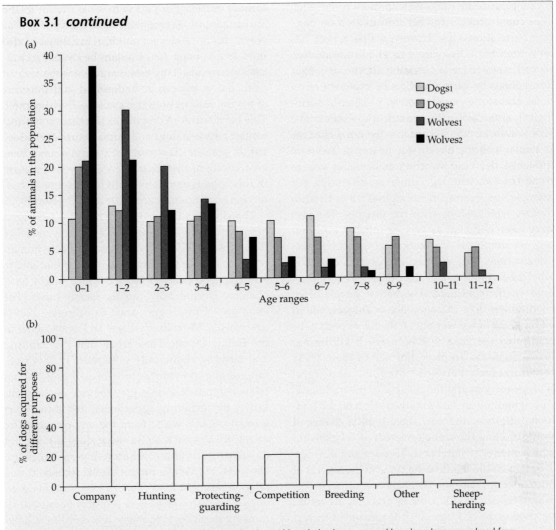

Figure to Box 3.1 (a) The age distribution of dogs and wolves. Although the data presented here have been reproduced from different sources they indicate marked differences in the age structures of the two species. Wolves1: collected for the 1991 report on the southern Yukon wolf population (radio-collared or killed wolves) (Hayes *et al.* 1991); Wolves2: live captured wolves in the Denali National Park, 1986–94 (Mech *et al.* 1998); Dogs1: based on a representative sample of the Swedish population (Egenvall *et al.* 1999); Dogs2: based on a sample of dogs presented at veterinary clinics in Canada (Guy *et al.* 2001c). (b) Purpose of acquiring dogs in Sweden. Dogs have to fulfil various social and working roles (data from Egenvall *et al.* 1999).

treated as members of the group or family, entitled to the same obsequies as humans.

The other extreme of the relationship might be represented by the use of dogs as a source of food (Podberscek 2007). In the archaeological record broken bones, bones with gnawed ends, and cut marks are usually regarded as evidence for butchery. Accordingly, dogs were part of the human diet for, example, in prehistoric central Europe until the Bronze Age (Bartosiewicz 1994), in the historic Maya culture of Mexico (Clutton-Brock and Hammond 1994), among the Maoris of New Zealand (Clark 1997), and also in Australia (Megitt 1965).

Comparative investigations in more recent societies could provide further information on dog–human relationships. However, this aspect has only rarely been the focus of cross-cultural studies, so such knowledge is based on shorter or longer descriptions by early travellers or explorers or on notes and stories mentioned in passing in sociological, anthropological, or cultural studies. The lack of such work is sad because the rapid changes in human culture, mostly due to strong 'Western' influence, decrease the chances of being able to reconstruct ancient dog–human relationships. For example, the Australian Aboriginal tribes lived in diverse relationships with the dingoes. Dingoes were eaten and kept as pets, or utilized for hunting, or simply as blankets during cold nights (Megitt 1965). This situation changed dramatically after the Europeans and their dogs gained a foothold on the continent. The native people often choose these dogs in preference to dingoes, and in addition the dingo was put on the list of pests to be eradicated (because it is accused of killing too many domestic animals, but see Corbett 1995). Hybridization between dingoes and feral (European) dogs and the collapse of the traditional Aboriginal culture had a marked effect on the traditional lifestyle, and now there is little chance of reconstructing the complex forms of relationship which once existed between humans and dingoes.

Recent cultures reflect the three main aspects of human–dog relations. (1) Dogs are utilized like other domestic or wild animals and they are raised for food or pelts. Dogs are still consumed in east Asia (Podberscek 2006) but have been part of the diet in many other parts of the world up to recent times. Dogs living in social relationship with humans are either (2) working companions and/or (3) provide emotional and social support ('pets'). In some communities cultural and religious customs forbid close association between dogs and humans, and people are rather passive in the relationship. But even here individuals (especially children) form strong working or emotional relationships with dogs. For example, on many Polynesian islands dogs are nursed like children, and then given to a child. The dog's soul is said to protect the infant and if the child dies the dog is often buried with it (Fisher 1983). The dog may also be used as a

form of currency or part of magical rituals. In the Turkana (north Kenya) dogs have the same role as playmates or nurses for children but they are also used as a 'sponge' for cleaning the child if it defecates or vomits. This may seem a strange way of using a dog, but can be understood in the context of having little or no water available (Nelson 1990). This habit survived despite the fact that such direct contact between dogs and humans carries a heavy risk of parasite (*Echinococcus*) transmission. There are indications that the incidence of hydatic disease in this tribe is associated with the amount of contact between humans and their dogs (Nelson 1990).

This diversity in dog–human relations has urged many researchers to search for a primary model of domestication. However, present-day dog–human relations may be the result of various evolutionary, ecological, or cultural factors, which might have changed periodically during recent times. For example, Coppinger and Coppinger (2001) described a 'Mesolithic village' on Pemba Island in the Indian Ocean. They argue that this hunting and farming community with more or less free-ranging dogs provides a model of early dog–human relationship where dogs play the role of commensalists by removing superfluous and dangerous human organic waste from the environment (by eating it). People tolerate these dogs but do not develop individual relationship with them. However, the Pemba people are Muslims and this religion strongly discourages close relations with dogs. Dogs are seen as evil, probably because they transmit parasites to humans. It is very likely, however, that this distancing between people and dogs has been a secondary development. Indeed it might be that these 'laws' or taboos were needed to deter people from showing their natural affection for dogs in order to prevent the spread of disease in the population where other preventive measures are not possible. This is also supported by anecdotal reports that some people like these dogs and even pet them if unobserved (Coppinger and Coppinger 2001).

Others argue that the dog's way into our society was paved by our devotion to all kinds of animals, and the hobby of pet-keeping. Actually, keeping pets (not only dogs, but the offspring of other species as well) was perhaps useful for people in learning about animals, which could have been advantageous in hunting societies (Savishinsky 1983), and might have contributed to their success. The traditional view of dog domestication emphasizes their role in hunting (Clutton-Brock 1984), by arguing that many hunting tribes keep various pets, including wolves or dogs. However, this does not provide direct evidence for the sequence of events, that is, that hunting with dogs developed from pet-keeping.

Given the limited amount of truly comparative data on the dog–human relationship in different cultures, it seems too difficult to select the primary model for ancient dog–human societies. Both archaeological evidence and present cross-cultural comparisons suggest that this association was very diverse from the beginning, and depended on the ecological conditions, as well as on the social and cultural organization of human societies (Chapter 5). Importantly, the role of dogs was not immune to changes during the course of human history. Recent history provides strong support for this, for example hunting or sledge-pulling dogs 'becoming' pets.

In the absence of comprehensive research on dogs in human populations worldwide, the following discussion is largely based on those societies where dogs are kept mainly as pets (including dogs in a working relationship) in a family setting, but we should not forget that the formation of other populations is also possible. In these societies dogs typically belong to a human family and/or have an owner who provides regular care and shelter and contributes in various other ways to the well-being of the dog. A high proportion of these owned dogs receive regular veterinary care (e.g. vaccination) and/or are registered with the local authority (if the law requires it), and special organizations are devoted to different aspects of dogs in the society (e.g. kennel clubs, association of dog trainers, etc.).

In a series of papers, Patronek and co-workers have provided a descriptive model for dog populations that cohabit with humans (Patronek and Glickman 1994, Patronek and Rowan 1995). The central unit of this model is the household, which provides the physical and social environment for the dogs. The number of dog-owning households varies considerably across countries; for example, it

is estimated to be around 40% in Australia (Marston and Bennett 2003) but only 14% in Austria (Kotrschal *et al.* 2004). The size of the dog population living in human households depends on many factors, such as the level of urbanization, historical traditions, or the current economic state of the country. In any case it is assumed that most dogs are associated with families (see below) and only a smaller portion of the total population live as free-ranging dogs (Chapter 4.3.2, p. 86) without individualized human contact ('owner'). The introduction of animal shelters aimed to reduce the population of free-ranging dogs which can cause economic damage (attacking domestic stock) or health problems (transmitting disease), and can be harmful to wildlife. Although many think of animal shelters as necessary institutions for regulating dog populations, people may be reluctant to give their dogs to shelters and release them into the wild instead. This practice is dangerous and could be considered inhumane ('incanine'?), but can be understood considering the fate and quality of life of many dogs in shelters (see below). In many countries a considerable proportion of dogs live (and die) in shelters, which should however be regarded as a necessity, and not a solution to the problem of ownerless dogs (Box 3.2).

3.3 Interactions between dogs and people in public

Living in the same society, both dogs and humans have to take their part in forming groups which can function under extreme situations, even if the actual group structure is different from the original one. Naturally both dogs and humans live in more or less stable family groups and are territorial. However, at present the social and physical dynamics of humans and their groups is radically different. People occupy overlapping and/or physically discontinuous territories, they are members of different groups at the same time, show tolerance to strangers, and form short-lived associations with groups varying in size. Thus dogs should be able to express similar social attitudes in behaviour in order to become integrated into human society. Most of these challenges can be overcome by an appropriate socialization process (Chapter 9.3.3, p. 207),

Box 3.2 A model of the dog population

Patronek and Glickman (1994) introduced a population model for dogs by analysing data for the USA. In principle this model could easily be generalized to other countries, and provides a useful tool for between-country comparisons. If such data were supplied (or collected) continuously, it could also show changes over time. It could also be used for forecasting, helping people managing dogs (breeders, veterinarians, shelter managers) and regulators. But even in its present state the model highlights some important problems. For example, 1 in every 10 dogs comes from a pet store, which is probably a high rate (0.5 million dogs) and is probably more typical for the USA than for Europe. There are more dogs surrendered to shelters (1.4 million) than shelter dogs finding new homes (1 million). The actual number of dogs in US shelters is

10 times the world's total estimated wolf population! Based on US surveys, Patronek and Rowan (1995) estimated an approximately 12% birth and death rate in dogs, which indicates that every 8th dog in the population is replaced yearly.

Country	Dogs in % of households	Estimated dog population (million)
Australia	40	4
Austria	14	0.6
Germany	20	8.8
Great Britain	15	6.8
Sweden	15	0.8
Switzerland	15	0.5
United States	34	52

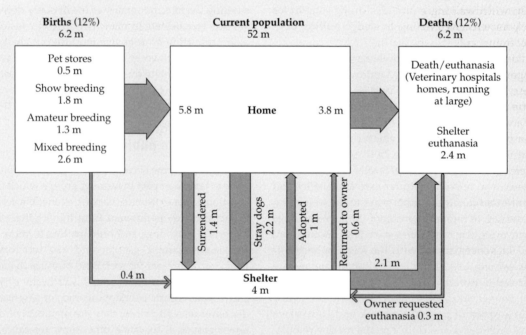

Figure to Box 3.2 Schematic model of dog population based on data from the USA (redrawn and modified from Patronek and Rowan 1995).

but not realizing the dynamic nature of modern human living could cause a lot of problems for inexperienced dogs.

Surprisingly little is known about the behaviour of dogs in public. Although there is a strong tendency to constrain the free movement of dogs in public places by making it obligatory to use the leash, Bekoff and Meaney (1997) found that in general off-leash dogs induce a manageable amount of problems to 'off-leash' humans. This emerged from the responses of both dog owners and non-owners, and from observation of the interaction between dogs and people. Most dog–dog (81%) and dog–human (85%) contacts were friendly or neutral, and only a smaller proportion of dog–dog encounters were described as aggressive. The presence of dogs in public places also facilitates interaction between people, and often led to conversation among strangers. One experimental study investigated the reaction of passers-by towards a human who walked with various 'things'. Not unexpectedly, when walking an adult or puppy Labrador the person received frequent visual or verbal attention from strangers who initiated social contact by looking, smiling, stroking the dog, or conversing. Importantly, inanimate objects (e.g. a teddy bear) were much less useful for this job, and similarly little interest was evoked by a Rottweiler dog (Wells 2004). These observations provide evidence that people are very sensitive to the image of dogs and find them generally attractive. However, this veers round if owners are seen with very large-sized dogs, or a dog that belongs to (or is similar to) breeds that have a "bad" reputation.

Instead of more constraining and alienating laws, more emphasis on the education of people and dogs could have a liberating effect on both species, leaving more space for free social interactions and experience.

3.4 Dogs in the family

Many people assume that dogs can easily adapt to live in human families because their ancestors also lived in similar social structures. Although it is true that the composition of a wolf pack and a human family have much in common, there are also large differences (Box 8.1). One of the major difference between wolves and their domesticated relatives is that, based on genetic factors and everyday social experience, dogs but not wolves are able to learn how to become integrated members of a human group. Similarities and differences between the dog and human family life lead to a lot of confusion, but here we restrict ourselves to demographical and some psychological aspects of dogs in the family (see Chapter 8.2, p. 166, Box 4.6).

We look at the family as the minimal social unit from the dog's perspective. Thus sharing its life with a person constitutes a 'family', just as two canids could be said to form a 'pack'. The function and role of dogs in the family have been investigated mostly by the use of questionnaires asking people about their pet-keeping habits, opinions about their pet's mental abilities, and their perceived relationship to the animals in the context of economic and social variables (Albert and Bulcroft 1987, 1988). Many studies suggest that dogs are still the most popular pets, thus their relationship with humans should be regarded as typical. It was recognized very early on that dogs play an important role in family life and are organic members of these groups (Cain 1985, Cox 1993). This is also reflected by the answers of family members to such questions: about 65–80% of the respondents regard their dogs as family members (Cain 1985).

Most studies agree that dogs are acquired for two main reasons. There is a general belief that dogs make good companions for older children (Albert and Bulcroft 1987, Edenburg *et al.* 1994), and there is both direct and indirect evidence that people in need of emotional support are also more likely to own a dog. This complements findings showing that people who have cared for a dog when young are more likely to have dogs in their family. This is also reflected in the motives for acquiring dogs, because the presence of older children and the lack of companionship are the foremost reasons (Edenburg *et al.* 1994, Arkow and Dow 1984). Similarly, Katcher and Beck (1983) assume that dogs (and pets) can provide certain emotional aspects of a social relationship for humans who do not receive this from their fellows.

Box 3.3 Do we like dogs or not?

People's reactions to dogs can be very different, and often depend on circumstances. Often dogs help in making people attractive to other people. This 'catalysing' effect of dogs has proved to be important as an additional benefit to people who need to rely on dogs for help. People living with disabilities are at disadvantage in society because of their limited physical abilities and the ignorance of others. Although modern technology can offer a lot of practical help for people living with disabilities, it seems that helper dogs have the additional advantages of catalysing the interaction between their owners and other members of the community and supporting emotional well-being.

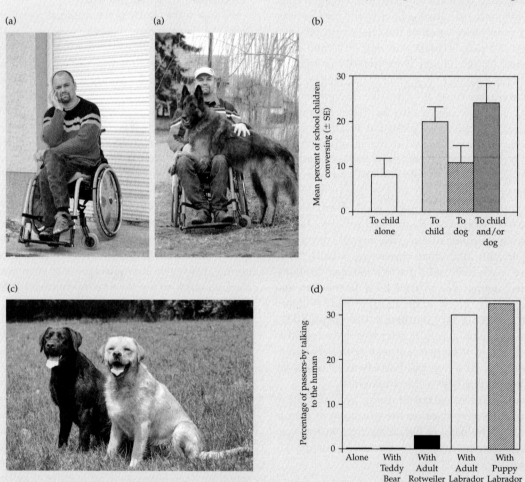

Figures to Box 3.3 (a) People with dogs are usually found to be more attractive which can be explained by the social facilitation effect of the animal. (b) Mader *et al.* (1989) have found that schoolchildren in wheelchairs were addressed more often verbally (direct social interaction) and experienced more friendly glances and smiles (indirect social interaction) from members of their social group. (c) The appearance of the dog also plays a role in the facilitating effect. People expressed clear preferences for dogs having long blond hair, show a tendency to approach them, or play (Wells and Hepper 1992). Most of such preferences are probably learnt and are strongly influenced by fashion trends and individual experience. (d) People with dogs and dog puppies are to be seen as more approachable and passers-by contact them more often directly (conversing) and indirectly (look, smile). Interestingly, the dogs belonging to breeds having a 'bad' public reputation do not have this effect (Wells 2004).

Thus it is not surprising to find that in the USA dogs are most likely to be present in families with children of preschool or school age (Albert and Bulcroft 1987), and about one fifth of these families have at least one other dog. Economic analysis showed that these families tended to have a higher income, but the willingness to spend money on a dog does not seem to be related to the amount earned. Importantly, in these families there was some trade-off between having infants and the presence of dogs. Dogs were relatively rare in families with very young children, in comparison to families before the birth of children or after children left home. The emotional bond between dogs and adult family members is weakest when older children are present in the family, indicating that during these years the main role of dogs is to be playmates. Studies indicate that dogs have a positive impact on the sociability and self-esteem of older children, although such correlative findings should be treated with care (Covert et al. 1985). Similar findings have been reported on dogs as emotional support (e.g. Salmon and Salmon 1983). Thus the effect and importance of dogs changes with the life cycle of the family.

The inclusion of dogs in the family network of relationships (Furman and Burhmester 1985) provided further support for their significant role. Bonas et al. (2000) asked people to quantify different aspects of the inter-individual relationships (e.g. companionship, intimacy, conflict, alliance, etc.) in the family. They found that dogs had been integrated into the web of family relations. Dog–human relationships showed higher scores for companionship, nurture, and reliance than human–human relationships. The opposite tendency was true in the case of affection and admiration. Generally, the negative aspects of relationships obtained lower scores for the dog–human than the human–human relationship. Thus the relationship with dogs often plays a compensatory role, that is, people establish a close relationship with dogs to compensate for low satisfaction they get from other family members (Bonas et al. 2000). Based on such observations some sort of anthropomorphism towards dogs is to be expected, and indeed there is evidence (from questionnaire studies) that a considerable proportion of dogs sleep on their owner's bed (35%), are allowed on the furniture (55%), get food from the table (20%), are talked to (30%) and enjoy a birthday party (30%) (Voith et al. 1992).

The fact that dogs are regarded as family members is also reflected in the negative aspects of the relationship (see also Hart 1995, Podberscek 2006). Aggressive interactions between humans and dogs can cause conflicting situations because people are attached to the animals but at the same time they are concerned about the future of the relationship (see below). The death of a dog can release emotional outbursts which are comparable to the loss of a human friend (e.g. Steward 1983).

The life of a mixed-species family also depends on the environment. One questionnaire study found that both the dogs themselves and their relationships with family members differed according to whether they lived in cities or rural areas of the Czech Republic (Baranyiova et al. 2005). Urban dogs tended to be smaller and more fearful, growled more often at family members, and showed more frequent mounting behaviour. They were allowed to sleep in beds, enjoyed vacations with the family, and had birthdays more typically than was the case for rural dogs. Urban people who regarded them as companions had more intense contact with their pets. It seems that in urban environments people may be more tolerant towards their dogs and attune themselves more to the behaviour of dogs; however, this attitude can also lead to problems.

The role of dogs in human families is emphasized by exceptional cases when people with little chance of joining a human family establish a social relationship by voluntarily adopting a dog. A preliminary study of homeless people in Cambridge (United Kingdom) indicated that these people took on a dog despite the fact that they gained little if any advantage from this relationship (Taylor et al. 2004), and more often the presence of the dog made their life harder. There is little evidence that the companionship of these dogs increases donations, although they can be useful as a night guard. However, there are also costs associated with such pet-keeping because homeless dog owners are not allowed into community shelters or hospitals with their animals.

Most of our present knowledge of the life of dogs in families is based on studies using questionnaires or other interviewing methods. Although these are well-developed methods for gathering certain type of information, much of this remains of doubtful value unless the results are supported by direct behavioural observations (Chapter 2.2, p. 28). A pioneering study revealed behavioural differences between dogs and cats by observing them in family settings (Miller and Lago 1990). The dogs interacted more frequently with their owner in the presence of strangers, and they initiated more contact with the strangers. Dog owners also gave more orders to their dogs. The frequency and kind of interaction between dogs and dog owners (in comparison to cats and cat owners) might actually underpin differences in attachment levels of humans towards their respective pet. Although precise behavioural observations are difficult to carry out, they seem to be a necessary complement to questionnaire studies.

3.5 Dogs at work

According to most theories of dog domestication, the working relationship between dogs and humans is present from the beginning (Clutton-Brock 1984). Even if we do not have indisputable evidence, hunting or guarding work was probably part of the lives of many dogs 8000–10 000 years ago. The tasks of dogs became more diverse in agricultural societies, and there are indications that dogs were specially bred for hunting, herding, guarding, or acting as war-dogs (Brewer *et al.* 2001). The actual economic value provided by these animals is difficult to judge, but using dogs in herding large groups of sheep or cows could save considerable human work. This period was followed by further diversification with the improvement of human hunting techniques, although by this time hunting had become more of a sport or a hobby than a necessity for sustenance. Modern societies have developed many novel roles for dogs. They cooperate in law enforcement (police dogs, border patrol dogs), help in search and rescue work, or assist people living with various disabilities. Some dogs provide emotional support for lonely people, or assist as mediators or catalysts in psychotherapy, especially for children (Hart 1995, Mader *et al.* 1989, Wells 2004, Prothmann *et al.* 2006).

Many books have been written on how to breed, socialize, and train dogs for these tasks, but in fact very little is known about the life of these animals. Not only are demographic data difficult to find but there is also a lack of observational studies. Adams and Johnson (1995) shed some light on the average days and nights of guard dogs. They observed interactions between dogs and people and also described the behavioural patterns of the dogs during their duties. Owners of premises equipped with guard dogs suffered less damage, so the dogs seemed to fulfil their deterrent role. Behavioural observations showed that this effect can be explained by the mere presence of these relatively large dogs (e.g. German shepherds, Rottweilers) and not because they behaved aggressively towards people. Although these dogs protected their territories against other dogs, they were more likely to back off if approached by human strangers. There was also a difference between dogs living continuously on the site and those working there only for a given period. The former were more likely to regard their working place as their territory, and showed more intense defence behaviours. Most dogs were more active during the day but they were generally very alert, and responded to various stimuli during the night, including barks of the other 'colleagues'. There are many aspects of guard dog life that have been not revealed by this study, but similarly studies on herding or hunting dogs are curiously lacking.

3.6 Social roles of dogs in human groups

Although people have been aware of the advantageous effects of dogs on individuals for a long time, research has not supplied supporting evidence (for a review see Hart 1995). But then interesting insights emerged from two different aspects of dog–human interaction. Levinson (1969) was among the first to suggest that dogs might be a useful medium for treating emotionally disturbed children and adults. Studying the survival rates of patients with coronary heart disease, Friedmann *et al.* (1980) found that dog owners (as well as pet owners in general) were more likely to be alive after 1 year. Both studies initiated research into the issue on direct and indirect health benefits of dogs. Such benefits can be categorized either on the basis of their nature or on the duration of the effect; for example, Hart (1995) distinguishes physiological and psychological benefits and effects on general health (see also Friedmann 1995). An alternative, perhaps more ethological, view would emphasize the role of the dog as a social stimulus. Thus direct social ('beneficial') effects (whether short- or long-term) could be related to the presence or absence of a group mate. Contacts with dogs can either revive deteriorating social relationships or increase the intensity and richness of existing social contacts. This also includes particular cases when dogs assist in developing or healing malformed social behaviours which either did not form in the first place or were retarded (e.g. therapy dogs for people living with autism).

Often, dogs replace some aspect of a typical social relationship. The effect of these companions is based on the same mechanism whether dogs are playing with children who have little access to pets (Bryant 1990), or are brought into contact with elderly people who have restricted human social relationships (Bernstein et al. 2000). Basically, a similar mechanism is at work when dogs act as a kind of catalyst between a group of people and lonely individuals. Dogs facilitate disabled children or adults becoming part of a social group; and the animal places them immediately into the attentional focus of others (Mader et al. 1989) (see Box 3.3).

Viewing the effect of dogs as an enrichment of social contacts also draws our attention to the fact that in order to be stable and supportive over time, social relationships need to be constantly reinforced by both parties. This could be a problem if the person concerned has little or no control over the means to express and support continuous interest in the dog. In such cases long-term effects can only be maintained by constant reinforcement of the relationship, which must be supported by outsiders such as parents, nurses, or therapists if the participant fails to do so. The lack of such help leads to rapid habituation, and the socialization effect evaporates (Banks and Banks 2005).

Social contact with or separation from group mates is often accompanied by physiological changes underlying emotional behaviour. The presence of dogs often has a calming effect which is also reflected in lowered blood pressure, heart rate, and skin conductance (Friedmann 1995, Wilson 1991, Allen et al. 1991). Thus dogs (like some other pets, or humans) exert their effect on people through mechanisms which control stress and alertness. It is not surprising that in certain situations members of a social species feel less stressed when enjoying the companionship of familiar group members. Being in a group also reduces the need for vigilance, which also leads to lower levels of stress. Interestingly, in the case of humans and dogs these effects are symmetrical to some extent; that is, humans have a similar stress-reducing effect on dogs (indicated by decreased heart rate), especially if the social contact is reinforced by tactile stimulation such as patting (McGreevy et al. 2005). Measuring the levels of cortisol, Tuber et al. (1996) found similar stress-reducing effect of humans in shelter dogs.

Indirect effects are those which could in principle be replaced by other means. For example, dogs are often reported to improve the health of their owners by 'forcing' them to do more physical exercise (Cutt et al. 2006). Dogs may well cause owners to take more exercise, although there are other means to the same end, such as gardening or jogging.

3.7 Social competition in dog–human groups and their consequences

Social competition is a natural way of distributing resources among group members. Importantly, aggressive behaviour is aimed at getting access to valuable items, or preventing the access of others. An individual may also act aggressively if it perceives a social situation as threatening its integrity. Aggressive behaviour consists mainly of ritualized behavioural units which evolved for signalling the inner state and physical potential of the contester, and does not aim at causing damage in the other. Nevertheless in many species aggressive behaviour includes elements that may cause physical pain (body hitting) or lead to injuries and wounds (e.g. clawing, biting).

Aggressive interactions are part of the everyday life of social animals, including mixed-species groups of dogs and humans (see also Chapter 8.3, p. 170). Although this situation seems to be quite natural for an ethologist, the enhanced media interest in 'dangerous dogs', pro- and anti-dog lobbies, and the contradictions in the scientific literature make this field very problematic (Beaver 2001, Overall and Love 2001).

3.7.1 Aggression and the human family

Human ethologists argue that the human family represents one of the most peaceful associations of individuals in the animal kingdom. This seems to be an evolutionary trend, because humans also show markedly reduced aggressive behaviour towards other group mates in comparison to our (living) primate ancestors. Many assume that this change also enhanced our possibilities for forming complex alliances, and engaging in sophisticated collaborative activities. This means that humans are very sensitive to any kind of aggression which could seriously disrupt group activities.

On this basis we can assume that during the domestication of dogs humans ensured that the animals displayed similarly peaceful attitudes, and dogs probably underwent selection for reduced aggression towards human companions (see Chapter 8.3.3, p. 173). Thus it is not surprising that aggressive behaviour by dogs has a strong negative influence on the human–animal relationship, and is the leading complaint in dog-owning families (Riegger and Guntzelman 1990).

Dog aggression is also seen as potentially dangerous because the patterns of human and dog behaviour are not fully compatible; that is, there is only limited overlap between the two species-specific sets of behavioural signals and action patterns that cause physical injuries and pain. Humans (especially children) may have innate tendencies for judging the 'meaning' of growling or persistent gazing, but they may not understand the signal indicated by erect tails and ears. Biting is only the last resort when it comes to aggressive interaction among humans, who prefer to use hitting as a form of physical deterrent. In contrast, the hitting element is missing from the repertoire of most dogs, but biting occurs relative often. In addition the mostly (or originally) thick fur of dogs provides some protection against the effects of a bite which can cause unexpectedly dangerous injuries in furless humans. The behaviour of dogs

could also vary depending on whether they perceive the situation as being social or predatory. Predatory behaviour is not signalled and is aimed at destroying the opponent, so such attacks could be even more serious. (Strictly speaking, predatory behaviour should not be categorized as aggression.)

With regard to aggression, the human–dog relationship is based on 'unconditional trust' (just like the human–human relationship). However, if this trust is lost for any reason, the original relationship will be difficult to reinstate. Thus serious aggressive interactions result in fatal outcomes for both the attacker and the victim. Physical pain and suffering might be accompanied by emotional disturbance (e.g. fear of dogs, see below) in humans, and the dog's fate is often dismissal from the group and death (euthanasia).

3.7.2 Studying the 'biting dog' phenomenon

Not only do dog bites cause physical and emotional suffering, but the associated medical care costs society many millions of dollars (Overall and Love 2001). In the last few years many epidemiological studies have been performed in different countries in order to assess the risk factors and suggest possible preventive measures (Beaver 2001). However, problems in collecting the data and interpretation of the results make generalization difficult.

Most problems relate to sampling methods. Data on dog bites can be collected from a sample that is representative either for the dogs or the humans (or ideally both). Interestingly, the neglect of sampling representative of human populations shows a bias towards the assumption that dogs are responsible for this situation, which is only half of the story. Often samples of the affected dogs are compared to some other reference populations, such as dogs registered with kennel clubs. However, this could also be misleading because many dogs (e.g. mongrels) are not registered.

Some studies collect data from volunteer respondents (e.g. Podberscek and Blackshaw 1993), others either ask some well-defined group of people (e.g. people visiting vets, e.g. Guy *et al.* 2001a) or ask victims directly. Studies also differ in whether dog owners or veterinary or medical personnel are questioned.

The different ways of categorizing aggressive behaviour also complicate the situation. Some categories are derived from the function of aggressive behaviour (i.e. territorial aggression), others are based on the assumed mechanism ('learned aggression'). A recent multivariate analysis suggested three basic categories such as 'dominance aggression', 'conflict aggression', and 'territorial aggression', which seem to focus on the functional aspect (Houpt 2006).

3.7.3 Identifying risks

Whether a social dispute develops into a serious contest between group mates depends on the biological characteristics of the participants (companion (dog and human)-related risk), the social experience or inexperience of the participants (socialization-related risk), and finally the particularities of the actual situation (situational risk). It should be stressed that all three types of risk can and should be identified for both humans and dogs, although there is a bias in the literature emphasizing the dog's side of companion-related risks (which is then easily codified by uninformed lawyers in the form of 'dangerous dog' legislation). Such three-way separation of risks might provide a useful framework, but one should expect interaction between these factors; for example, the relative risk related to socialization might depend on the biological features of the companions (Overall and Love 2001).

Companion-related risk
Companion-related risks have been often identified for dogs with regard to breed, size, age, gender (including the effect of neutering), and health status. Most debates surround the problem of whether there are breeds that are over-represented in the population of 'biting dogs'. Setting aside the problem of what constitutes a breed, studies provide a mixed picture. Reviewing 11 studies from 1970–96 in the USA, Overall and Love (2001) did not find a clear trend for the same breeds to come top of the listing of the three most affected breeds. The only breed that is indicated in 8 out of these 11 studies is the German shepherd, but even this does not provide evidence for a breed effect, partly because each study used a different way to calculate the relative

risk involved. In a recent Canadian sample, Guy *et al.* (2001a) do not list German shepherds among the three breeds that caused most bites (Labradors are at the top of their list) (Box 3.4).

Most studies also agree that large dogs cause more injuries, which could reflect problems with the sampling because people might not take bites delivered by smaller dogs so seriously (Guy *et al.* 2001b). Many studies find that younger dogs bite more often, indicating the role of social experience. Male dogs display more aggressive behaviour in general (e.g. Podberscek and Blackshaw 1993, Guy *et al.* 2001a, Horisberger *et al.* 2005) but there are also exceptions (e.g. Guy *et al.* 2001b). Even more contradictory are the effects of neutering. This factor is also problematic because the operation can take place either before or after the aggressive act, which is often not taken into account. Supporting evidence for a positive effect (less aggression) in males is weak, and there are indications that neutering increases aggression in female dogs (Wright and Nesselrote 1987, Guy *et al.* 2001a). Thus neutering has no unequivocally decreasing effect on the frequency of aggressive behaviour.

The human side suggests a somewhat clearer picture. There is an overall agreement that most dog bites happen in the family setting at home or in familiar places and involve members of the family (Guy *et al.* 2001b). This is to be expected, because dogs and humans interact most frequently in these situations where dispute over resources could take place. Most studies find that children get bitten more often than expected from their proportion in the population (Overall and Love 2001). This might be explained by assuming that there are more frequent social contacts between children and (their) dogs, there is more competition for the same resources (e.g. toys, resting place), and children have smaller resource-holding potential than adults (see Chapter. 8.3, p. 171), which means that dogs might be more willing to initiate agonistic interactions towards them. Moreover, in the case of improperly socialized dogs children might be perceived as a potential prey. In addition, young teenagers (Guy *et al.* 2001b; Horisberger *et al.* 2005) as well as male adults (e.g. Podberscek and Blackshaw 1993, Maragliano *et al.* 2006) have a much greater risk of being bitten.

Box 3.4 Dangerous dogs: retrievers, German shepherds and Rottweilers

In recent years many countries have implemented 'dangerous dog' legislation with the aim of reducing the frequency of dog attacks and biting incidents. In most cases some special event triggered this move by lawmakers, with the backing of the general public. In contrast, dog owners and other supporters protested against these changes, which hit owners of some specific breeds regarded as 'dangerous' especially hard. The issue of the epidemiology of dog bites is now receiving more attention, but old beliefs still persist. Recently various demographic investigations have been published, but differences in the methodology make comparisons difficult. Guy *et al.* (2001a) and Horisberger *et al.* (2004) present comparable data on three similar-sized breeds (Labrador and Golden retrievers analysed together, German shepherds, Rottweilers), which will be used as an example to highlight the difficulties in the analysis.

From the dog's point of view the data provided by Guy *et al.* (2001a) reinforce the view that in Canada Rottweilers are more 'dangerous' because every fifth animal that visited the clinic bit somebody. However, percentage data can be partly misleading because the number of biting Rottweilers is only a quarter of the number of retrievers. Thus in absolute terms retrievers have a greater impact on society in terms of biting incidents. Van den Berg *et al.* (2003) assume that genetic factors might contribute to this unwanted behaviour in retrievers.

From the human's point of view, German shepherds cause the most problems in Switzerland (Horisberger *et al.* 2004). Every fourth person visiting a doctor is bitten by this breed, whereas injuries by retrievers and Rottweilers are less common. Nevertheless, projecting the frequency of biting dogs onto the reference dog population we find that Rottweilers and German shepherds bite more often than expected.

In conclusion, this little comparison shows that there are no 'dangerous' dog breeds in general. Most breeds that seem to bite more often than expected make up only a small part of the whole dog population. In the end more bites by dog breeds with a small population roughly equals the number of bites by dog breeds with a large population. Thus the problem of reducing dog aggression is truly breed-specific and may include genetic selection, problems of socialization, and education of the public (see also Collier 2006).

Study 1 (based on data from Guy *et al.* 2001a)

Reference population: dogs visiting one of 20 veterinary clinics in Canada for any reason during a period of 15 months

Dog	No. visited clinic	No. bitten by dog	%
Retrievers	383	54	14
German shepherd	166	23	14
Rottweiler	55	12	21

Study 2 (based on data from Horisberger *et al.* 2004)

Reference population: humans visiting family practitioners or accident and emergency departments in Switzerland for treatment of a bite injury during a period of 12 months.

Biting dog	N (total = 299)	%	% of dog breed in the reference population
Retrievers	24	8	12.1
German shepherd	72	25	12.8
Rottweiler	20	6.7	2.1

continues

Box 3.4 *continued*

(a)   (b)   (c)

Figure to Box 3.4 Which of them will bite? (a) Labrador retriever (photo: Enikö Kubinyi) (b) German shepherd (c) Rottweiler. Depending on the statistics used, arguments for 'dangerousness' can be put forward for all three breeds.

Socialization-related risks

These involve the lack of appropriate early socialization of dogs and problems in the 'interpersonal' or hierarchical relationships in the group. Many people assume that uncertainties in the rank order of the group or anthropomorphism on the part of the owner are the causal factors for the aggressive behaviour. Some people believe that certain social situations may increase the dominant tendencies in dogs, resulting in a higher frequency of attacks. Thus letting a dog go ahead of you, feeding it before the human mealtime, allowing it to sleep on the bed or in the bedroom, or allowing it to win in tug-of-war games is expected to increase aggressiveness. Questionnaire studies on large samples have had variable success in finding support for such associations (e.g. Jagoe and Serpell 1996, Podberscek and Serpell 1997, Guy *et al.* 2001c, Rooney and Bradshaw 2003). The main problem with most of these results is (as the authors themselves acknowledge) that these associations say nothing about cause and effect. Finding that a dog sleeping in its owner's bedroom is more aggressive could indicate that either close contact during the night or sharing the resting place leads to more intense competition, or that a dog with higher assertive tendencies fights out its 'right' to sleep

with the owner. It is more likely that such situations reflect the lack of proper and consistent socialization of the dog during development, which is the normal time to acquire the rules and forms of social interaction.

Improper or inadequate 'socialization' of children (or adults) to dogs can also be a causal factor, although this is often neglected.

Situational risk

Situational risk factors are perhaps the most difficult to identify because respondents may not remember the circumstance of the event exactly or are less willing to cooperate in revealing the problem. Many bites occur when the dog is in the possession of food or toy, in the course of play (Horisberger *et al.* 2005), or suffering from unrelated pain or stress, such as a skin problem (Guy *et al.* 2001c). Very often the problem relates to one party misunderstanding the behaviour of the other. Thus children (but also inexperienced adults) are more likely to fail to recognize behavioural signals indicating higher levels of tension in the dog, but at the same time a dog could also misread human behaviours if they do not conform to the habitual forms. As expected, most situational risk factors can be reduced by paying more attention to the socialization process in general, but this is true for both dogs and humans.

There is a strong and often neglected relationship between fear and aggression. Fear can often cause agonistic interactions, and at the same time it can also be an unfortunate outcome of such contests. Recent surveys suggest a positive relationship between increased aggressiveness and both asocial fear (e.g. loud sounds) and social fear in dogs (Podberscek and Serpell 1997, Guy *et al.* 2001c). Similarly, fearful humans (both children and adults) may more easily become victims of dog attacks. Nevertheless, early and gradual exposure to social stimuli may have a moderating effect on the later development of fear. This can be especially advantageous in the case of young children (Doogan and Thomas 1992). Moreover, in humans early exposure to dogs can be a preventive measure against the development of fear of dogs in case one suffers a dog attack at some later time. Early and regular experience with dogs in the nursery or at primary school (as a part of the curriculum) could have a positive effect. Similarly, exposing pups to humans, especially children, could decrease fear. There are only a few studies dealing with fear of dogs in adults and children. A recent survey on a random adult human population revealed that 43% of the respondents fear dogs (Boyd *et al.* 2004). Interestingly, a large proportion of fearful people expressed fondness for dogs, and their fear was mainly the result of negative experience of having been attacked, threatened, or witnessing an attack. The prevention of development of fear in humans towards dogs (and vice versa) could also decrease the frequency of dog bites (Box 3.4).

Overall and Love (2001) argue that to increase our understanding of dog bites there is a need for (1) more detailed description of the biological features of the attacker, (2) identification of the risks provided by canine and human behaviour, (3) development of behaviour profiles for biting dogs, and (4) more detailed descriptions of the situations. In addition there is a need for long-term, longitudinal questionnaire studies which should be supplemented with direct behavioural observations (Netto and Planta 1997, van den Berg *et al.* 2003).

3.8 Outcast dogs: life in animal shelters

Dog shelters are relatively novel innovations, developed to provide housing for 'unwanted' animals. Over the years the role of shelters has increased because of the growing number of dogs that are relinquished by their owners, and there is also a greater demand to put free-ranging dogs into shelters. Recent publications suggest that at any time 5–10% of the total dog population might live in shelters if such facilities are made available (Patronek and Rowan 1995, Marston *et al.* 2004). In the USA this could mean around 4–5 million dogs. Apart from managing a substantial part of the dog population, shelters also have an important role in reintroducing dogs to the human community.

However, shelters also face immense problems. Although they offer a valuable service for the community, they often do not have the financial means to provide the dogs with an appropriate environment. The management of dogs is also bound by regulations, some of which actually decrease the well-being of the dogs living in the shelter.

Most dogs admitted to shelters experience a big change in their life by losing all former social contacts. This can be very detrimental in the case of family dogs, where social deprivation is also accompanied by an altered physical environment. In many shelters dogs are housed singly (or sometimes in pairs) in a small kennel (4 m²) (Wells and Hepper 1992, Hennessy *et al.* 1998, Marston *et al.* 2005b). Note that the EU recommends 4 m² floor space for pair-housed dogs below 20 kg and 8 m² floor space for dogs over 20 kg. Although this type of housing is preferred because it decreases the likelihood of spreading disease, it is detrimental for a social animal. Dogs that spend a considerable time in a social group (monitored by the staff of the shelter) retain much of their social nature and are more likely to adapt to their new homes if adopted (Mertens and Unshelm 1996). Although environmental enrichment can help to some extent (visual access to another dog, increased visual access to visitors, or provision of novel olfactory, auditory, and visual stimuli) (Wells and Hepper 1998, 2000, Wells 2004), ultimately no stimulation can replace direct social contact (Marston and Bennett 2003).

There are arguments that this deprivation is only short term and therefore does not reduce well-being. Indeed, some shelters report that dogs spend on average less than 1 week in the shelter before being re-homed or put down (Wells and Hepper 1998, Marston *et al.* 2005b) but this is apparently not so at many other shelters and some dogs spend up to 5 years there (Wells *et al.* 2002). One study did not find major change in the behaviour for over 6 days after entering a shelter (Wells and Hepper 1992), but longer-term housing for months or years can have a negative effect on the welfare of dogs (Wells *et al.* 2002). This could be especially problematic in countries that have introduced 'no euthanasia' rules (e.g. Italy) because some dogs (especially older ones) stayed for more than 6 months on average.

The critical effect of being introduced to a shelter was revealed by measuring increased levels of the stress hormone cortisol during the first 5 days, in comparison to control pet dogs that stayed with their owners (Hennessy *et al.* 1997). Such abnormally high stress levels can be markedly reduced by human petting, which provides further support for the need of direct social contact for shelter dogs (Hennessy *et al.* 1998). Shelter dogs also very rapidly develop an attachment relationship with a human (Gácsi *et al.* 2001; Chapter 8.2, p. 168). Thus from the animal welfare point of view regular access to daily social experience might be obligatory for these dogs. In a more recent study Wells *et al.* (2002) found that activity of the dogs was related to the time they spent in the shelter, and marked changes occurred some time between 2 and 12 months. In order to avoid these problems, in many countries volunteers have developed so-called 'temporary adoption programmes' for providing homes for the unwanted dogs (Normando *et al.* 2006; Box 3.5).

Shelter dogs are not representative of the dog population because people are more likely to relinquish dogs that show behavioural problems (e.g. aggressiveness or distractive behaviour). In addition, free-ranging dogs coming to shelters are often poorly socialized and thus experience difficulties in developing a natural relationship with humans. The reintroduction of these dogs to human families is more successful if each dog receives individual attention. Making a behavioural profile of the dog by utilizing standard behavioural tests might also help in finding a

---

Box 3.5 Dog shelters: hostels, homes, or retraining centres?

Ideally, a dog shelter should be a place where dogs who are found without a human partner, or unwanted companion animals, can be kept for a short time until they find a new welcoming home. Recent research has started to collect data on the dogs that enter shelters, and on their fate both at the shelter and in their new homes. The main problem is that the number of dogs introduced to the shelter is higher than the number adopted. Although it may be unrealistic to expect all shelter dogs to get a second chance to join a human family, the shelter environment should increase this possibility.

Leaving a dog at a shelter is clearly the saddest aspect of dog–human relationship, 'a tie that does not bind' (Arkow and Dow 1984). There are many reasons for separating from a companion but the same reasons could cause problems for the prospective adopters as well.

This table below suggests that the relationship is broken more often by humans than by dogs. The most frequently reported behavioural problem causing relinquishment was aggression, followed by the tendency to escape and hyperactivity. After adoption, owners reported more than one behavioural problem in their dog. The most frequent problem was fear and hyperactivity, and we cannot exclude that the shelter environment contributed to the emergence of these unwanted behaviours. Since the shelter may induce novel problems in dogs, there is an increased need for continuing socialization (Mertens and Unshelm 1996) and for behavioural rehabilitation (Orihel *et al.* 2005). Standardized questionnaires for relinquished and adopted dogs can also help to identify the problems.

continues

Box 3.5 *continued*

Reasons for relinquishment	Problem with adopted dog within a month[a]		
	% (N = 3123) (Marston *et al.* 2004)	% (N = 62) (Marston *et al.* 2005b)	% (N = 556) (Wells and Hepper 2000)
Owner factor (e.g. moving, financial, health)	32		
Dog behaviour (total)	14		
Escape	2.6	22.3	13.4
Hyperactive	2.2	61.1	37.4
Barking	1.1	24.7	11.3
Predatory	0.9	24.1	–
Aggression (dogs & humans)	3.2	18.7	12 (approx.)
Fear	–	32.2	53.4
For euthanasia	7.9		

[a] overlapping categories.

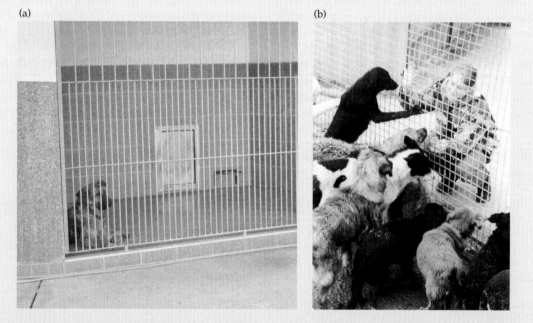

(a)

(b)

Figure to Box 3.5 At the moment there seems to be a trade-off between recommendations for 'healthy' and 'happy' environments. (a) In many shelters dogs spend most of their time alone or in pairs in a barren environment. (b) Enjoying group life with peers could enhance transmission of disease. Are there not better options? (Photo: Enikö Kubinyi)

matching human companion (Marston and Bennett 2003, De Palma *et al.* 2005). The chances of adoption can be enhanced by subjecting dogs to some corrective behavioural training if it seems necessary (Orihel *et al.* 2005). Unfortunately, such measures are just being introduced at some shelters around the world, and return rates of dogs are still relatively high, ranging from 8% to 50% for different shelters. In the long term it might be better to view shelters not as transient sanctuaries for a couple of days but rather as rehabilitation centres for dogs that have lost contact with human society.

3.9 Conclusions for the future

For any in-depth research there is a clear need for the collection of comparative data on the dog populations living in various regions. Such demographic surveys should include information on the population biology of dogs, cultural differences in the dog–human relationship, and the living environment. If possible, data collection should take place at the international level using standardized instruments.

More data are also needed on the life of dogs that work for humankind. General behavioural observations are lacking, and in most cases methods have not been developed to measure efficiency of working performance or monitoring welfare.

The dark side of human–dog relationships needs also more attention. Although dogs can physically hurt humans by biting, we also hurt them if they are left to suffer in shelters. Clearly, research on dog biting needs to be advanced in areas including the identification of risk factors (separately for human populations and dog breeds), the development of behavioural testing (Netto and Planta 1997), and the provision of recommendations for dog-breeders.

Further reading

Many issues of human–dog relations have been discussed in recent books (e.g. Podberscek *et al.* 2000) including the contribution of dogs to human health (Robinson 1995, Wilson and Turner 1998).

CHAPTER 4

A comparative approach to *Canis*

4.1 Introduction

The ancestry of dogs seems to be settled. Geneticists have provided convincing data showing that the wolf is the nearest living relative of the dogs, although there is some doubt that the extant wolf is the ancestor of dogs. Coppinger and Coppinger (2001) stressed that we should speak of a common ancestor of dogs and wolves, and dogs originated probably from a special wolf-like ecological variant. Thus instead of looking for the direct phylogenetic ancestor(s), which might have died out, a wider comparative perspective on *Canis* species could be more helpful.

First, there are 'adaptive stories' to explain why the wolf was the only possible species to choose, but from a wider perspective these arguments are less convincing. In principle other species of *Canis* (such as coyotes or jackals) might also have, or have had, the potential to become domesticated; however, the wolves were the only ones 'lucky' enough to be at the right place at the right time. Once some groups of humans got over the first hurdle and dogs emerged, there was no incentive to domesticate others. Some support for this view comes from the fox-selection experiment (Chapter 5.6, p. 132) which clearly shows that directed selection for 'tameness' results in a few generations in dog-like behaviour and looks (Belyaev 1978).

Second, with respect to their ecology and behaviour some recent species or populations could more directly resemble some ancestor wolf-like populations that provided the evolutionary 'material' for dog domestication (see also Koler-Matznick 2002) independent of their genetic relationship to present-day dogs.

Third, another aspect of comparative investigations should aim in particular to reveal diversity within wolves. It seems that this species actually covers the whole range of traits which are present in a more restricted and isolated form in the other species of the genus *Canis*. Although there has recently been an immense development in wolf research, this knowledge finds its way very slowly into the dog literature, and more importantly, secondary sources actually present an unrealistic (or untrue) picture. Thus it is important that for comparative reasons we obtain a relatively broad perspective on the wolves, although we will restrict ourselves to only a few main points, as other volumes dedicated to this topic are available (Mech 1970, Harrington and Paquet 1982, Mech and Boitani 2003).

4.2 Putting things into perspective: an overview of *Canis*

4.2.1 Systematic relationships and geographic distribution

The Canidae consists of 15 genera, one of which is the *Canis* genus, which consists of 7 wild species and the domestic dog (Sheldon, 1988). It is interesting that both the family and the genus got their name (*canis*) from the youngest and probably least typical member of the group. Based on chromosome number, recent classifications refer to a group of 'wolf-like canids' that include the dhole (*Cuon alpinus*) and the African wild dogs (*Lycaon pictus*) (e.g. Wayne 1993).

Apart from the wolf (and the dog), which will be discussed in detail below, six further species are categorized in the genus. The jackals, which are probably the descendants of extinct *C. arnensis*, represent the most southerly species. The side-striped jackal (*C. adustus*) occurs from the north of South Africa to Ethiopia; the present habitat of the golden jackal (*C. aureus*) covers mainly North Africa but it can also be found in southern and middle Europe; the black-backed jackal (*C. mesomelas*) is most typical in East Africa (Uganda, Tanzania); the Ethiopian jackal (*C. simensis;* often referred to as the Ethiopian wolf) is mainly confined to the mountain regions of Ethiopia. Coyotes (*C. latrans*) live in expanding populations in North America, and the red wolf (*C. rufus*) now has recognized species status (Nowak 2003) (Box 4.1).

4.2.2 The evolution of *Canis*

Paleozoologists agree that in the history of the Carnivores the Canidae family is represented by two extinct subfamilies (Hesperocyoninae and Borophaginae) and one living one (Caninae) (for a more detailed review see Wang *et al.* 2004). Species belonging to these subfamilies originated 40 million

Box 4.1 Present-day distribution of the wolf and other canids

Wolves are clearly the most widely distributed *Canis* species. Unfortunately, expanding human populations have driven them to extinction in many locations. Thus the wolf has largely disappeared from Mexico and the USA, although in recent years some population growth has been reported in the USA, and there are attempts to rescue the Mexican population. Once wolves inhabited the whole of Europe; now, mostly due to protection in some countries, local wolf populations of 5–200 individuals are surviving or even increasing at a few locations. The total wolf population was estimated at *c.*300 000 by Ginsberg and MacDonald (1990) and *c.*150 000 individuals by Boitani (2003). (In comparison, there are 52 million dogs in the USA alone.)

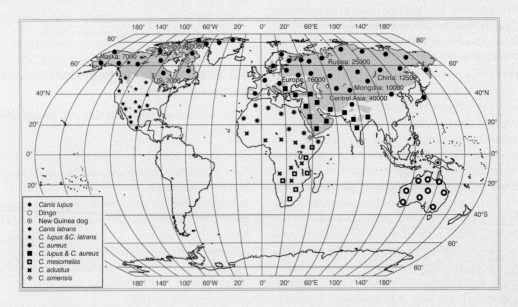

Figure to Box 4.1: Distribution of wolves and other *Canis* species. The numbers on the map refer to estimated wolf numbers given by Boitani (2003). The drawing is based on Clutton-Brock (1984), Mech and Boitani (2003).

years ago and evolved in North America. Many species of the Hesperocyoninae and Borophaginae can be detected in the fossil record up to 2 million years ago, and throughout their history these subfamilies remained endemic to their continent of origin. In contrast, species belonging to the Caninae subfamily crossed over to Eurasia approximately 7–8 million years ago, and rapidly radiated to most parts of the Old World (see below). One very intriguing characteristic of the Canidae is the range of their feeding habits. Both hypocarnivory and hypercarnivory occur, with the former showing signs of a more omnivorous diet (extending size of the molars: increased grinding ability); in contrast, the increased size of the carnassial at the expense of the molar (increased shearing ability) suggest obligatory meat eaters often specialized in eating big game. More importantly, the change at the level of different species emerges frequently and independently in these subfamilies, probably reflecting actual environmental constraints (parallelism, see Box 1.3.).

The first recognized member of the Caninea subfamily, the fox-sized *Leptocyon*, lived in the early Oligocene (32–30 million years ago) (Box 4.2). Later, in the medial Miocene (10–12 million years ago), a jackal-sized canid emerged. *Eucyon*'s most characteristic feature is the presence of the frontal sinus, which is retained in the descendants of this clade. *Eucyon* colonized Europe by the end of the Miocene (5–6 million years ago) and was evidently present in Asia in the early Pliocene (4 million years ago).

Another significant parallel event was the evolution of the *Vulpini* around 9–10 million years ago (late Miocene). All extant foxes are the descendants of this clade. One difference between the fox and dog clades is that recent species of the former group are more resistant to displaying complex social behaviour.

During the transitional period from the Miocene to Pliocene (5–6 million years ago), North America gave rise to canids which are regarded as the first members of the *Canis* genus (Wang *et al.* 2004). These mostly jackal-sized species display evidence for hypercarnivory. In the early Pliocene they arrived in Europe and radiated throughout the Old World. The exact order of events then becomes very hard to follow because of the huge areas potentially covered by various species and the possibility of crossing to and fro between Eurasia and America. The situation is even more complex because significant climate changes often caused expansions, as well as reductions or extinctions, affecting a range of species.

Today's coyotes (*Canis latrans*) represent the only surviving endemic species in the New World, originating from the extinct *Canis lepophagus* about 1.8–2.5 million years ago (Nowak 2003) or 1 million years ago (Kurtén and Anderson (1980) (for the importance of this date see also Chapter 5.3.2, p. 109). In contrast, *Canis* species diverged in the Old World during the late Pliocene and Pleistocene (1.5–2 million years ago), colonizing Europe, Asia,

---

Box 4.2 Phylogenetic relations based on palaeontological findings

The reconstruction of the evolution of wolf-like canids is complicated because most species were very mobile and dispersed over large areas, sometimes two or three continents. It appears that although the *Leptocyon*, *Eucyon*, and *Canis* genera all emerged in North America they rapidly crossed to Eurasia. Especially in the case of *Canis*, there is evidence that both lines have surviving species. Palaeologists assume that the American *Canis* is the ancestor of the recent coyotes while the African and Asian dogs (jackals, wild dogs, cuon) originated from the Eurasian branch. The last large 'natural' migration occurred around 100 000 years ago when *lupus* populations crossed the Bering Strait for the last time before the two continents separated. However, dogs have found a way to solve this problem and make sure that dispersion of *Canis* goes on despite geographical barriers: they have joined humans on their migration routes.

continues

Box 4.2 *continued*

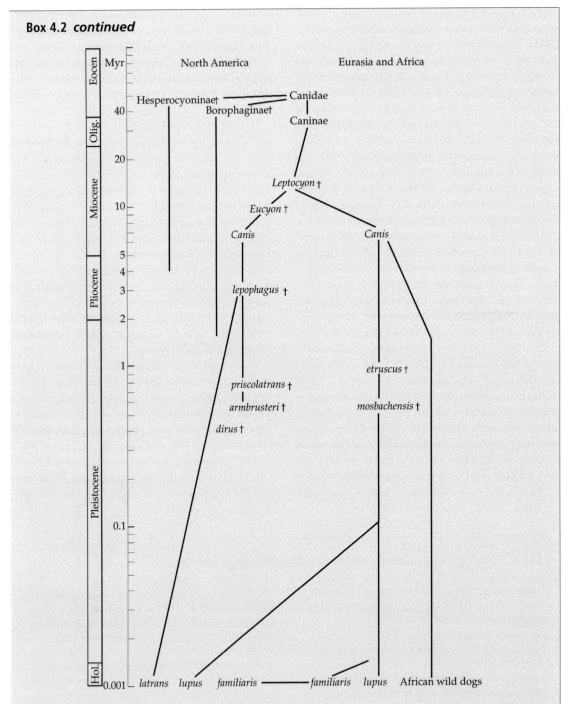

Figure to Box 4.2 Phylogenetic tree of Canidae branches which led to the emergence of extant *Canis* species. A cross indicates extinct genus. Note logarithamic scaling of time. (Based on Wang *et al.* 2004 and Nowak 2003).

and Africa, and this radiation gave rise to canid forms such as wolves, dholes, and wild dogs. The Eurasian *Canis etruscus* and a further descendant form (*Canis mosbachensis*) are regarded as the ancestors of the grey wolves (*Canis lupus*), the dholes, and African wild dogs. This larger radiation took place in Eurasia and Africa where wolves emerged by 130 000–300 000 years ago and extended their habitat to North America by crossing at the Bering Strait 100 000 years ago (Nowak 2003, Wang *et al.* 2004). During glacial periods populations survived south of the ice sheet in middle zones of the continent. Importantly both wolves and coyotes proved to be very resistant species, and according to the archaeological records they have remained virtually unchanged morphologically up to our times (Olsen 1985), excluding variation in size and probably also in behaviour. The conservative nature of canids is also evident on a longer time scale; Radinsky (1973) found only a slight relative increase in brain size over a period of 15–30 million years.

The overall phylogenetic relations are supported by the comparative analysis of DNA samples of extant species, although the relationship among closely related species shows some ambiguity. Phylogenic trees generated on mitochondrial DNA (2001 bp protein coding region) (Wayne *et al.* 1997) and nuclear DNA (both exons and introns representing variable regions) (Lindblad-Toh *et al.* 2005) agree on the close relation between wolf (dog) and coyotes and indicate an African origin for this clade, but show differences with regard to the relationships among jackals, the Ethiopian Jackal, and the dhole (Box 4.3).

4.2.3 The ecology and dynamics of group living in some canids

In many respects Canidae (including *Canis* species) represent an odd group within the carnivores. They are not strictly carnivorous, and have a strong tendency to form and live in groups (Kleiman and Eisenberg 1973, Gittleman 1986). In addition, these differences vary not only across species but also among populations. Although there have been attempts to categorize Canidae species according to their social structure (Fox 1975), there are more exceptions to the rule, and local long-term ecological factors and selective pressures often push some populations towards extremes. The comparative study of extant species is also made difficult because human activity often has marked effects on the ecological conditions; for example, human activity has provided new food sources (rubbish dumps, water, domestic animals), but has destroyed habitats or aimed at extermination of canid populations. Evolution of Canidae has

---

Box 4.3 Evolutionary relationships within wolf-like canids

With the advances in molecular genetic techniques, the comparison of DNA sequences offers an alternative way to construct phylogenetic trees. The power of such comparisons depends crucially on the DNA which is used. At the beginning the sequencing of DNA was complicated and expensive, so only short sequences of well-known genes were compared (A: cytochrome B, 736 bp (base pairs); Wayne 1993). Later studies included more genes which provided longer sequences (B: TRSP and RPPH1, 673 bp and 684 bp respectively, Bardeleben *et al.* 2005). Lindblad-Toh *et al.* (2005) used a much longer sequence of 15 000 bp (C) obtained from several locations on the genome (both introns and exons were included). Other investigations were based on the comparison of mtDNA which is inherited only from the mother (D: 2001 bp, Wayne *et al.* 1997). Despite the differences in methods used, the overall picture is very similar. As expected, dogs and wolves show the smallest divergence, which indicates a close relationship. From the wolf's perspective the next relative species is the coyote, followed by the golden jackal. Similarly, at the base of the tree we find two African species: the African wild dog and side-striped jackal. Based on this observation Lindblad-Toh *et al.* (2005) argued for an African origin of recent *Canis*.

continues

Box 4.3 *continued*

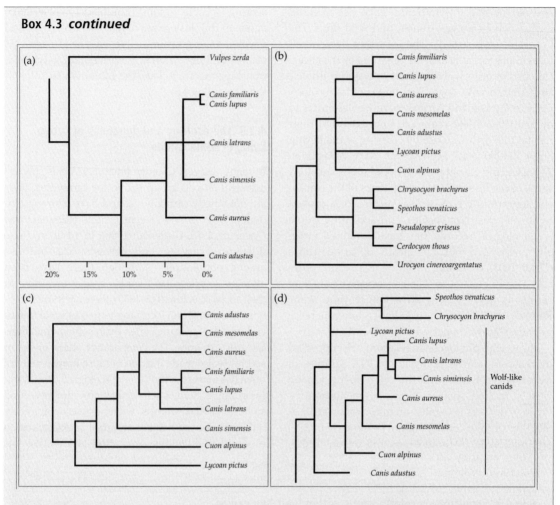

Figure to Box 4.3 (a) Cytochrome b; maximum parsimony tree; (b) TRSP, RPPH1 DNA strict consensus maximum parsimony tree; (c) 15 kbp genes; maximum parsimony tree; (d) 2001 bp mtDNA consensus tree. (redrawn based on references above).

already shown that these species are highly adaptive to a wide range of ecological conditions, and therefore it is not surprising that intensive human interventions contributed to increased variation in the canid social structure (Table 4.1).

In fact a careful overview of these related species suggests that it is very difficult to pinpoint skills that are confined to only one species and never emerge in others. In line with this, Macdonald (1983) argued that the early evolutionary factors were the same for all canids, whether fox or wolf, and this common heritage is retained in recent species, combined with a flexible (mostly behav-

ioural) capacity to adapt to local ecological factors related to feeding or predation.

In answering questions on why most Canidae express some level of sociality ranging from long-term pair bonds to extended family packs, arguments have usually focused on collaborative hunting, the defence against other predators, or increased reproductive success of the larger family. Without denying the importance of these factors, Macdonald (1983) proposed that in an evolutionary perspective the concentrated distribution of some food resources could have selected for communal feeding in canids (and other carnivores), and this

Table 4.1. Comparative summary of *Canis* species based on Sheldon (1988)

Species	Shoulder height (cm)	Weight (kg)	Diet	Gestation	Social organization	Home range (km²)
Side-striped jackal (*Canis adustus*)	41–50	6.5–14	Omnivorous; carrion, small animals plants/fruits	8–10 weeks (max. 7 offspring)	Pair + offspring	c.1.1
Golden jackal (*Canis aureus*)	38–50	7–15	Carrion, small animals; coop. hunting	63 days (max. 9 offspring); biparental, alloparental	Very variable, pair + offspring (+ yearlings)	Hunting range 2.5–20
Black-backed jackal (*Canis mesomelas*)	38–48	6–13.5	Carrion, coop. hunting plants/fruits	61 days (max 9 offspring); biparental, alloparental	Pair + offspring	c.18
Ethiopian jackal (*Canis simensis*)	53–62		Rodents; hunts alone		Pair (+ offspring)	
Grey wolf (*Canis lupus*)	45–80	18–60	Carnivorous; carrion, plants/fruits coop. hunting	62–65 days (max. 13 offspring); biparental, alloparental	Very variable, pair + offspring + yearlings	18–13 000
Coyote (*Canis latrans*)	45–53	7–20	Carnivorous; carrion, plants/fruits (coop. hunting)	c.60 days (max 12 offspring); biparental, (alloparental)	Very variable, pair + offspring (+ yearlings)	1–100
Red wolf (*Canis rufus*)	66–79	16–41	Small animals, carrion, plants	60–62 days (max 8 offspring); biparental	Very variable, pair + offspring (+ yearlings)	40–80

could have led to the emergence of secondary social characteristics, such as joint hunting and defence of the territory or alloparental behaviour. Interestingly, Kleiman and Eisenberg (1973) also note that in contrast to felids, canids are notable for 'peaceful communal feeding'; that is, they are relatively tolerant of the presence of others at the food source (e.g. at the kill).

The *Canis* species that live under similar ecological conditions show many morphological and behavioural parallelisms. Many regard the coyote as an ecological equivalent to the jackal, and populations of small wolves (living in western or eastern Asia) also show similar adjustment to the environment. They all live in small families, juveniles stay with the parents for 1 or 2 years, and they display a range of feeding behaviours from scavenging and solitary hunting to organized attacks by a group of subadults and adults.

Thus it seems that during the evolution of *Canis* both size variation and the adjustment of social behaviour were key factors in adaptation to the local niche (Box 4.4). Modifications could be achieved by varying the strength of association between group mates, which resulted in a varying pattern of dispersal from the pack. In contrast to foxes (where young normally leave within 6–10 months after birth) (Baker *et al.* 1998), *Canis* offspring stay usually at least until the next breeding season or, more frequently, for the next 1–2 years. Loyalty is greater if the animals of the next generation are not involved in sexual competition. This can be achieved by delaying maturation for 1–2 years, which is more likely to happen in species with a larger body size. Thus *Canis* species represent a finely tuned series with a considerable amount of overlapping variation in terms of their morphology and behaviour. However, if environmental factors push the species in one direction then differences can emerge. An example of this is the well-organized group hunting behaviour in wolves, in which all or most members of the pack participate, independent of sex and age (Chapter 4.3.3, p. 79).

Box 4.4 Size diversity in Canidae

The *Canis* species are often categorized on the basis of size, measured as body weight, shoulder height, body, or skull length. Detailed morphological examinations reveal that jackals, coyotes, and wolves are nearly isomorphic, that is, the size relations of their body are constant (Wayne 1986a, b, Morey 1992). In short, wolves have bigger heads because they have a larger body, but if shrunk they would just look like coyotes or jackals. Importantly, such nearly isometric relationships are not only present between the body size and skull length but also remain constant between different dimensions of the skull, including for example width vs length (Box 5.5).

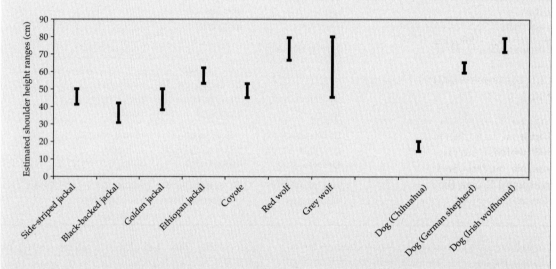

Figure to Box 4.4 Variation in shoulder height in *Canis*. Although the literature provides no possibility for statistical comparisons, the obtained values (smallest and largest) and the size of the ranges show a considerable overall similarity among the species except the wolf and the dog. In the wolf the wide range of shoulder height is represented by different subspecies. In the case of the dogs the range of shoulder height is even wider, but the species is 'divided' into breeds.

4.3 An overview of wolves

Any argument on wolves (*Canis lupus*) as ancestors of dogs should rely on detailed knowledge of the species. Often the picture of wolves is oversimplified, hindering our understanding and interpretation of dog behaviour. Just as there is no such thing as 'the dog', there may be no such thing as 'the wolf'. We would argue that the range (variability) of the wolf phenotype covers many of the features that can be found compartmentalized (in a mosaic pattern) in other species of *Canis*.

Up to the beginning of the last century wolves could be found everywhere in the northern hemisphere, in contrast to much more localized *Canis* species such as jackals or coyotes. Wolves must have a very adaptable genetic system in order to survive in such different environments. Repeated cyclic changes in their environment (e.g. ice ages) might have selected for a very plastic phenotype in wolves, and this plasticity could became significant at the time when they met humans.

This phenotypic plasticity makes comparative evolutionary investigations very difficult. Some see homologous relationships on the basis of some phenotypic similarity between dogs and wolves living in some recent populations, but this can be either a case for convergence (Chapter 1.4, p. 14) or an adaptation to a particular environment which emerges only during the time when the population is exposed to these environmental variables. Thus it is difficult to argue that dogs are the direct descendants of one recent wolf population, solely on the basis of phenotypic similarity. For example, there have been assumptions that dogs originate from southern wolf populations (e.g. *C. l. pallipes*) because these wolves are relatively small (e.g. Hemmer 1990); however, at the time when dogs evolved (and if we assume that small size is at all important in this respect) there might have been small wolves in various other places depending on the particular ecological conditions.

4.3.1 Geographic distribution and systematic relationships

Until 1800 the wolf was dispersed across Europe apart from the British Isles. Now, large populations (500 wolves) survive only in Spain, Poland, Romania, Bulgaria, Serbia, the Baltic states, Ukraine, and central Russia (Boitani 2003). There are crude estimates of *c.*65 000 wolves living east of the Urals and in Asia, and probably a further 2000 living in Asia Minor and Egypt. The population in the Americas is judged to be about 60 000 individuals, of which only 10% are in the USA. Thus, based on estimates by Boitani (2003), there might be about 160 000 wolves living in the Holarctic. In contrast, Ginsberg and Macdonald (1990) estimated around 300 000 wolves, and it is thought that wolves have lost more than 50% of their original habitat during the last few hundred years (Box 4.1).

The Grey wolf has always provided a lot of work for taxonomists. Some of the problems stem from the uncertainties surrounding the species concept. The situation is made even more complicated by the wolf's complex relation to various forms of domesticated and feral dogs. There is limited evidence that all *Canis* species can interbreed, with fertile offspring. Genetic studies revealed wolf–dog hybrids in Italy (Randi *et al.* 1993, Randi and Lucchini 2002), but they occur elsewhere too. Hybridization also takes place between wolf and coyote (Lehman *et al.* 1991) producing fertile offspring (see also Wilson *et al.* 2000). Thus according to the classical definition of a species, all *Canis*

could be considered as a single species. However, the revised biological definition of species is based on interbreeding natural populations that are separated from other similar groups (Mayr 1963). This separates wolves from coyotes (or jackals), even if there is some limited evidence of hybridization between these species if their habitats overlap (Wayne and Vilá 2001). This would argue for categorizing wolves and dogs into separate species. However, some taxonomists now seem to disagree whether the classic Linnaean categories (*lupus* and *familiaris*) are still valid. This has led to the unfortunate and confusing situation that many European zoologists, behavioural scientists, and geneticists over the world still refer to the dog as a separate species, while in many papers published by North American authors dogs are categorized as a subspecies of wolves (*C.l. familiaris*). The 'lumpers' argue that dogs and wolves are not differentiated enough to qualify for species-level discrimination (e.g. Wayne 1986a, b). However, the ecological species concept takes Mayr's definition even further by saying that species are adapted to a specific niche in their environment as a consequence of an evolutionary/ecological process. Thus if such a niche and a set of particular adaptations can be identified in a population, then a species-level categorization might be justified. This logic was applied by Coppinger and Coppinger (2001) and others when they argued that dogs show specific adaptive traits for living in an anthropogenic niche (Box 5.1). Since both the population-based and the ecological definition seem to be fulfilled by dogs, we will retain the original labels used by Linnaeus.

Similar problems at a different level emerged in the taxonomy of the *lupus* subspecies. Based mainly on the distribution of populations and morphological traits, wolves were categorized into various subspecies. For example, based on Hall and Kelson (1959), Mech (1970) listed 24 subspecies in North America, which were collapsed into 5 subspecies based on a detailed morphological analysis (Nowak 2003). Thus the present list includes the Arctic wolf *C.l. arctos*, Mexican wolf *C.l. baileyi*, Eastern wolf *C.l. lycaon*, Plains wolf *C.l. nubilus*, and Northwestern wolf *C.l. occidentalis*. According to Nowak (2003) there are 9 living subspecies in Eurasia: Arctic wolf

C.l. albus, Arab wolf *C.l. arabs*, north-central wolf *C.l. communis*, *C.l. cubanensis*, Italian wolf: *C.l. italicus*, *C.l. lupaster*, common wolf *C.l. lupus*, and Indian wolf *C.l. pallipes*, but only 7 were listed in Mech (1970). However, there are problems with the present system too: *C.l. chanco* (originally described from China and Mongolia) is not mentioned by either source, which presents a problem because this subspecies has often been referred to in connection with the domestication process. Genetic analysis (see below) seems not to support the distinction between Italian and other European wolves. Moreover, the two 'arctic' wolves (*arctos* and *albus*) might create some confusion in the literature (Nowak 2003). Finally, if we define the wolf subspecies on the basis of geographical distribution then this does not fit with the idea of including the dog as an additional subspecies which is distributed all over the world.

It seems that the wolves escape our classic notion of species and subspecies. In suggesting a way forward, Wayne and Vilá (2001) argue that instead of trying to categorize extant populations of wolves, we should regard them as a series of intergrading populations—a concept that is also supported by the genetic evidence.

4.3.2 Evolution of the wolf

Today the wolf is recognized as a top predator throughout the northern hemisphere, but the situation was quite different even a few hundred thousand years ago (Wang *et al.* 2004). At that time herbivorous species were controlled by much larger predators on both continents. This was probably the result of a runaway evolutionary process in which there was a trend for increasing size in carnivore predators to outwit competitors. Their larger body size could be only sustained by a strongly carnivorous diet (Carbone *et al.* 1999), and these species (e.g. dire wolf, sabertooth cat) became increasingly dependent on the amount of meat available. The ancestors of today's wolf had to share their habitats with at least 11 other predators of the megafauna (most of which were bigger), and thus occupied a lower rank in the food chain as a mesopredator (Wang *et al.* 2004).

However, the fate of the wolf seems to have taken an unexpected turn. Starting sometime during the

middle Pleistocene (500 000 years ago) in Eurasia, and culminating at the end of this period (10 000 years ago) in North America, those large mammals 'suddenly' disappear from the fauna. The reasons for this are still debated; some emphasize climate changes, while other suspect that the successfully hunting humans had a catastrophic effect on the ungulate prey populations of the dire wolf (*C. dirus*) and others. This situation (especially towards the end of the Pleistocene after the end of the last glacial maximum at 18 000 years ago) gave the wolf a unique chance to fill a vacant niche (Wang *et al.* 2004). The large dire wolf became extinct in America by 10 000 years ago, and wolves probably were just about to (re-)colonize the Old World when they first crossed to the New World around 50 000–100 000 years ago. By the time humans begun migrating to the New World (15 000–20 000 years ago) wolves had probably established their position of being one of the few top predators (Fig. 5.1).

During the Pleistocene wolves had to survive either relatively warm or cold climates, including the advance and retreat of the ice sheet. These changes probably caused a set of phenotypic changes including overall morphology and behaviour. During unfavourable periods, e.g. when the temperature decreased, surviving wolves retreated into safer environments (refuges) and thus smaller or larger parts of the wolf population were separated from each other for a period of several thousand years. During glacial periods wolves might have been pushed far to the south of North America or Asia, whereas in interglacial times they could regain territories far into the Arctic. The need for periodic adaptation to the local environments and subsequent dispersal over large areas, paralleled by hybridization with wolves from other refuges, renders the evolution of wolves very difficult if not impossible to determine. For example, archaeological records have revealed that the size of wolves both followed changes in the local climate and differed according to geographical regions (Kurtén 1968) (Box 4.5).

Recently, researchers have collected both extinct and extant wolf mitochrondrial DNA (mtDNA) sequences (see also Chapter 5.3.2, p. 110) over the entire geographic area inhabited by this species in order to reconstruct wolf evolution by phylogenetic means (for a review see Wayne and Vilá 2001). The genetic comparison differentiated fewer major groupings of wolves (in contrast to the 5+9 subspecies listed above) that might attain a subspecies status. The presently available collection of wolf mtDNA indicates that North American and Eurasian wolves do not share haplotypes, although the differences are relatively small. This might show that wolves originating from Asia migrated repeatedly to North America, or that an early invading population was very diverse (Vilá *et al.* 1999). Recent work indicates that the (nearly extinct) wolves in Mexico might represent an ancestral population which migrated very early from Asia, and then was repeatedly driven southwards during glacial periods but often had the chance expand into the plains of North America (Wayne and Vilá 2001, Leonard *et al.* 2005). Another separate wolf population (*pallipes* subspecies) inhabits lowlands in India and regions in western Asia that seem to have separated very early (estimated 400 000 years ago) from the other wolves (Sharma *et al.* 2003). A further differentiated grouping was found among wolves living in the southern Himalayas and Tibet (*chanco* subspecies). Interestingly, neither former population seems to share mtDNA haplotypes with relatives that were among the ancestors for the domesticated dog. This suggests that neither population contributed to the dog's gene pool, although wolves living in this region were among those that could have been in very early contact with dispersing humans (Chapter 5.2, p. 97). More importantly, however, other wolves, which are also currently identified as 'chanco', seem to carry mtDNA that is very closely related to dog haplotypes. This could mean that only some of those populations, which are all recognized currently as representing the *chanco* subspecies, participated in the domestication process. The clear separation of mtDNA sequences between 'native' Indian breeds, local pariah dogs, and *pallipes* wolves provides evidence for a strong reproductive barrier between the two species (Sharma *et al.* 2003).

The overall diversity of the mtDNA is not as large as might be expected from a species distributed over the entire Holartic (sequence divergence within species: wolf–wolf = 2.9%; coyote–coyote = 4.2%; between species: wolf–coyote = 9.6%). This could be

---

Box 4.5 Wolf phenotypic plasticity

One reason why wolves may have been successful as the ancestor of dogs could be their phenotypic plasticity. Evolving and living in the temperate zone and surviving many glacial periods could have led to a species which has the means to adapt relatively rapidly to changing environments. To illustrate morphological and behavioural plasticity in wolves we combine data from various authors partially reported or cited by Mech and Boitani (2003).

• Recent wolves follow the Bergman rule, thus in general their size decreases from north to south. Here we use skull length as a measure because it correlates with body size but is less dependent on the actual state of the wolf (in some cases estimates based on the condylobasal length was used). Wolf skulls show a very marked increase in length (approximately 30%), and a clear sexual dimorphism (a).

• There is also a relationship between territory size and latitude in North American wolves which is partially attributable to the change in biomass (Fuller *et al.* 2003). From the behavioural point of view this means that wolves can adapt to areas where they have to travel long distances. This provides also indirect support for the rapid dispersion of any wolf sub-species, especially in the northern regions of Eurasia and America (b).

• Comparative data suggest that pack size increases in relation to prey size: the mean size of wolf packs hunting on bison may be twice as large as wolf packs for which white-tailed deer are the main prey (Mech and Boitani 2003). Naturally, pack size depends on many other environmental factors but this comparison shows that in certain environments wolves can be under selective pressure to maintain larger packs (c).

continues

Box 4.5 *continued*

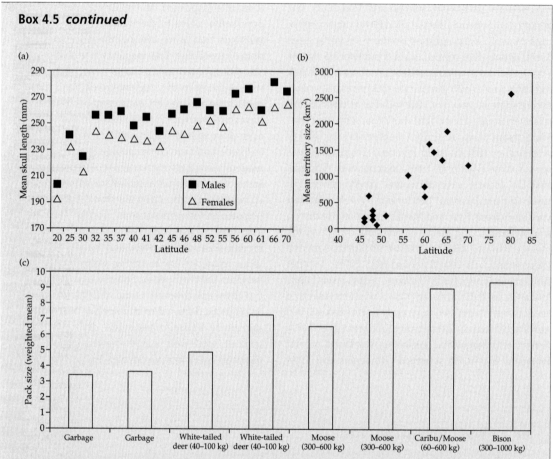

Figure to Box 4.5 (a) Mean skull length suggests the operation of the Bergman rule. Data for the lower latitudes come from Eurasia (Europe and Asia Minor) (Mendelsohn 1982, Okarma and Buchalczyk 1993, and other references cited herein); North American skull lengths have been obtained from Pederson (1982). (b) Territory size increases with latitude (based on data from Mech and Boitani 2003). (c) The relationship between prey size and (weighted mean) pack size (based on data reported by Mech and Boitani 2003). Prey weights refer to the smallest (female) and largest (male) values for the species reported and should be regarded only as approximate. In the case of 'garbage', 'white deer', and 'moose' the results of two independent studies are reported.

explained by the extraordinary mobility of wolves, which is revealed by cases when individuals from as far away as Portugal and Turkey share identical haplotypes. Local extinctions during glacial periods and, more importantly, more recent extermination of wolves in many parts of the world have had a major effect on the genetic diversity of the living population. In some respects recent wolves can be regarded as experiencing a 'glacial' period when in many locations wolves are forced to retreat to restricted locations. The good news for conservationists is that during evolution they survived many such situations;

thus, given appropriate environmental conditions, wolves could recolonize lost territories (Wayne and Vilá 2001).

According to Vilá *et al.* (1999), the genetic diversity found in wolves would predict a minimum of 1 million extant females in contrast to the actual 160 000–300 000 living animals. According to phylogenetic modelling, wolves have lost a considerable part of their genetic diversity. Based on the same data set the historical breeding female population size was estimated being around 5 million, of which 6% or even less have living relatives today.

The great mobility of wolves makes it not unexpected that there is no relationship between haplotypes and geographic distance, even at the continental level (Vilá *et al.* 1999, Verginelli *et al.* 2005). Wolves living in neighbouring countries or areas have often only distantly related mtDNA, but similar haplotypes may be shared by wolves living many thousands of kilometres away. The statistical comparison of more than 200 wolf mtDNA sequences indicated that Asian haplotypes might reflect the ancient condition, indicating the location for species evolution. Nevertheless, in view of historical fluctuations in population size and distribution during glacial and interglacial periods, manifested by local extinctions and hybridization, any direct phylogenetic connection between extant and extinct populations is doubtful.

4.3.3 Behavioural ecological aspects

Wolf research has been pursued in two directions. Large, undisturbed populations of wolves in the USA and Canada have become preferred objects of extensive field research, providing data on population and behavioural ecology of the species. However, the researchers have had to overcome many difficulties. Perhaps the most problematic thing is to get the wolves into the observer's visual range. Many populations avoid humans, live over vast areas, and move swiftly for long distances. Individuals migrate even further when leaving the pack. The xenophobic wolves do not tolerate the presence of others, and years can pass before zoologists are 'allowed' in the vicinity of the group.

Many ethologists and zoologists choose to observe wolf groups living in captivity in order to gain a detailed description of their behaviour. Although the lack of such data from the field made such investigations indispensable, there has, not surprisingly, been some disagreement about how such data should be interpreted (Packard 2003). First, there were arguments that the captive wolves were confined to a small space and had no chance to disperse over a larger area. Submissive individuals are prohibited from "leaving" the pack for shorter or longer periods in order to get out of sight of the more dominant companions. This could be problematic as the pack gets older, because under natural conditions wolves

more than 3 years old leave the group. The stress caused by reduced inter-individual distance and other disturbing environmental factors (such as the regular presence of researchers and other visitors) could result in behavioural abnormalities. Second, the composition (e.g. relatedness) of captive wolf packs is often arbitrary, and the structure does not correspond to that observed under natural circumstance. Third, captive wolves reported in different studies originated from different geographic regions (not always made clear in the published reports) which could be reflected in the observed behavioural variation. Thus studies on captive wolf packs are better viewed as modelling the potential forms of social behaviour which can happen in the wild, and one must be cautious in using such data to generate a behavioural model of the wolf pack (Packard 2003).

Territorial behaviour
According to Mech and Boitani (2003), wolf packs defend the area they inhabit, so for them home range and territory have the same meaning. The determination of territory size in wolves provides a great challenge because they travel a lot (up to 14 km per day; Mech 1966) and often cover huge areas. Field work utilizing various methods has provided evidence for exclusive use of areas by wolf packs, with very little overlap at the edges. This does not exclude the facts that some wolves (e.g. at dispersal) travel great distances, or some packs follow migrating prey (e.g. caribou, Sharp 1978), and that wolves cross into each others' territory when food becomes scarce.

The size of the territory might vary according to prey abundance. Territories become smaller with increasing amount of prey (biomass). This is probably also reflected in the relationship between latitude and territory size, hence wolves occupy a smaller area in the southern regions of their distribution (Mech and Boitani 2003). The largest home ranges can be found in northern Canada and Alaska (1000–1500 km²); European wolves (often living in natural reserves) usually inhabit much smaller home ranges (80–150 km²) (Okarma *et al.* 1998).

Pack size
The number of pack members can vary over the years. Wolves can have 1–6 offspring per breeding

season, but juvenile wolves leave the pack at the age of 9–36 months. Counting the actual number of individuals belonging to a pack is made complicated by lone wolves. Some of these have been expelled, but might be allowed to join again. In addition, wolf packs often split and reunite especially during the winter, and are generally smaller in the summer. The formation of larger packs is often constrained by environmental factors or simply because of the lack of offspring in a dwindling population (Pullianen 1965).

The size of the wolf pack can be anything between 2 and 42 individuals, but Fuller *et al.* (2003), after reviewing more than a dozen field studies, found the average pack in North America to consist of around 8 wolves. Average pack size in Europe is probably somewhat smaller (5–6 wolves) (Okarma *et al.* 1998). In some regions lone wolves could make up 90% of the population (Pullianen 1965).

Although a single wolf can seize an adult male deer or even an adult moose (Mech and Boitani 2003), wolves typically hunt in packs when foraging for larger game. Accordingly, it is often assumed that there is a relationship between the size of wolf pack and prey size because there is an optimum number at which the group can maximize net energy gain of hunting (Macdonald 1983). Pack size might be determined by their most frequent (or preferred) prey. Compiling a set of studies from North America, Mech and Boitani (2003) showed that there is a tendency for larger packs to coexist with larger prey (Box 4.5, Fig(c)). In areas where the white deer is the primary prey wolves live in packs of 5, while packs preying mainly on moose or caribou tend to reach the size of 9 individuals. In Poland the most frequently observed packs consisted of 4–6 individuals preying mainly on red deer. Jedrzejewski *et al.* (2002) explained this by the fact that such packs consume the kill at a sitting.

Changes in pack size also take place when the main prey varies according to season. Decease in size can, however, be the result of different confounding factors, such as increased mortality by the end of the winter or increased dispersal. During food shortage the number of individuals expelled from the pack increases (Jordan *et al.* 1967).

Bigger packs have a higher killing rate (Schmidt and Mech 1997), although the latter also depends on the availability of prey animals and the size of the last meal. Both American (Mech 1970) and European wolves (Jedrzejewski *et al.* 2002) hunt on average every second day.

More recent investigations emphasize that competition from scavengers, such as ravens, could mean that bigger packs are more successful in defending killed prey (Vucetich *et al.* 2004). There is probably also an optimal size for the actual hunting team. This is supported by the frequent observation that bigger packs break up before hunting, and the hunting teams are usually assembled from 4–6 wolves (Mech 1970). Derix *et al.* (1993) argue that cooperative hunting and defending prey strengthens the bond between males.

The flexibility of pack size in wolves may be critical to their success in inhabiting a range of very different environments. As shown above, actual pack size depends on the presence and interaction of many different factors, including prey size, optimal number of the hunting team, consuming the kill at once, defending the kill from scavengers, food availability and density (Mech and Boitani 2003, Okarma and Buchalczyk 1993). Trends for pack size at one locality may not hold true for other regions.

Feeding habits
The feeding habits of wolves vary according to their habitats, which were probably not so markedly different during prehistoric times when the habitats were less fragmented and prey animals could also disperse over vast areas (although they might have experienced increased competition from larger predators, see above). At present wolves in North America and Canada still have the chance to focus only on large herbivorous prey, whereas their Eurasian companions, especially in Europe and west–south Asia, have to maintain a much more varied diet (Fuller *et al.* 2003). The main prey of North American wolves consists of caribou, moose, and reindeer, although they also forage for smaller prey, particularly in the summer. In contrast, European wolves feed on red deer, wild boar, and roe deer but their diet more often includes smaller prey such as hare, ground squirrel, or mice (Jedrzejewski *et al.* 2000). Wolves also prey on domesticated animals (most often on sheep; not on adult cattle, but only on their young) but this occurs

more frequently in regions where there is less opportunity to hunt in the wild. Once wolves habituate to the presence of humans, which often happens in Europe and western Asia, they also visit refuse dumps, as found in the case of Italian and Israeli wolves (Boitani 1982, Mendelsohn 1982). In extreme cases eating garbage could account for 60–70% of their food intake.

Although wolves have a broad diet, it is interesting to note that in most cases the two prey species most often consumed amount to 80% of the total food consumption (Mech 1970). This suggests some form of specialization or preference for particular species. In Poland, Jedrzejewski *et al.* (2002) found that wolf predation affected mainly the number of red deer in the Białowieża forest. There was no close correlation between number of wolves and size of the deer population, but the presence of wolves in this area slowed down the rate of deer reproduction. Wolf killing amounted to 40% of the annual increase in red deer, and was responsible for 40% of mortality. In contrast, no such effect was observed in the sympatric wild boar, roe deer, and moose populations.

Wolves also optimize their prey preference so that they choose the easier alternative if possible. If large prey of different sizes is available, then wolves take the smaller one (Mech 1970), but such an effect can be explained partially by the wolves themselves being relatively small. Peterson *et al.* (1984) found that in Alaska larger wolves tend to hunt on larger game. Smaller wolves in south-east Alaska hunt mainly deer, whereas much larger individuals living in the interior of Alaska prey mostly on moose. They argued that the hunter, as an individual, needs also to have a certain weight (strength) to be effective. This can also explain why wolves living in disturbed southern areas do not prey on large wild herbivores, and develop a preference for human waste or domestic animals (e.g. Mendelsohn 1982). Another case of such specialization was reported by Darimont *et al.* (2003) who described wolves preying on salmon, but eating only the head of the fish. This preference could reflect avoidance of parasites in the body, or a preference for the more nutritious head; in any case it would be interesting to know how wolves acquired this habit.

4.3.4 Social relationships between and within wolf packs

In contrast to morphological and genetic research, comparative research on behavioural differences among wolf subspecies is lacking. Discussions on social behaviour always refer to 'the wolf' in general. However, the dynamic and variable social system of wolves probably played an important role in their survival in a range of different ecosystems. This ability to establish various group structures might have arisen as a consequence of being exposed to diverse environmental conditions during glacial and interglacial periods during the Pleistocene, when the changing climate affected many aspects of their habits. The wolf gene pool might have gained certain features that allow for a flexible phenotype and most of the morphological and behavioural features that separates their population are signs of developmental plasticity (Chapter 5.5.4, p. 125), and are not 'adaptations' in the strict sense. Nevertheless, until further research is done genetic differences in social behaviour cannot be excluded.

Inter-pack relations
Wolf populations inhabiting diverse geographic locations should be viewed as a complex network, which is maintained by the dynamic relationships among packs. The number of wolves and their distribution in this network probably depend on two main factors involving food supply and diverse social factors (Packard and Mech 1980). Both seem to have an effect on the population size, although large variations have been observed. In some cases population size does not follow increasing availability of food resources and seem to stabilize at a lower level (Mech 1970), but in other instances rapid population growth was recorded (Wabakken *et al.* 2001). Similarly, mortality can affect wolf populations to a varying degree. A survey of studies on wolves suffering only little human disturbance indicated an annual average mortality around 25%, more than half of which consisted of the death of starving cubs (Fuller *et al.* 2003).

Inter-pack relations are influenced by three main factors: dispersal of young, territorial defence, and acceptance of unrelated individuals in the pack.

Under natural circumstances the rule is that both male and female young wolves leave their native pack. The proximate causes for this behaviour might involve food and/or mate competition, but the avoidance of inbreeding can also play a role. Dispersal is a gradual process; some individuals might return for a shorter or longer period to the pack before leaving for ever. Based on 75 dispersed juvenile wolves from north-eastern Minnesota (USA), Gese and Mech (1991) reported that most individuals left the pack at 11–12 months of age (26%) and most of the departing wolves migrated before their second birthday (79%). The majority (67%) of older wolves (up to 3 years) succeeded in finding a denning place; in comparison, only 25% of the younger wolves (less then 1 year old) were able to establish an independent life. The condition (weight) of the departing wolf did not seem to affect its chance. Both sexes left the pack at the same frequency but females remained nearer their original pack than males. In general juveniles migrated further than adults. The extent of dispersal ranged between 8 and 432 km. Dispersal seems to be based on individual decision (although animals are often 'forced' to leave the pack), as only single animals left the pack despite the obvious hazards associated with this behaviour. The success of the dispersers depends on various factors, such as finding a suitable mate and the number of available territories. Observations showed that the rate of dispersal is lower under both favourable and poor food conditions and becomes more variable in intermediate conditions.

Wolves do not tolerate strangers on their territory, which often leads to fierce fights if neighbours encounter each other at the edge of the territory. The behavioural rules of territorial aggression are different from those in the pack; thus, in contrast to within-pack clashes, wolves are often killed in these situations, but are generally not eaten. Similarly, packs behave aggressively towards lone wolves who often follow them at a distance (Mech and Boitani 2003). In some exceptional cases, usually if a pack has lost breeding individuals, wolves might also 'invite' strangers to join. Younger wolves have a better chance of being accepted. Stahler *et al.* (2002) reported a pack that lacked a dominant animal and allowed a breeding male wolf to join.

It has long been believed that the dispersal behaviour of wolves increases genetic diversity between adjacent packs; in contrast, close within-family ties result in higher levels of inbreeding, hence less divergence within a pack. Observations and genetic analysis suggest a more complex situation within populations. Relatedness between packs decreases with distance, probably because after a pack splits wolves usually stay in neighbouring territories, and most dispersing wolves join packs living nearby. However, the genetic difference between packs is actually smaller than was thought previously (Lehman *et al.* 1992). This also suggests that wolves are quite successful in joining neighbouring packs. One might assume that if a former family member had already been accepted into a pack, newcomers from the same pack might have a better chance, as in the case of packs where all members are strangers. As noted earlier, successfully dispersing wolves often establish kinship between geographically distant populations, thus wolves can be related over a wide distance ranging from Alaska to eastern Canada and southern Minnesota (Roy *et al.* 1994).

Intra-pack relationships

Our assumptions about social relationships in a wolf pack have undergone significant changes over the last few years. Today most zoologists agree that the wolf pack should be regarded as an extended family, which consists of a breeding pair and their offspring (Mech 1999, Gadbois 2002, Packard 2003). Most of the problems were rooted in the disagreement between field and captive studies on the social structure and hierarchical relationships within wolf packs. Observers of wolves living in captivity (often characterized by restricted range and unnatural pack composition) witnessed a heightened level of agonistic interactions and the development and stabilization of strictly hierarchical rank relationships. This provided the basis for a model that described the social system in wolves as linearly hierarchical. Others (e.g. Zimen 1982, Fentress *et al.* 1987, Derix *et al.* 1993) were biased in favour of a separate hierarchy for males and females with the position of the wolf being strongly determined by its age (sex/age graded hierarchy). Such a social system is often characterized by agonistic tensions which are caused by either harassment and suppression of (younger) subordinates, or the repeated challenges and provocation of the dominants (see Packard 2003 for a review). Mech (1999) argued against separate male and female hierarchies because in wild packs males dominate females, and breeding males never submit to females, but the reverse often happens. However, the relatively small sexual dimorphism in wolves does not seem to support a forceful maintenance of hierarchy. Thus ethologists watching wolves slowly became convinced that the model described above overestimates behavioural enforcement of wolf hierarchy by aggressive behaviour.

A significant conceptual change occurred when Mech (1999), Packard (2003), and others suggested that the wolf pack should be viewed as an extended family (Gadbois 2002). They argued that in most cases a pack is formed by two young wolves that are strangers to one another, and develop into an extended family by sharing their life with companions 1–3 years old that are their offspring.

The oldest and most experienced wolves in the pack are the parents (the founding breeding pair) which share the leadership role and both have greater rights to make decisions in the group. In most cases this leadership role is focused on the same-sex companions, but the female seems to assume a leading role when there are pups to be raised, while the male is primarily involved in organizing foraging and provisioning. According to Packard (2003) this view of the wolf pack is still very deterministic; she argued for two-directional relationships between parents and their offspring. The family model of the wolf pack suggests a more flexible hierarchy and that the behaviour of the offspring has also an influence on the decision-making process in the pack (Box 4.6).

The family concept does not exclude hierarchical/dominant relationships. It is natural that parents have more chance to exert control over their offspring because of their advantage in both

---

Box 4.6 Modelling the social structure of wolves

In recent years researchers have begun to revise the original social model of the wolf pack which was based on a behaviourally enforced strict linear hierarchy (a). This model assumed that all wolves aim for the dominant position because this is the only way to ensure the propagation of their genes. This view was changed on the basis of field observations which showed that although most packs raise only a single litter, pack members belong to the same family and young wolves leave the pack between 1 and 3 years of age (Gese and Mech 1991). This provides the wolves with an alternative tactic to ensure reproduction. In addition, detailed observations failed to find statistical support for a linear hierarchy (Lockwood 1979).

One alternative model is a sex/age graded hierarchy (e.g. Zimen 1982) which is based on observations that males may dominate females and that parents more often show dominant behaviour towards offspring (b), but at the same time this model stresses separate hierarchies for

males and females (e.g. Fentress *et al.* 1987). This view was challenged by Lockwood (1979) and Packard (2003), partly because sex and age confound assumptions about dominance.

Thus Packard (2003) advances a family model of the wolf pack (c). The main difference in the family model is that besides recognizing the agonistic aspect of inter-individual relationships, it stresses that the dominant presence of mutual affiliative and attentive behaviours ensures 'peaceful' social life in the pack for most of the time. In her terms the 'dominant' behaviour of the parent and the 'submissive' behaviour of the offspring might be viewed as 'parental aggression' for executing behavioural control. In parallel, younger wolves might display 'exploratory aggression' for finding out the limits of parental indulgence on the part of the pups. Lockwood (1979) suggested that the wolf social system could be described as one in which animals switch from one social role to another as they get older.

continues

Box 4.6 *continued*

The 'hierarchy' and the 'family' models have many common elements. However, while the former model refers to wolves as 'alpha, beta, . . . omega animals' or 'dominants', the family model prefers categories such as 'leaders' or 'breeders'. This propagation of new categories has created some confusion in the literature and it would be useful to settle for one unified nomenclature. In any case, however, these changes in our understanding of the wolf social system should be also a warning for those who apply these concepts uncritically to dogs.

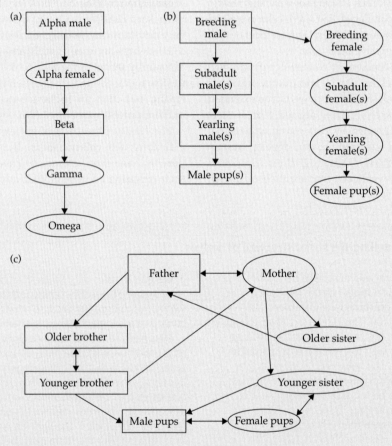

Figure to Box 4.6 Various models of the social system of a wolf pack (redrawn from Packard 2003).

physical strength and experience. Thus as a default in most packs parents play the role of leaders, controlling pack movements and taking other decisions. Peterson *et al.* (2002) reported that breeding parents (who also did most of the scent marking) were more likely to lead the pack during travel or pursuit of prey, and they seemed to share this role, apart from the period when the female had cubs. Lower-ranking wolves provided leadership only shortly before their dispersal, or when they were members of larger packs. However, even in such cases the behaviour of the dominant wolf often influenced the pack's activity. If age and experience are important for leading a group of wolves, packs with such animals may be at advantage. Such knowledgeable individuals could know more about the availability of food or the optimal movements across the territory.

In line with the family concept, both Ginsburg (1987) and Packard (2003) emphasize the emotional aspects of inter-individual relationships within wolf packs, in which cohesive and agonistic forces work in parallel and their balance determines the social stability of the pack. Accordingly, the relationship between wolves is influenced not only by dominance and rank order but also by affective behaviours individuals display towards their companions. This would suggest that the craving for a higher social status is counteracted by the need to maintain close emotional ties. Affective relationships might develop during puberty, when maturing individuals are slowly integrated into the structure of the pack. Observations on wolves indicate that the social stability of the pack is most important, and all members display a tendency to show appeasing behaviour apparently in order to reduce tension (Schenkel 1947, 1967, Fentress *et al.* 1987, Packard 2003). Zimen (1982) reports that for captive packs, lower-ranking males often assume 'pup-mimicry' possibly in order to avoid male aggression; similarly, lower-ranking females try to be as cryptic as possible in order to avoid attacks by the ranking female.

Social relations and mating in wolves
The mating season starts in midwinter, and wolves court and mate from January up to beginning of April. It seems that this is the critical time of the year when agonistic social interactions intensify mainly intrasexually. Packs typically produce a single litter per year, and sampling a range of 3–16 wolf packs in Denali (USA) over 7 years Mech *et al.* (1998) reported 0.7–5 cubs per pack, averaging 3.8 cubs raised in a pack per year. Field observations suggest that most courtship activity is confined to the breeding pair, and the dominant male interferes with any attempts by lower-ranking males to approach his mate (Harrington and Paquet 1982, Mech 1999). The male and female of the breeding pair follow different tactics to prevent mating between other pack members, which influences the temporal pattern of agonistic interactions in the group (Derix *et al.* 1993). Breeding males concentrate their intervention efforts on the period of mating. They try to prevent male–female sexual interactions, especially if their mate is involved. In

parallel, they are aggressive towards other males. In contrast, there is a lower level of intrasexual aggression among females but the breeding female assures that this is maintained during the whole year in various contexts, including feeding or group howling. (Such prevalence of intrasexual aggression could actually bias towards the view of a separate dominance hierarchy.)

According to Packard (2003) the development of a monogamous relationship is more likely in packs in which there is a stronger attraction between the breeding pair, offspring is reproductively premature, and parents are successful in intervening in all courtship attempts in the pack. In general the breeding pair has a greater chance of raising their offspring in a larger pack. Thus breeding animals have to balance between making the pack a comfortable place to stay for the yearlings or older non-breeding animals and preventing their mating opportunities by force.

In certain conditions the structure of the wolf packs deviates from a family unit; such groups are larger in number and have a more complex pack structure involving many unrelated wolves and multiple breeders. It is conceivable that these bigger groups are organized more hierarchically which is enforced by a dominant male (the alpha male). It is more likely that such hierarchical relations are less confounded by factors of age and relatedness. Although such packs form less frequently, their regular occurrence suggests that wolves are able to live in flexible social hierarchical systems.

In large packs complex mating patterns might emerge: the second-ranking male often succeeds in mating with the breeding female, or the breeding male ties with a lower-ranking female. The occurrence of multiple litters suggests that the presence of the breeding female does not physiologically suppress the lower-ranking females, and they retain the potential to reproduce throughout the mating season (Packard *et al.* 1985). Although most authorities agree that a typical wolf pack produces one litter per year, the rare presence of multiple litters indicates that the reproduction is affected by different and sometimes opposing factors. It could be in the interest of the breeding pair to restrict the production of a litter to themselves, but good environmental conditions could favour multiple litters. There are assumptions that the variability in number of litters in wolf packs reflects variation in food availability. So far infanticide has not been reported from the wild (at least it is not mentioned in Mech and Boitani 2003) but it is not exceptional in captive packs (see Packard 2003).

In some cases females can suppress or speed up their maturation. Normally female wolves are sexually mature in their second or third year, but in captivity female wolves can reach maturity within a year (e.g. Medjo and Mech 1976). This suggests that the timing of maturity may be under influence of environmental factors such as food availability or social suppression by other females.

Social relations and food sharing
The harmony of the wolf pack may come also under threat in cases of sharing food (Packard 2003). The sharing of prey depends on its size (Mech 1999), and generally breeders control food distribution. Inter-individual relationships often influence dyadic tolerance, and appeasing individuals have a chance to gain some meat (Packard 2003). In the case of large prey (e.g. adult moose) there are only minor disputes and everyone is allowed to eat. If the prey is small (e.g. musk ox calf), dominant animals eat first. Quarrels are more intense between juveniles if parents are not present. Mech (1999) described an 'ownership zone' around the mouth of an individual, by observing that once a wolf succeeded in securing a piece of meat in his mouth (or within lungeing distance) this is 'respected' by the others. Lower-ranking animals may carry a small piece of food in front of the dominants 'provocatively' with raised tail and head.

A special case of food sharing occurs during whelping. Protection and feeding of the cubs generally involves a collaborative effort from the breeding pair and partly from older offspring in the pack. Although many assume that alloparental behaviour of young adults or juveniles from previous years is an important contribution to the maintenance of the pack, field observations suggest (Mech *et al.* 1999) that staying with the pack could be in their own individual interest. In general, breeding wolves control the amount of redistributed food; however, hunting yearlings might

also regurgitate food for their younger brothers and sisters, although at other times they compete with them for food from returning parents. The parents are more likely to give food away when it is scarce, and at such times it might be more advantageous for the parent to feed yearlings than cubs (Mech 1999).

In captive wolves having freely available food Fentress and Ryon (1982) observed selective feeding. Adult wolves fed both yearlings and cubs, and mothers mainly got food from male adults (Paquet and Harrington 1982).

4.3.5 A comparison: social organization in free-ranging dogs

What are 'free-ranging dogs'?
The cohabitation of dogs and humans is a dynamic process, and a considerable part of the dog population has lost contact with humans for shorter or longer periods. Such incidents have often happened during the (at least) 15 000 years of domestication and also take place daily in our own time. There are arguments that such 'free-ranging' dogs provide a natural situation in which one can investigate dog behaviour ethologically. Bradshaw and Nott (1995) complain that the complex interaction and influence of humans prohibits researchers from being able to observe the species-specific aspect of social behaviour in dogs. Others maintain that these free-ranging dogs are good models for the ancestral canid populations prior to domestication (Coppinger and Coppinger 2001, Koler-Matznick 2002).

Unfortunately both the systematic and ecological categorization of free-ranging dogs is the subject of considerable discussions. The main reason is that researchers have found it difficult to separate genotypic and phenotypic effects on these dogs, and are often using a confusing set of definitions. Taking into account the arguments in the modern literature we suggest the following distinctions (see also Box 4.7).

Feral dogs differ from their domestic companions because they have not been exposed to close human contact early in their life (lack of socialization) but they have a gene pool that is typical for domesticated dogs (i.e. these dogs have not been exposed to selection by the natural environment) (Daniels and Bekoff 1989, Boitani and Ciucci 1995, Boitani

Box 4.7 Wolves and dogs in the anthropogenic environment: socialization, feralization, genetic changes

There is often a misunderstanding in the use of categories and the labelling of the processes which differentiate wolves from various populations of dogs (see also Boitani and Ciucci 1995). Wolves kept in captivity with little human contact can be regarded as *habituated*. After more direct human contact wolves can be *tamed*, especially in the case of young individuals. Human foster parents can *socialize* a wolf if they replace the parent just after birth, maintain close contact with the wolf, and exclude conspecifics. *Domestication* is the result of a genetic change; however, dogs become socialized only if they are raised in a human social environment (owned dogs). Some dogs lead a

relatively free life despite being socialized to some extent. These dogs have or can establish a social relation with human(s) and may be fed and sheltered regularly (stray dogs, village dogs). Dogs are regarded as *feral* if they have not been socialized and therefore have no individualized contact with humans. Feral dogs spend most of their time away from human settlements. They can revert to being stray dogs or owned dog if socialized to humans, but stray dogs or owned dogs can also infiltrate their society. Finally, if dog populations experience no influx from other dog populations for many generations genetic changes might stabilize. Dingoes are one example of such dogs.

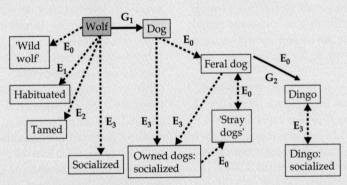

Figure to Box 4.7 A conceptual model of environmental (developmental) and genetic effects on dogs and wolves. G_1, domestication; G_2, genetic changes after isolation; E_0, no humans present in the environment; E_1, E_2, various levels of human exposure; E_3, early (and extensive) socialization.

et al. 1995). Not only were the ancestors typical dogs living in the anthropogenic environment, but these populations receive continuous influx from dogs that share their life with humans (Beck 1973). Thus the lack of genotypic change makes it possible that these dogs can be 'rescued' by early (during the socialization phase, see Chapter 9) socialization to humans. Adult socialized offspring from feral dogs should be indistinguishable from other dogs living in human families. Note that in this sense feralization is the opposite process to socialization and not to domestication, which was often implied in earlier writings (Kretchmer and Fox 1975, Price 1984).

If separation of some dog populations from humans occurred long ago, and there was no chance of further

genetic influx from domestic populations that were under continuous selection by humans, then genetic changes might have taken place. These could involve the realization of a founder effect, genetic drift, or various forms of selective directional changes (Chapter 5.5, p. 119). It is assumed that if these genetic changes affected systems involved in the original domestication process, then despite the exposure to humans (socialization) these animals will deviate from the domesticated (and socialized) phenotype of dogs.

So far there is no direct evidence that some dog populations have undergone evolutionary changes (because there are no controlled experiments which attempted to socialize these animals, although Australian Aborigines practised such activities, Chapter 3.2, p. 50) but there are strong indications that dog populations in Australia (dingoes) (Corbett 1995) and New Guinea (singing dogs) (Koler-Matznick 2002) would fit into this category. The crucial point here is that continuous influx from other dog populations disrupts genetic changes like those listed above. However, in the case of these island populations a separation for many thousands of years was probably enough for the stabilization of genetic changes, some of which might include adaptive changes to their actual environment. In addition it might have been not accidental that such 'dingoes' evolved on islands that lacked competing carnivores, e.g. wolves.

We should see here that just as in the case of domestication, genetic isolation from the other population is a prerequisite for genetic divergence. In the absence of a proper term the process of genetic isolation that takes place over many thousand generations in the wild could be referred to as *dingalization*. It is unfortunate that today the genetic isolation has broken down and ('pure') dingoes have increased chance to hybridize with (feral) dogs (Corbett 1995).

Thus both feral dogs and dingoes can be regarded as 'free-ranging', but in the case of the former changes in the phenotype are caused mainly by being exposed to a different developmental environment, while in the case of the latter both phenotypic and genotypic effects are responsible.

Feral dogs

The comparison of feral dog populations with wolves can reveal how domestication affected the organization of social behaviour because both forms have the chance to express their behaviour in the same environment (such populations were studied in Italy by Boitani 1983, Boitani and Ciucci 1995). If feral dogs show some phenotypic similarities to wolves then it is less likely these aspects of the behaviour have been subject to genetic changes.

Long-term detailed observations point to a divergent nature of feral dog society (e.g. Daniels and Bekoff 1983, Boitani and Ciucci 1995, Boitani *et al.* 1995, Macdonald and Carr 1995). The organization of group size and social structure depends on the habitat, the food supply, and the kin relations in the group. Reports on European populations suggest that feral dog populations are not self-sustaining, and their survival depends on a steady influx from

other dog populations. However, this does not mean that feral dogs are not able to survive in the wild. Successful establishment of a population depends among other things probably on the food supply, competitors, human extermination, or increased sensitivity to disease and parasites.

Some behaviour patterns typical for domestic dogs make feral dogs vulnerable to seasonal changes (Boitani *et al.* 1995). The low survival rate of feral dog pups is often attributed to the lack of paternal care (typical for all other *Canis* species) and alloparental behaviour which leaves the care of pups exclusively to the female. Feral dog mothers often rear their young at some distance from the group (Daniels and Bekoff 1989). Feral dogs also maintain a dioestrus cycle, which means that one litter is often born at a less favourable time of year (e.g. late autumn or winter) and thus the female has less time to provision the first litter before the next whelping.

Interestingly, observations on West Indian feral dogs show a different picture. Pal and colleagues (1998, 2003, 2004) provided evidence that these dogs have only a single oestrus cycle, and the males show parental behaviour. They stayed with the pups at the den and protected them against intruders, and one male was observed to feed the pups by regurgitation. Biparental care did not reduce high (63%) mortality in these dogs, although this rate falls within those observed in wolves (Fuller *et al.* 2003).

If food is abundant and there is no human interference, feral dog populations have the potential to survive and propagate because they are organized around one or more monogamous breeding pairs and group members are relatively tolerant of each other (Macdonald and Carr 1995, Pal *et al.* 1998). In comparison to wolves multiple litters are more frequent, and mothers are usually not exposed to harassment from companions, and have sometimes been observed to feed each others' young (Pal 2004). However, most juveniles disperse by the end of their first year, earlier then observed in wolves (Pal *et al.* 1998). Thus at many locations when no feral dog control is implemented and where the climate puts less stress on breeding (e.g. food availability), feral dog populations are on increase.

At some sites feral dogs have been observed to hunt and kill larger prey, both wild and domesticated animals (e.g. Jhala and Giles 1991); however, in general they prefer scavenging or hunting on small prey if given a choice (Butler *et al.* 2004). They hunt mostly alone; group hunts are rarely observed. Organized pack hunts have not been described for feral dogs.

Similarities between wolf and feral dog groups depend on ecological factors. Like wolves, feral dog packs are territorial, maintain a home range of variable area, and show a similar pattern of daily activity. Although there is apparently large variation among feral dog populations, and few observations on undisturbed packs have been published, some researchers still doubt whether the social organization described for feral dog groups meets the criteria of a canid pack.

Dingoes
Similarities between wolves and dingoes in social traits could suggest that early domestication (*c.*5000 years ago) did not affect these features of behaviour (although they can be the result of parallelism). In general the social structure of Australian dingoes is more similar to that of the wolves than to that of feral dogs. Both Corbett (1995) and Thompson *et al.* (1992) report dingo packs made up of relatives, who jointly defend their territories, hunt together, and participate in the nursing of young animals. This means that in contrast to feral dogs, dingoes form 'real' packs. Dingo packs are also comparable to wolf packs in numbers and home range area, although variations can be large when different regions and habitats are compared (Corbett and Newsome 1975, Thompson *et al.* 1992). Thompson *et al.* (1992) observed packs ranging from 2 to 13 individuals (21% of all sightings were solitary animals; mean pack size was 2 on territories covering 40–110 km^2).

Dingoes live in a hierarchy, in which the breeding male is usually observed as the leader during travel, when eating prey, or when approaching drinking holes for the first time. In contrast to feral dogs, large mammals usually comprise a relatively large part of the dingo's diet (*c.*20%). They hunt for various species including reptiles, birds, marsupials, and domesticated animals, and in some areas dingoes can cause major losses of domesticated stock (Corbett 1995).

Although little information is available, dingoes living in packs are described as being less aggressive to each other then wolves. This is also the case in the mating season, when the dominant breeding pair does not seem to inhibit mating of other pack members. For captive dingoes Corbett (1988) reported that most packs raise only a single litter because after whelping the dominant female kills the offspring of other mothers. The male dingoes take part in raising the offspring by providing food and social experience. Having lost their pups, subordinate females contribute to feeding the dominant female's offspring. Corbett (1995) suggested that this behaviour in dingoes might reflect a behavioural adaptation to the extreme ecological conditions, because it assures that at least a single litter (with many alloparents from the pack) survives when food and water is scarce. Although loss of pups due to cannibalism was also reported in wild dingoes (Thompson *et al.* 1992), some packs actually raised multiple litters. Thus the role of infanticide as a means of population regulation under natural conditions remains uncertain.

The similarities between dingoes and wolves suggest that at the time of their separation domesticated populations had not lost the behavioural potential for paternal (males) and alloparental (yearlings) care, although the possibility that dingoes 'reinvented' this behaviour cannot be excluded. The increased individual tolerance during mating periods could be the result of early domestication reducing aggressive behaviour towards group mates (low levels of aggressive interactions have been also noted in feral dog groups) or an adaptation to environmental challenges. The emergence of infanticide in dingoes could be viewed as an example for the operation of Dollo's rule ('evolution is not reversible'). The reduced intra-group aggression towards mates selected an alternative solution to group size regulation: while wolves suppress multiple litters by within-sex aggression, the same outcome is achieved in dingoes by infanticide.

4.4 Wolf and dog: similarities and differences

Historically, scientists have tried to identify morphological or behavioural (more recently genetic) features which would help in the objective identification of wolves and dogs. Such categorization has turned out to be very difficult. Although molecular genetic work has found molecular markers that distinguish reliably between wolf and dog (Vilá *et al.* 2003), phenotypic markers are difficult to establish.

The problem of describing categorical differences between dogs and wolves is rooted in the fact that despite their ecological separation, the two species share most of their phenotypic traits and qualitative differences (traits that are present in only one of the species) are rare. In reality most differences are quantitative, and there is a large overlap between the species-specific variations. In addition, most of these quantitative traits have never been examined in detail and compared across species.

4.4.1 Morphological traits

It is clear that by looking at their morphological and anatomical features wolves and dogs can be easily told apart, especially if the dog belongs to some specially selected breed. The situation becomes more difficult if one compares wolves with 'wolf-like' breeds like the German shepherd or the malamute, or if only a smaller set of morphological evidence is available (such as a tooth or a long bone).

Although there is little conclusive evidence, there are indications that dogs and wolves might be distinguished on the basis of a few qualitative traits. Such discrimination is usually based on features that are missing from the wolf but may be present in the dog. It follows that these features are useless if the dog does not show them. Linnaeus himself noted the sickle-shaped tail of dogs. Such a tail shape has not been observed in any wolf; similarly, wolves never have the drooping ears which emerge in some dogs (but not all) (Clutton-Brock 1995).

In the case of quantitative variables the categorization is based on statistical methods, which make the process very complicated because usually one phenotypic variable is not enough. For example, wolves and dogs have an overlapping variability in the length of the humerus (Casinos *et al.* 1986). The Irish wolfhound probably has a longer humerus than most wolves. Thus dogs and wolves cannot be told apart on the basis of humerus length.

Measuring the diameter of this bone, it turns out that wolves have a thinner humerus than dogs. Statistical methods (linear regression) reveal this difference between the two species. However, some dog breeds, e.g. Afghan hound, have a similar length/diameter ratio to the wolf. Thus upon finding a humerus one cannot be certain whether it belonged to a wolf or a dog, and only the inclusion of further phenotypic variables allow successful identification (Wayne 1986a, b) (Box 4.8).

4.4.2 Behavioural comparisons

Over the years ethologists have compiled a long list of behavioural elements (an ethogram) which characterize the wolf (e.g. Schenkel 1947, Fox 1970, Frank and Frank 1982, Feddersen-Petersen 1991, Packard 2003) and researchers raising wolves and dogs have often reported on the observed behavioural differences between individual animals (e.g. Fentress 1967).

However, comparable ethograms including quantitative data for dogs are lacking with a few exceptions (e.g. Bradshaw and Nott 1995, Goodwin *et al.* 1997). General behavioural observations on various dog breeds, mongrels, or feral dogs suggest that they represent certain 'mosaic' constructions of the ancestral wolf pattern. Thus any given dog population displays only a restricted subset of the wolf ethogram (e.g. Coppinger *et al.* 1987, Goodwin *et al.* 1997). Many observe a large individual variability in the behaviour of dogs, which makes them less predictable than wolves (Fox 1970, Ginsburg and Hiestand 1992).

Fox (1978) lists four possible sources of quantitative behavioural difference between dogs and wolves, of which barking provides a good example (see Coren and Fox 1976, Schassburger 1993, Pongrácz *et al.* 2005). Both wolves and dogs bark (see Chapter 8.4.2, p. 185) but it seems that in (many) dogs the threshold for barking is lower (*threshold change*). The pattern of barking in dogs differs, as they emit this vocalization in long bursts and combine it with other vocalizations (*sequential changes, omission*). Wolves bark in special social contexts ('warning and protesting') whereas in dogs different types of barks are emitted in various social situations (*ritualization*). Dogs can be taught to bark (or withhold barking) in response to some external stimuli (*ontogenetic modification*: learning, training).

Some behavioural differences might be secondary—associated with alteration of other morphological features, sensory ability, hormone levels, etc.—or might be the result of phenotypic plasticity and do not indicate genetic changes (e.g. wagging of hind end of the body in the absence of a tail; Fox 1970). The greeting pattern in dogs might be different because of the absence of certain glands (e.g. the supracaudal gland) used for olfactory

signalling (Bradshaw and Nott 1995), or the lack of movable ears or tails could cause changes in the communicative behaviour.

According to Fox (1970) wolf-like grinning is used by dogs (lips are retracted vertically and horizontally exposing the teeth) only towards humans. To many this resembles a human grin, whereas others describe it as 'smiling'. The use of this signal might provide a case for ontogenetic ritualization (Chapter 8.5, p. 190).

Finally, it is interesting to note that the New Guinea singing dog, which is genetically a close

---

Box 4.8 Comparisons between wolf and dog

Over the years scientists have compiled lists of features that can be used for identifying wolves and dogs. Unfortunately, most such lists are based on qualitative comparisons and provide very general statements. Wolf and dog population-level comparisons are lacking.

There are some features of the skull that could be typical for one species on the basis of relative comparisons. For example, a tooth could indicate the species if found in a mandible, but not alone. For most such measures there is a need for some sort of scale along which the individual data could be categorized.

Morphological traits

Some suggested differential morphological traits that have been regarded by many authorities as distinguishing wolves and dogs:

• *Dew claws*: Wolves never develop dew claws (first digit: hallux) but they are also missing in most dog breeds (Clutton-Brock 1995).
• *Tail*: Wolves never have a sickle-shaped or tightly curled tail, but this is also lacking in most dog breeds (Clutton-Brock 1995).
• *Ears*: Wolves' ears are always erect and never drooping (but many dogs also have erect ears).
• *Tail glands*: The supracaudal gland is absent or reduced in dogs (Fox 1971, Clutton-Brock 1995).
• *Lower jaw*: Turned-back apex on the lower jaw in dogs (which is present only in some wolf subspecies, Chinese wolf (*C. lupus chanco*) (Olsen and Olsen 1977).

Relative differences in the skull

Some suggested differential morphological traits in which relative differences in the skull are indicative of the species (most references from Clutton-Brock 1995 if not stated otherwise):

• *Skull and body*: Skulls of dogs are shorter and smaller (volume) for the same body weight (Kruska 2005).
• *Skull and teeth*: Teeth are smaller in relation to the skull (Wayne 1986b, Morey 1992).
• *Skull length and width*: The muzzle is wide relative to its length; in the skull the palate and maxillary region became shorter and wider, in relation to skull length (this is why the dog appears to have a shorter nose) (Box 5.5).
• *Skull and sinuses*: Frontal sinuses are enlarged in dogs.
• *Skull and bullae*: The auditory (tympanic) bullae are smaller and flatter in dogs.
• *Skull and forehead*: The angle of the forehead ('stop') tends to be larger in dogs.
• *Skull and orbit*: In the dog the shape of the orbit is more rounded, and the eyes look more directly forwards.
• *Mandible and teeth*: The upper tooth row is more bowed and the angle of the mandible deeper with the ventral edge more convex, mandible deeper in wolves.
• *Mandible and teeth*: Teeth in dog are often more compacted, especially in the premolar region.

continues

Box 4.8 *continued*

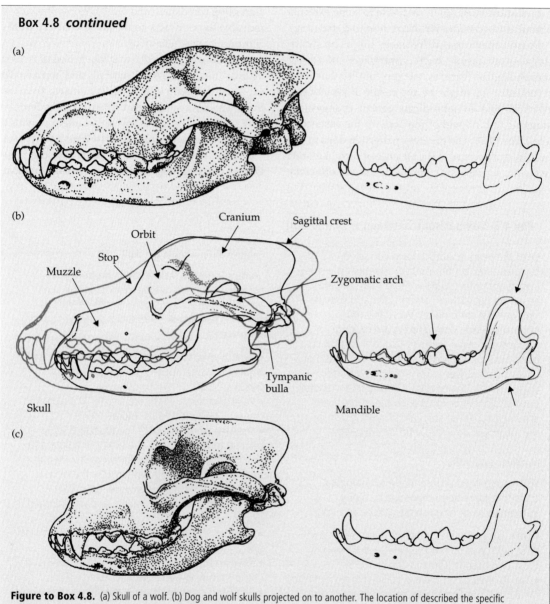

Figure to Box 4.8. (a) Skull of a wolf. (b) Dog and wolf skulls projected on to another. The location of described the specific difference are indicated by arrows, see the test for details. (c) Skull of dog. (Based on Clutton-Brock 1995).

relative of the Australian dingo, apparently shows many peculiar behavioural traits that have not been described for the dog or the wolf. These behaviours are mainly associated with inter-individual communication and sexual behaviour (Koler-Matznick *et al.* 2000, 2003). From the data currently available, it seems that these dogs may represent a special case of changes associated with living under particular environmental conditions,

and probably originating from a small population (founder effect, Chapter 5.4, p. 117).

4.5 Conclusions for the future

Species of the *Canis* genus represent a very successful group of animals. On the whole they are more similar to each other than otherwise, which is also underlined by the fact that despite their relatively long evolutionary separation they can still hybridize with each other. The *Canis* genome may function like a Swiss army knife which can easily be adjusted to any challenges represented in the actual environment.

In the lack of evidence to the contrary we cannot exclude that any *Canis* species had (or has) the potential to become a 'dog'. Increased sociality is the main argument in favour of the wolf, but this can be selected for in a few generations. Targeted socialization experiments involving various *Canis* species and subspecies might reveal similarities and differences in behaviour towards humans.

There is much to be learned about behavioural variation in the wolf, and whether this has a genetic and/or environmental basis. The evolutionary history of wolves might have resulted in a genotype with increased phenotypic flexibility. In parallel, there is also a need for quantitative description of phenotypic variability for present-day dogs.

Further reading

The volume edited by Macdonald and Sillero-Zubiri (2003) offers a wide perspective on the comparative biology of canids, and a similar approach has been adopted by Mech and Boitani (2003) who focus on the wolf. It is interesting that the thought-provoking study on feral dogs by Beck (1973) has not found its followers.

CHAPTER 5

Domestication

5.1 Introduction

The term 'domestication' is often used in two different contexts. The first meaning of the word designates a historic (often including prehistoric) period during which some 'wild' animals and plants were transformed by humans. This view emphasizes the contribution and role of domestic animals in human history. Accordingly, domestication is described as a set of technological innovations such as keeping animals in captivity, breeding them, and selecting them.

Biologists prefer to study domestication in the context of evolution. For example, Price (1984) defines domestication as an evolutionary 'process by which a population of animals becomes adapted to man and to the captive environment by genetic changes'. Thus domestication is a Darwinian process including forms of selection that are present in natural populations. As a consequence, domesticated animals found their way into a specific environment (*niche*) created by humans. It is an interesting question whether in the case of domesticates there is one such niche or many. But in any case we assume that the human-created (anthropogenic) niche differs in many respects from the natural ones, and this is most obvious in the case of the dog.

Providing an evolutionary realistic framework for dog domestication is difficult. The first task is to reconstruct the actual selective environment, including the possible actors—the 'ancestors' of both dogs and humans—and the particular causal factors. Usually such reconstructions take into consideration both abiotic factors such as possible geological events (e.g. glaciation, continental movements, ambient temperature), and biotic elements of the environment such as the presence of possible food sources, other competitors, or potential predators. The evolution of dogs is closely linked to the emergence and spreading of humans (*Homo sapiens*), so some knowledge of the latest phase of human evolution (the last 50 000 years) is required. This also means that changing views of human evolution can affect our understanding of dog domestication.

5.2 Human perspective on dog domestication

In recent years we have witnessed an increased interest in theories that aim to explain the evolutionary events that led to the domestication of dogs. Most ideas are non-exclusive and use different type of arguments. There is no disagreement among researchers that the history of dogs and humans is tightly interwoven, but views of the role that humans played in this process vary (Box 5.1). As a first approximation it may be useful to look at the last 50 000 years of the two species in parallel.

The human colonization of extra-African regions involved four major phases (Finlayson 2005) (see Figure 5.1). First, older members of the *Homo* genus (now described as *H. erectus, H. heidelbergiensis,* and *H. neanderthalis*) left Africa around 300 000–400 000 years ago (Finlayson 2005), and probably encountered wolves along their journey. By this time wolves were the main predators in the Holarctic (Chapter 4.3.2, p. 76); in addition, some species of wolves (and/or jackals) inhabited the north-east part of Africa, so it is very likely that humans had cohabited with wolf-like canids long before leaving Africa. This means that at least three species of *Homo* lived for over 400 000 years alongside wolf populations over a vast area ranging from the Atlantic Ocean to eastern China. Note that as far we know, no change in wolf populations took

Box 5.1 Non-exclusive theories of domestication

Over the years many different theories have been proposed which are summarized here with respect to the evolutionary mechanisms. Each theory is important in explaining a particular aspect of the process, so all five together probably give the most plausible account of the sequence of events (see also Figure 5.2).

1 Individual-based selection

Humans regularly picked wolf cubs from the den, and after socialization in human groups, individuals showing the 'right' temperament and/or affiliative tendencies were selected for over many generations (e.g. Lorenz 1950, Clutton-Brock 1984, Paxton 2000). This idea is supported by observations that pups of wild canids show very distinct and diverse characteristics in behaviour towards humans (MacDonald and Ginsburg 1981). However, it is likely that such individual selection occurred not at the start but only at the end of domestication (when breeds were selected for).

2 Population-based selection

Dogs are the descendants of a scavenging canid population by either of two processes:

A The activity of humans induced changes in the environment by providing a novel, easy-to-exploit food source. This food source was utilized by (some) wolf population(s) that in parallel underwent morphological, physiological ('protodomestication': Crockford 2006), and behavioural changes and finally, isolated themselves from the rest of the 'wild' population. This novel niche was provided by human hunters or appeared in the form of human settlements (Coppinger and Coppinger 2001).
B An already existing population of wolf-like canids leading a scavenger lifestyle associated itself with human communities, and exploited food provided by human activities. As the production of food waste by human groups grew, the animals became more dependent and an exclusive relationship evolved (Koler-Matznick 2002).

Although feasible, version A of the theory has problems in explaining why domestication started only at a few locations, and there is very little factual evidence for version B.

3 Dog–human co-evolution

Co-evolution is defined as an interaction between two species that results in adaptive changes which correspond to some function fulfilled by the other species. Accordingly, both dogs and humans have changed in functional (adaptive) ways because of their evolutionary relationship. Paxton (2000) suggests that dogs have taken over the job of orienting in the environment (because their superior smelling ability) and this allowed for selective changes in human facial (nasal and oral) structures for more skilled production of speech sounds (see Bekoff 2000 for critique).

4 Human group selection

Some traits emerging at the group level can be favoured by selection. Critically, group selection works only if individuals are faithful to their group, which might have been the case during periods in human evolution (Sober and Wilson 1998). Human groups building their culture on their relationship with dogs could also have experienced some advantage if dogs contributed to increased fitness of humans. Preference for observing wolves might aid in the development of hunting or establishing settlements (Sharp 1978, Schleidt and Shaller 2003), and human groups could also show variability in tolerating wolves or dogs around them.

Little factual support is available for this theory.

continues

Box 5.1 *continued*

5 Cultural–technological evolution

Diversification of dogs runs in parallel with cultural–technological evolution. At the beginning dogs had a restricted role as work aids (perhaps also as a food source), and humans could have developed a ritual relationship with dogs (Morey 2006). Marked diversification occurred when humans found ways to use dogs for different tasks involving herding, guarding, pulling sledges (Morey and Aaris-Sorensen 2002), or recently e.g. in assisting handicapped humans. Such diversification occurred repeatedly during human history and dogs seem to mirror the increasing complexity of objects with the advance of culture.

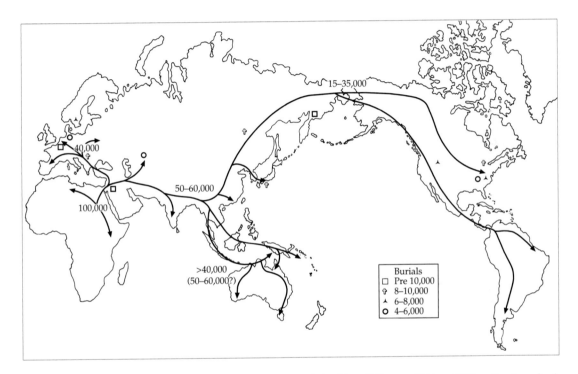

Figure 5.1 Current view of early human (*Homo sapiens*) migrations 'out of Africa' (Cavalli-Sforza and Feldman 2003) and locations of early dog remains based on archaeological dates reported by Crockford 2006 and Morey 2006.

place during this time that could be related to the presence of humans (but see Olsen 1985), although in principle these human hunting groups could have produced some surplus food, which would have attracted local wolves.

The second phase took place when the ancestors of modern humans (*Homo sapiens*) left Africa. This was a very turbulent process, involving numerous populations, many of which died out before they could establish a strong presence in east Asia.

Archaeologists and evolutionary geneticists seem to converge on the idea that humans colonized east Asia in several waves between 45 000 and 120 000 years ago, but they were often forced into refugia when the climate became colder (Finlayson 2005). This date would fit with suggestions that recent dogs emerged as a consequence of the encounter between modern humans and some wolf-like wild canids around 100 000 years ago (Vilá *et al.* 1997, see Chapter 5.3.2, p. 114).

If dogs had evolved soon after their encounter with humans (around 50 000 years ago) one would expect dogs to have joined human groups on their migration routes from the beginning. Unfortunately, at present there is no evidence for such early association between humans and wolf-like creatures. Thus the population-based theory of dog domestication based on a novel, food-rich anthropogenic niche faces problems when it has to explain the apparent lack of any detectable change during a very long period of cohabitation between humans and wolves. It might be the case that during these times human hunters did not produce enough waste food to sustain large groups of wolves around their camps (Box 5.2). The amount of food could be important here, because if the animals had to complement their diet by additional hunting on their own then they could come into contact with conspecifics, which would jeopardize the isolation of the 'wild' and 'anthropogenic' populations. However, we may assume that humans hunting especially on large game (e.g. horses) did produce surplus food potentially available to wolves (and other scavengers) over a very long period. Indeed, in central Europe some authors

find indications for change in local wolf populations showing signs of domestication (Musil 2000) dated to around 12 000 years ago. It is important to note that hunters were mobile, so it was not necessary for the wolves utilizing food remains to come into close contact with the people. The animals could have visited the places of the kill or butchery after humans had already left.

Authors committed to the population-based view (for example, Tchernov and Horowitz 1991, Coppinger and Coppinger 2001, Crockford 2006) come to the conclusion (based on different lines of argument) that the first step to exploitation of the human-provided food source was a marked reduction in size. Although this idea is basically supported by the archaeological record, nobody seems to consider the possibility that wolves might have had competitors in exploiting this novel food source. In contrast, in the absence of contradictory data it seems plausible that all the way from Africa human hunters have been followed by other small carnivores like the Golden jackals that are still distributed over most parts of south Asia today. These were the 'right' size and probably had most behavioural adaptations to scavenge on surplus food left over by humans. It is important to note that

---

Box 5.2 How much meat keeps a wolf going?

A review of various studies suggests that a free-ranging adult wolf might need more than 5 kg of meat per day (Peterson and Ciucci 2003). Henshaw (1982) estimated 1–1.5 kg meat per day based on the basal (resting) metabolic rate (BMR), but other calculations yield a minimum of about 0.55 kg for an inactive animal. Based on arguments provided by Peterson and Ciucci (2003), the relationship between body weight (W) and energy requirements can be described (following Kleiber 1961) as

$$BMR \ (kcal/day) = 70 \ W^{0.75}$$

(Replacing the constant 70 by 12.19 gives the result in kJ/h.)

Coppinger and Coppinger (2001) argued that the reduction of size during domestication was important because early dogs had to survive on food with a smaller energy content. Indeed, if we assume that a wolf survives on 1 kg meat per day (because it is fed by people) then an average pack would need about 6 kg meat per day, which is about 180 kg per month. This means that the humans would have needed to hunt about three deer (each weighing approximately 50 kg) each month just to keep the animals going. Thus eating alternative food including human leftovers, as well as decreasing body size (including a relatively smaller brain), could be advantageous in survival. A further possibility would be to select

continues

Box 5.2 *continued*

for dogs with reduced basal metabolic rate, partially because the basal metabolic rate of the wolf is higher than predicted for carnivores generally (Kreeger 2003). Unfortunately, present data are difficult to compare because the basal (resting) metabolism can be measured by different

methods. Providing early dogs with food was probably a critical condition of domestication, which took place at locations where people could afford and find ways to maintain packs of these canids.

---

because of the colder climate contemporary wolves were much larger than the extant subspecies inhabiting south Asia, thus wolves had to decrease their size first before outwitting other canids. But more likely jackals did not follow humans migrating to the north, and thus wolves had a better chance to invade this niche there. Alternatively, wolves have out-competed smaller canids around the human food sources.

The third phase started after the end of the last glacial maximum, around 20 000 years ago, when human populations showed rapid expansion and began to move in several waves to east–central Asia, Siberia, and from there north-westwards to Europe and eastwards over the Bering Strait into North America. The 'exact' dates are less interesting; it is more important to note that by 10 000–15 000 years ago most continents had some human occupants (Australia was reached along the coast relatively early around 40 000–45 000 years ago); perhaps Patagonia was one of the last territories to be discovered.

This phase includes the transition from hunting-gathering to agriculture, which was actually not a smooth, one-way process. Agriculture has emerged independently in several places (Smith 1998) and it was often accompanied by switching back and forth between hunting and farming. In some places both activities were practised in parallel for many generations. For example, in the Near East an early period characterized by the evolution of farming around 14 000 years ago was followed by a period of 1000–2000 years where humans reverted to hunting, possibly because of marked changes in climate that made early and vulnerable agriculture impossible to maintain (Goring-Morris and Belfer-Cohen 1998). Such changes in human activities could have influenced an already established relationship with wild canids. The critical issue here is whether a genetic

separation between the 'wild' and 'human-associated' populations could be maintained. At the moment it is less clear how mobile hunting humans could prevent these wolf-like populations from mixing. An alternative explanation probably relating to this phase was suggested by Koler-Matznick (2002) who argues that the domestication was based on a scavenger canid living in east Asia. Accordingly, domestication began only when human populations reached this part of the world 20 000 years ago (Fig. 5.1). This would explain the lack of earlier findings along the route of humans, and explain why there are no transient wolves showing decrements in size. The argument could be supported by the observation that in Asia there are no other smaller *Canis* species showing a scavenger lifestyle, in contrast to America (coyote) and Africa (jackals). However, if domestication was based on a scavenger (sub)species, why did it not start with jackals?

The fourth phase started when humans established permanent settlements. Coppinger and Coppinger (2001) suggest that the enduring human territories in the form of villages provided a natural barrier between wild and anthropogenic populations because in order to get food scavengers needed to spend time near humans. If early settlements provided a permanent habitat for dogs, then these newly evolved creatures might have accompanied humans if they decided to return to hunting.

In fact, not human activity *per se* but the changing/switching lifestyles might have speeded up the domestication. Humans could be in a better position to realize the beneficial potential of these animals if they practised both farming and hunting. There might have been only a handful of places where humans developed such a balanced method of food provision. As soon as animals with dog-like characters

Box 5.3 Where did dogs originate?

The analysis of mtDNA for 466 dogs collected from various continents suggests that dogs originated from one or a few ancestral populations and that east Asia was the central place for the early events of domestication (Savolainen *et al.* 2002). The available mtDNA sequences were categorized into six groups known as clades A–F. If the distribution of these clades is plotted in relation to major geographic areas, then the proportion of dogs belonging to each clade is very similar in most cases (a). The presence of clade D in Europe and south-west Asia might indicate limited local hybridization events with other wolf-like canids. The analysed

sample revealed 70 unique mtDNA sequences (sequences that are present only in one geographic region). Assuming that the dispersal after domestication led to reduced variability in dogs in the newly inhabited areas, the largest variability should indicate the centre of domestication events. Most unique samples were found in east Asia, followed by Europe and south-west Asia (b). If local hybridization events had contributed to a large extent to the mtDNA of the surviving dog population then we would not expect such difference in the distribution of unique sequences. (see also p. 111)

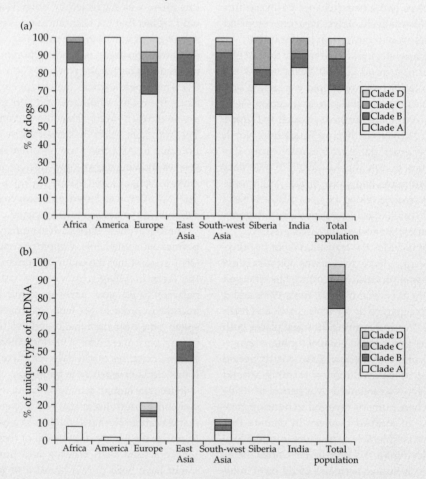

Figure to Box 5.3 (a) Distribution of different mtDNA sequences (haplotypes) and (b) unique sequences from clades A, B, C, and D across geographic areas (expressed as a percentage of the dogs associated with that geographic location). Data from Savolainen *et al.* (2002); clades E and F are omitted for simplicity.

emerged, trading humans and fortunate dispersal events could rapidly widen the spread of these animals. Once they showed a preference to stay with humans, dogs were very likely to be easily adopted by other communities of exclusive hunters or farmers. This could explain why dogs appear relatively rapidly at western and northern European sites around 12 000 years ago, and accompany humans crossing to North America probably with the second or later waves around 10 000 years ago.

Thus these wild canids were not only able to gather food in their new niche but in order to survive they had to be able to follow rapid changes in human lifestyles. Importantly, by this time a uniquely strong social bond seemed to have evolved between humans and dogs, as suggested by early dog burials (Morey 2006), but this was not accompanied by any marked diversification of dogs during the next 4000–6000 years. It may be the varying and often unforeseeable life of humans which inhibited the development of specific forms of dogs; or, on the contrary, dogs had a special function to play, either ritual or practical. It is likely that the diversification of dogs is associated with rapid technical changes during the *Neolithic revolution* when around 5000–7000 years ago humans started to select dogs for various working roles; this resulted in the development of 'breed-based' dog populations, some of which showed characteristic sets of morphological and behavioural traits. However, it is likely that most of these early dog breeds do not have any direct phylogenetic descendants in recent populations, and most of them died out during famines or wars. It is very likely that even if some recent breeds look similar to old dog drawings, they have been partially recreated relatively recently (Box 5.4.). A new process started around 200–400 years ago, when dog breeds were developed and maintained in strict reproductive isolation. Thus the present breeds represent a new 'cocktail' of the wild canid genome (Figure 5.2).

5.3 Archaeology faces phylogenetics

For many years the reconstruction of the origin of dogs has been based on the fossils and remains of wolf- and dog-like creatures. Although dog domestication is perhaps not the main focus of archaeozoology, the collection of remains has increased, and technical advances have permitted a more precise determination of temporal and spatial relationships. In contrast, the genetic analysis of phylogenetic connections has started only recently, mostly based on material collected from living specimens.

In principle, zooarchaeological and phylogenetic models of dog domestication should not differ; however, given the fact that the data have a very different nature, they might tell a different side of the same story. In the case of fossils the date and location seem to be fixed and the task is to reconstruct the evolutionary relationship, while in the case of genetic data, we assume or predict (by using statistical methods) ancient events and their relationships based on genetic similarity (DNA sequence) in living organisms. Thus these approaches are often complementary, or determine each other, as in the case of calibrating molecular clocks on the basis of fossil ages (see Chapter 5.3.2, p. 110).

5.3.1 The archaeologists' story: looking at archaeological evidence

Two related but different kinds of evidence are usually collected to describe the process of dog domestication. When the interest is in the evolutionary aspect the emphasis is on the skeletal remains, but otherwise researchers look for possible indications of the relationship between humans and canids (Morey 2006).

Most comparative archaeozoologists agree that in general dogs can be discriminated from wolves on the basis of their generally reduced body size, shorter snout and facial part of the skull, and relatively small (often crowded) teeth in relation to the maxillae (e.g. Musil 2000, Box 4.8). Note that most of the listed characters are quantitative and express metric relationships between different parts of the bones. This means that any kind of conclusion rests critically on complex statistical comparisons.

The archaeologists' task is to separate three different types of events. The first event is related to the divergence of the ancient canid population, giving rise to the ancestors of today's dogs. Such divergence could have taken place potentially at many geographical locations where wild canids

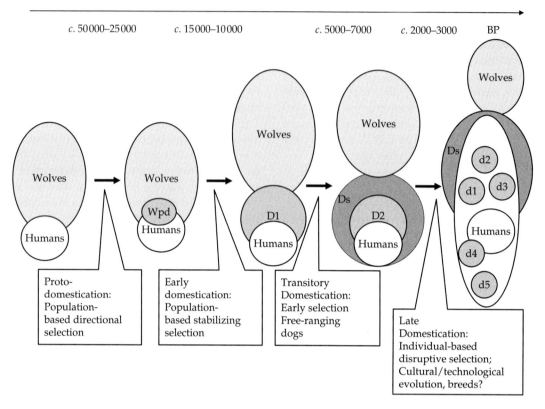

Figure 5.2 Key steps in dog domestication. The combination of recent theories gives a relatively straightforward evolutionary description of the domestication process. Protodomestication (Crockford 2006) and early domestication was based on wolf-like populations, but during the transitory and late domestication period there was a tendency to rely on individual selection. Early domestication was characterized by the emergence of a smaller dog-like canid in many places, and during transitory domestication morphologically distinct categories of dogs emerged. Late domestication produced typical dog breeds, perhaps repeatedly in different locations and historical periods. Importantly, the type of selection processes changed also during domestication BP–before present. For simplicity the hypothetical effects on humans are not included here. Wpd, protodomesticated wolves; D1, D2, early dog population(s); d1–d4, dog breeds; hatched area, stray/feral dogs (d5). (Chapter 5.4.2, p. 119)

and humans shared the same habitat, and at very different times. The separation of these ancient dogs was probably paralleled by changes in morphological characteristics. Although it is likely that behavioural changes preceded morphological alterations, this delay could have been relatively short, taking only a few generations.

The second type of event concerns the variation within the ancient wolf and dog populations. In general some variation is expected within any population (e.g. sexual dimorphism). There are arguments that domestication has in the long run produced a more variable population of dogs in comparison to wolves. However, this can only take

place if the reproductive barrier is maintained. In the case of dogs, increase in variation is often taken as evidence for diversification of function in human society.

The third type of event relates to the problems of what happened if the reproductive barriers disappeared after shorter or longer isolation (e.g. interbreeding). This involves events when hybridization took place between dogs and wild canids (mostly wolves) or different forms of dogs that had been separated for a long time. The introduction of modern European dogs into the New World after Columbus is a well-known example, when native dogs came into contact with European breeds after sharing a common ancestor perhaps more than 10 000 years ago. Such encounters could have occurred frequently when large human populations moved across continents accompanied by their dogs.

As more and more finds come to light, other circumstantial data can also help to clarify the process, such as the colonization of islands by humans. For example, Japan was first colonized by humans *c.*18 000 years ago, but dog remains are found only from *c.*9000–10 000 years ago. Because immigrations were probably regular during this period, this discrepancy might reflect important changes in the dog, which might have been taken on a sea trip only after it had established a close relationship with humans.

Other investigations place more emphasis on searching for clues indicative of the cultural aspect of the relationship. There is now evidence from all parts of the world that people practised ritual burial of dogs as soon as anatomical differences between dogs and wolves emerged (Morey 2006). Most of our earliest finds come from dogs that have been intentionally buried by humans. As this practice seems to be mostly restricted to dogs (other domesticated animals were buried much less frequently), domestication may have been paralleled by a spiritual relationship with dogs. There are, however, indications of less mystical relationships. In some cases dogs provided meat for humans, or served as carriers of loads.

The sequence of events as shown by the archaeological record

In order to present a non-exhaustive account of dog domestication we have arbitrarily divided the time into periods to allow for parallel presentation of the events at various geographic locations. Early dates are given in years before the present (BP). To give a rough estimate of the progress of changes, skull length (SL) and/or withers height (WH) will be indicated. The numbers either refer to an identified specimen or give ranges for the smallest and largest specimen reported. Nevertheless it should be kept in mind that such measurements actually have little relevance to the domestication process itself, although they might suggest a general overall trend.

Before 14 000 BP Although most researchers assume that dog domestication started much earlier than

this, convincing archaeological evidence is lacking. If we assume that behavioural changes had preceded any morphologically detectable change, a separation into two more-or-less permanently isolated populations should have produced morphological changes in a few generations (e.g. Trut 2001). On this basis there would be little reason to assume that the divergence within any wolf population had started much earlier.

14 000–12 000 BP Perhaps surprisingly, some of the earliest evidence comes from North Europe, near Oberkassel in Germany (around 13 000 BP). In 1979 Nobis described a small mandible found in a human grave. The missing two premolar teeth suggest that this specimen was a dog, because such an abnormality is very rare in wolves. More recently, two large ancient dogs (estimated WH 70 cm; SL 240, 256 mm) have been reported from the Bryansk region of Russia (Sablin and Khlopachev 2002). The early presence of these very large dogs contradicts the assumption that domesticated descendants become smaller. Alternatively, they might have been local wolves living in captivity, in close contact with humans the descendants of an even larger wolf subspecies, or hybrids of some sort. (But see Box 5.5). Archaeologists assume that these animals were playing an important role in the life of these hunter-gatherers by helping in the hunt or guarding the settlement.

12 000–10 000 BP At a northern Israeli site dating from the Natufian period, dated around 11 000 BP, a carnassial, a fragment of a mandible, and a skeleton of a puppy were found (Davis and Valla 1978). The skeleton was recovered from a human grave. Interestingly, the hand of the deceased human was positioned over the body of the puppy, suggesting an affectionate relationship. In order to determine whether the fossils belonged to a wolf or a dog the archaeologists compared the length of the two lower carnassial teeth (M_1) to both contemporaneous and recent wolves. The analysis showed that the teeth in question are smaller than the carnassials in recent (and relatively small) Israeli wolves, and much smaller than Pleistocene wolf teeth collected from the same region. A more recent find of two dog-like canids buried together with three humans shows similar difference in M_1 size when compared to both recent and extant wolves of the region (Tchernov and Valla 1997).

Investigating the skeletal remains of three locations in central western Germany, Musil (2000) reported the presence of relatively small wolf-like canids. These settlements (Kniegrotte, Teufelsbrücke, Oelnitz) were established by hunter-gatherers living in the Magdalenian culture who lived by hunting horses. This scenario offers a potential role for dogs to participate in hunting. Various measurements obtained from these maxillae fall below the range of wolves that lived at the same location 10 000–12 000 years earlier.

Chaix (2000) reported a more complete skeleton (described as a dog and dated from 10 000 years ago) from a cave in the French Alps (estimated WH 40 cm). The skull was exceptionally small (SL 149 mm) in comparison to both Paleolithic and Neolithic wolves (SL 240–276 mm), which suggests a size reduction of 38–46%. At present this constitutes the earliest dog-like find in western and central Europe.

It is likely that towards the end of this period the first dog-like canids accompanied hunter-gatherers who crossed the Bering Strait to America. Although humans had first migrated 20 000–35 000 years ago, later invasions could have been more successful because of the partnership with dogs.

10 000–8000 BP Apart from the debated mandible found in a cave at Palegawa (Iraq), dating from around 10 000 BP (Turnbull and Reed 1974), the earliest remains from Asia Minor were recovered from Jarmo (Iraqi Kurdistan; Lawrence and Reed 1983). The only skull and many jaws are clearly distinguishable from corresponding wolf bones but suggest robust specimens (from c.9000–7700 BP), and the simultaneously excavated figurines of dog-like animals (with curved tails) provide additional evidence of the early presence of such canids. The presence of dogs is also confirmed by wall drawings depicting hunting scenes from Catal Hüyük (Turkey), one of the first centres of agriculture.

In Europe the frequency of more dog-like remains increases, found in association with hunter-gatherer groups living at permanent settlements. Various skeletal remains from relatively uniform dogs have

been excavated at Star Carr and Seamer Carr (England) dating back to 9900–9500 BP (Clutton-Brock and Noe-Nygaard 1990, estimated WH 56 cm). Similarly small-bodied dogs were recovered at Bedburg-Königshoven (Germany; Street 1989), and ancient dogs from this period have also been described from Sweden, Denmark, Estonia (see references in Benecke 1992), and Siberia.

Further remains in central Europe on the banks of the Danube (Vlasac, Serbia) reveal the presence of small dog-like canids living along wolves (8500 BP) (Bökönyi 1974). Apparently these dogs belonged to fishing and hunting communities who ate these animals, as indicated by the high number of broken long bones and skulls. Interestingly, dog burials were also reported from nearby (Radovanovic 1999), suggesting a wide spectrum of dog–human relationship within the same time frame.

The first archaeological evidence that dog-like canids reached North America dates to around 9000 BP. Mandibles and skull fragments were recovered from the Danger Cave (Grayson 1988) in northwestern Utah (USA). Dog-like canids also appear in Japan at c.9300 years BP (Shigehara and Hongo 2000). Importantly, these specimens seem to have no direct relationship with the native (today extinct) Japanese wolf; they probably accompanied the settlers invading these islands.

8000–6000 BP Although after 6000 years one would expect some morphological changes emerging at sites providing the earliest evidence for domestication, little if any progress is revealed. Parts of skulls and other bones, which have been recovered from submerged settlements in the Mediterranean Sea off the Israeli coast (Atlit Yam, Kfar-Galim) show practically no difference in comparison to the much earlier specimens from the Natufian period (Dayan and Galili 2000). For example, the length of the two lower carnassials (M_1) is identical to those recovered from at least 2000 years earlier (Davis and Valla 1978). There is some circumstantial evidence that dogs were introduced from the Near East to Egypt, and later dispersed throughout northern Africa. Towards the end of this period, the first dog burials from Egypt are discovered in agricultural communities of

Merimde (6800 BP), suggesting an important role for canids in these cultures (Brewer *et al.* 2001).

Joint burials of dog-like canids and humans have been found at various places in south-eastern North America (e.g. Tennessee, Kentucky; see Morey 2006 for review). There is a pronounced tendency to bury dogs with people, which suggests an intimate relationship between native American hunters and dogs at least for the next 2000 years (Schwarz 2000). A detailed account based on the fragmented remains of two specimens from western Idaho (USA; 6600 BP) reveals that these canids had a relatively small skull (SL *c.*172 mm) and low withers height (WH *c.*47.7–52 cm) (Yohe and Pavesic 2000).

By the end of this period even smaller dogs had emerged which shared their life with people in Central America. The most widespread dog was the so-called Mesoamerican common dog (SL 160 mm, WH 40 cm), which is believed to be a direct descendant of the first dogs that arrived with humans to the central part of the American continent around 8000 BP. These dogs remained morphologically unchanged for the next 6000 years until the arrival of the first Europeans (Valdez 2000). Remains excavated in Patagonia (Chile) from before the end of this period indicate the end of the colonization of the Americas.

6000–4000 BP   The identification of dogs becomes much easier, partially because by now there are many independent clues, such as drawing of dogs or small sculptures. By this time the size variability of dogs surpasses the variation present in local wolves at any given time or location. This is the first period of breed diversification, and there are also indications of the presence of stray dogs that had no dedicated relationship with humans, and rapidly became a nuisance.

Dog remains recovered from different sites in Mesopotamia (Tepe Gawra, Eridu) show skeletal similarities to recent salukis or some greyhounds (Clutton-Brock, cited in Clark 2001). The presence of such dogs is also supported by representations of saluki-like dogs on pottery and seals towards the second half of this period in Mesopotamia (Tepe Gawra, or near Mosul). A depiction on a vase (from *c.*6000 BP) shows both a lone hunting wolf and a leashed dog hunting with humans on bezoar

goats. This indicates that the painter was aware of both the similarities and the differences between dog and wolf hunts.

On Egyptian pottery and in rock art (5700 BP) dogs look like sighthounds with slender body, erect ears, and curly tails. Most scenes involve hunting game, such as gazelles, but some dogs are depicted as being on the leash or lying under their owners' chair (Brewer *et al.* 2001). Another type of dog is more reminiscent of the modern saluki, with a shorter muzzle, lopped ears, and curved or sabre tail. There are also early representations of dogs with massive muzzles, long tails, and lop ears. However, there is some disagreement whether these drawings are representations of a mastiff-like type of dog or just represent a less skilful depiction of the dogs. Towards the end of this period there is also pictorial evidence of short-limbed dogs displaying erect ears and curved tail. Although a number of dog remains have been identified throughout the period of the Egyptian dynasties, only a very small part of this material has been subjected to careful analysis. Preliminary comparative analysis of withers height by Brewer *et al.* (2001) suggest that there had been at least one or possibly two forms of dogs that could be separated from the 'wild' or 'feral' dog population ('pariah dogs') of the day (WH 42.5–49 cm). One type of dog looked like a modern saluki (but somewhat smaller) and was possibly used for hunting (WH 47–57 cm); the other type was short-legged. Importantly, Egyptians discriminated their companion dogs from the pariah dogs. Favourite companion or hunting dogs were named, cared for, and often provided with a special burial; some had their own sarcophagus and their memory was perpetuated by the carving of statues.

Remains from various parts of Europe suggest a relatively uniform dog fauna, dominated by medium-sized dogs (Benecke 1992) (e.g. SL 135–175 mm in Switzerland, WH 47 cm in Hungary–49 cm in Germany), in comparison to the wolves of the day (SL 230–240 mm; WH 68 cm).

At the same time, dogs living in Armenia (SL 193–213 mm) approached the size of the wolf (Manaserian and Antonian 2000), and small dogs are apparently missing. Relatively large dogs (SL 192 mm; WH 50.5 cm) were also found in

Kazakhstan, living with horse hunters at Botai in 6300–5600 BP (Olsen 2001). Skeletal remains were recovered from pits in houses, suggesting close association between people and dogs. Apart from cooperation in hunting, dogs could have also played a role in guarding the house. Comparative analysis provided some evidence that Botai dogs are reminiscent of today's Samoyeds (SL 176 mm; WH 48 cm). Ancestors of this breed might have derived from the dogs of the Samoyedic people who migrated from this part of central Asia to northern Siberia accompanied by their dogs, but this suggestion is impossible to verify on the basis of osteological evidence. Botai dogs were somewhat heavier then recent Samoyeds, and so were probably better prepared for the cold climate and able to survive extreme cold temperatures.

By this time there are relatively small dogs in Japan (SL 151–157 mm) which would not fundamentally change their body conformation in the following 4000 years. It is believed that some of these dogs have survived in the form of the recent Shiba breed (Ishiguro et al. 2000).

Although the archaeological dating indicates a later period (3500–4000 BP), it is assumed that the first dogs had arrived in Australia during this period, and rapidly colonized the continent. Some animals or populations may have stayed with the Aboriginals for longer durations, possibly for some generations (Corbett 1995).

4000–3000 BP (2000–1000 BC) Dog remains from Italian sites show wide variability in size. By now there is a more than 60% difference between the skull length of the smallest dog (SL 127 mm, WH 36 cm) and the largest (SL 194 mm, WH 62 cm). But even the biggest dogs did not reach the size of the local wolf (Mazzorin and Tagliacozzo 2000). Similar large dogs have been reported from England (SL 176–202 mm, Harcourt 1974), and broad size ranges have been described from other sites in Switzerland and Germany. Although these dog skulls are markedly smaller, the qualitative analysis showed a considerable overall similarity to the wolf (Benecke 1992).

In parallel, only relatively large dog skulls were excavated at various sites in Armenia (SL max. 224 mm). Animal figures and rock carvings suggest that at least by the end of this period, dogs

were used in herding and also guarding the house. The drawings of dogs portray individuals of different sizes with curled tail and floppy ears (Manaserian and Antonian 2000).

Remains have been recovered from the eastern Arctic of dogs that lived with the Pre-dorset people (Morey and Aaris-Sorensen 2002). The systematic collection of bones indicates that before dogs became customary they repeatedly disappeared for long periods, and often had to be reintroduced. It is also not clear whether these dogs participated in transportation. Skeletal finds make it more likely that for a long period loads were placed directly on the dogs, and they were used for pulling vehicles only some 2000 years later, after the invention of modern sledges.

By this time dog burials in the north-eastern USA (Handley 2000) point to the existence of two types of dogs. In a sample collected from period of over 3000 years, dogs belonging to two size classes could be differentiated. Smaller dogs (SL 163 mm) may have looked like a recent spaniel, while larger ones (SL 213 mm) were more wolf-like, although they did not reach the size of the local wolf subspecies.

3000–2400 BP (1000–400 BC) At Pyrgi (Italy) a large dog skull (SL 213 mm), falling within the range of smaller wolves, (Mazzorin and Tagliacozzo 2000) suggests a trend towards larger dogs. This is supported by a find from the Durezza cave (Villah, Austria) which revealed a large set of dog bones in addition to human and other animal bones, possibly as result of dead bodies being collected at this place (Galik 2000). Based on multivariate analyses of skull measurements, dogs could be categorized into two groups. Although there were size differences (which could be partially subscribed to sex differences), most dogs seem to have medium to fairly long skulls (SL 195–255 mm, WH 49–63 cm) with a relatively wide palate. Qualitative features suggest an overall homogeneity in these dogs. This might be the first indication of selection for increased size in European dogs that resulted in some dogs approaching or surpassing wolf proportions (SL 230–240 mm; WH 68 cm).

In the Mesoamerican region the common dog was still by far the most widespread, but new forms began to appear. Although in general all dogs look

Box 5.4 Where do breeds come from?

The Mexican Xoloitzcuintli is regarded as one of the oldest American dog breeds (see references in Vilá *et al.* 1999). This hairless breed was thought to be a relative of another morphologically similar breed, the Chinese crested dog. However, analysis of the mtDNA sequences showed that the Mexican dog is neither a native American breed which was domesticated locally, nor it is in close genetic relationship with the Chinese breed. Vilá *et al.* (1999) found that the Xoloitzcuintli's mtDNA has a Eurasian origin, and the frequency of the haplotype also makes it unlikely that this breed was derived from the hairless Chinese crested dog. It seems more likely that the Xoloitzcuintli is a survivor of the dog population that accompanied early humans to the Americas and was only later developed into a dog breed showing these special morphological characteristics.

The resemblance between Egyptian dog paintings and sculptures and the recent Pharaoh hound breed has deceived many dog experts into thinking that these dogs originate from ancient Egypt. However, analysis of their DNA suggests that this breed has been relatively recently re-created by crossing other dog breeds. (Parker et al 2004) The result is a genetically modern dog with a look that is indistinguishable from the paintings in pyramid tombs many thousands of years old. Thus the similarity in appearance does not support an ancestral origin. The situation is probably similar in the case of other ancient dogs, such as salukis or mastiffs, which were also often depicted by old painters. It is not the breed (in genetic sense) that has a long history, but only the 'form'. Dogs defy the rules of biological evolution because after separation dog populations were isolated only for a short time before their genetic isolation was interrupted by human intervention. In dogs, behavioural (and morphological) similarities often represent a case for convergent evolution, so similarity is not evidence for a 'common ancestor' (homology) (Chapter 1, p 14). This provides a further argument for the genetic plasticity of the dog, that is, similar phenotypes can be selected for on the basis of different genetic material (e.g. Belgian and German shepherds belong to different clusters on the basis of their genetic make-up, despite their morphological and behavioural similarity; Fig. 5.3).

(a)

(b)

Figure to Box 5.4 Similarity does not support descent or close evolutionary relationship. (a) Although both the Xoloitzcuintli (on the left) and the Chinese crested dog (on the right) originate from the same domesticated population in general, they have been developed to breeds independently in geographically different locations. (b) The Pharaoh hound looks like the depiction (see next page) but was not the model for it. The present-day breed is a recent development from modern dogs, the resemblance is secondary. (c) A reproduction of a wall painting in Ptahhotep's tomb (5th dynasty, c.4500 BP).

continues

Box 5.4 *continued*

(c)

very uniform, the bones suggest the emergence of a smaller type (WH *c*.30 cm), the tlalchichi, which had somewhat shorter legs and spread from central Mexico towards the coastal areas. Based on a shorter face, Valdez (2000) describes the short-nosed Indian dog (WH *c*.35 cm) that lived during same time period, probably restricted to the territories of the Maya people. Fossils point to a novel type of dog at 2000 BP which is assumed to have looked just like the recent Xoloitzcuintli breed (Mexican hairless dog) (WH *c*.40 cm) (Box 5.4). Unfortunately, most native dogs disappeared shortly after the arrival of the Spaniards in Central America. Some researchers see a similarity between these early dog types and present-day feral dog

populations in Mexico, and assume that some genetic material of these extinct native dogs might have survived in the present-day feral populations (Valdez 2000).

2400–1500 BP (400 BC–500 AD) During the Roman period in Europe (Italy) large dogs are constantly present, although their skull length does not reach the size of the Pyrgi dog. The most interesting feature of Late Roman times is the appearance of very small dogs (SL 115 mm; WH 26 cm) suggesting the beginning of targeted selective breeding (Mazzorin and Tagliacozzo 2000). Small dogs (lapdogs) were possibly selected for their looks (and maybe also for their behaviour) and not for their value at work. The maintenance of very small animals needed special care and effort. Very large dogs (WH up to 72 cm), in the wolf size range, are also present (Bökönyi 1974).

Lapdogs were introduced to many Roman provinces, and they have been found in both the western (Britain) and eastern (Pannonia, Hungary) border areas of the empire. A survey of dog remains from the Roman town of Gorsium (Tác, Hungary) revealed dogs with very short long bones (WH 23–25 cm; Bökönyi 1974). Based on a qualitative and partly metric investigation of both the skulls and long bones from this site, Bökönyi (1974) concluded that the contemporary dog population might have comprised five different morphological forms. It is perhaps no coincidence that similar ranges are reported by Harcourt (1974) (SL 116–206 mm; WH 23–72 cm) on the basis of British finds, which leaves little doubt of the uniformity of the dog population under the Romans.

In the eastern part of the Roman Empire, the Danube provided a natural border to the Barbaricum. This offers the possibility of contrasting the dog populations of the Romans and the neighbouring Sarmatian people living to the east of the Danube. Statistical evaluation showed that Roman dogs were more variable in size for most measures of the skull (SL 138–220 mm) than the dogs of the neighbouring barbarians (SL 174–226 mm) (Bartosiewicz 2000). This difference could be accounted for by the presence of relatively small dogs in the Roman population, and it was suggested that the use of the dogs was partly different on the two sides of the Danube. It is likely that the Sarmatians preferred dogs that could be used in the management of other animals or for guarding, and for both roles animals with a certain size and strength were at an advantage.

Apart from dogs from Egypt and possibly from China, there is less evidence for such divergence in other parts of the world.

500 AD–present Although the collapse of the Roman Empire and the migration period brought changes to the dog population in Europe, this species retained its diversified character throughout the Middle Ages. Some measurements indicate differences between dogs living in towns and in the countryside, but there is no doubt that selective breeding was practised. The increasing distance between social classes, and the use of dogs for certain tasks or sports, contributed to the stabilization and perhaps increase of both morphological and behavioural differences. Because there was no artificially maintained reproductive barrier between these forms of dogs, new types could be created relatively rapidly by hybridization and selective breeding. Thus some types of dogs for a given task (e.g. herding) could be established locally if no other sources were available, or a few imported individuals could be hybridized with the representatives of the local populations. The final stage of this process began with the emergence of 'breeds', when 'pure' blood lines were maintained and hybridization was discouraged. According to kennel clubs (e.g. the American Kennel Club and the Fédération Cynologique Internationale) 400–500 breeds of dogs are registered, some of which are nearly identical genetically, while others differ to a greater degree. This unfortunate situation has probably slowed down dog evolution at present, especially because feral dogs are excluded from these breeding systems (although 'lucky' accidents may happen).

5.3.2 The geneticists' story: evolutionary genetic evidence

In the last 10 years there have been immense efforts to use modern evolutionary genetic tools in order to answer the questions left open by archaeological research. New data and some hypotheses advanced by the phylogeneticists have contradicted the picture presented by fossil dog remains in some respects, which has led to debates about the validity of these phylogenetic models (e.g. Coppinger and Coppinger 2001, Morey 2006).

Although the basic logic of the ideas is relatively simple, the actual modelling process is complex. Often extensive knowledge is necessary to understand the validity as well as the constraints of these models. The basic idea is that genetic variation changes over time and space, so tracing or modelling the history of these changes in genetic variation could lead to the reconstruction of the evolutionary process. Such modelling of genetic variation assumes a range of different processes like mutation, genetic drift, selection, population bottlenecks, or founding effects. Although there are exceptions, it is important to note that most of these reconstructions are based on extant species, from which DNA can be conveniently collected. However, this is also a limitation on the accuracy or resolving power of these models, because some important evolutionary events remain hidden if there are no survivors.

Studies differ in other respects which make a critical overview for the non-specialist difficult. For example, different forms of DNA are analysed which have a particular evolutionary fate. Mitochondrial DNA (mtDNA) and Y chromosome DNA do not recombine; the former can be inherited only from the mother and the latter only from the father. In contrast, autosomal DNA recombines during meiosis when the gametes are formed and is inherited from both parents. It is therefore to be expected that models based on different forms of DNA will vary, e.g. the mtDNA studies will not reveal the effects of hybridization events by male wolves. The sample sizes used in the studies are very variable, particularly the number and ratio of dogs and wolves. For example, using wolves as roots for a given tree could be tricky because even blind sampling across a wide geographic range could produce inequalities. For example, the three Russian wolves in Vilá et al. (1997) clustered very differently: one with Estonian/Finnish, the second with Greek, and the third with Arab wolves.

How does the 'molecular clock' work?
The idea of the molecular clock is based on the observation that mutations (changes in the DNA sequence) occur at some rate continuously over time. If there is a split from a hypothetical common ancestor than the number of mutations in the descendants could give an idea of the time that has passed since the divergence. However, as often happens with clocks, calibration is required. Such external reference is most often provided by archaeologists or palaeontologists who use independent methods (e.g. radiocarbon dating) for estimating time. The molecular clock of dog evolution is usually calibrated to the time when wolves and coyotes diverged (Chapter 4.2.2, p. 69).

This split is assumed to have occurred sometimes between 1 and 2 million years ago (Kurtén 1968). Based on a small part of the non-nuclear DNA (mtDNA) the genetic divergence between wolf and coyote was calculated to be in the range of 7.1–7.5% (Vilá et al. 1997, Savolainen et al. 2002) whereas divergence between wolf and dog was estimated at around 1%. Although most recent phylogenetic models use a more sophisticated approach of calculating evolutionary dates, for a moment assume a simple linear relationship between genetic divergence and time. If the divergence of 7.5% between wolves and coyotes has been realized in 1 million years since their split, then approximately 140 000 years are needed to obtain the 1% divergence between dogs and wolves (Vilá et al. 1997). However, the date of the dog–wolf split depends on both the accuracy of estimating the divergence between dogs and wolves (localizing actual changes in the DNA sequence) and the choice of date for the wolf–coyote split. Replacing 1 million years by 2 million years in the calculation, we arrive at 280 000 years for the domestication of the dog. Closer dates to the present for dog domestication can be calculated if we accept a later wolf–coyote split (700 000 years ago) for which there are also arguments in the literature (see Coppinger and Coppinger 2001).

The calculation of wolf–dog genetic divergence is also problematic. There are indications that recent wolf populations have undergone a rapid decline in the last 200 years and thus lost some of their genetic variability (Leonard et al. 2005). In dogs, the establishment of breeds in recent years probably also results in less variation than was the case even a few hundred years ago. If the same data had been collected in the Middle Ages, a smaller divergence between dogs and wolves might have indicated an earlier date for domestication.

In order to explain the discrepancies between dates of domestication proposed by the archaeologists and phylogeneticists, Ho and Larson (2005) suggested that molecular clocks should be adjusted when they are used for dating events that happened earlier then 2 million years ago. They argue that the mutation rates in present populations of the younger species are overestimations, because there has been little time for purifying selection to act. This means that in dogs there has been less time to select out those deleterious or slightly deleterious mutations which had disappeared from wolves during their 1 million year history. Consequently, calculations based on a smaller divergence between dogs and wolves would indicate a more recent date of domestication.

Finally, the molecular clock 'ticks' by generations; that is, any genetic change can manifest itself only when there are offspring to carry it. Generally this is not a problem because the generation time does not change between related species, but we know that at some point in time dogs switched to breeding twice a year. In addition, changes in the selective environment of dogs could also influence the observed mutation rate, because many assume that relaxed selection resulted in increased diversity (Björnerfeldt *et al.* 2006, Chapter 5.4.2, p. 118). Finally, as dogs spread out rapidly towards all parts of the Holarctic the rate of selection could be a function of the geographic location. At the moment it is difficult to judge how the outcome of these different processes affected the observed genetic divergence between dogs and wolves.

Genetic variation in space
One basic assumption of phylogenetic analysis is that the greatest genetic divergence present in the extant population of a species indicates the geographical centre of evolutionary changes. The logic behind this argument is that after populations radiate from this location there is an increased chance that the genetic material loses a considerable part of its variability because of genetic drift or founder effects. However, this idea rests on the (generally fulfilled) assumption that after separation the effect of hybridization between species is minimal (or non-existent) and that the populations are more or less localized, that is, they stay at or near the same place where they evolved. Although these conditions are possibly true for most wild among these populations and their relatives, there are indications that wolves and dogs defy these rules, and we should not uncritically assume that canids stay put at any given geographic location. Wolves migrated over thousands of kilometres, and in Eurasia there seems to be no east–west barrier for them. In line with this, no relationship was found between distance of the wolf populations and similarity of mtDNA at large

distances (Vilá *et al.* 1999, Verginelli *et al.* 2005). Thus the observed genetic similarity between certain dogs and wolf populations does not indicate that the dogs originated from the location where these wolves live today.

An even more serious problem is that in most phylogenetic models species are assumed either to occupy an end point of a tree (Box 4.3) or to represent a node of further divergence. This seems not to apply to dogs, however. First, researchers found evidence for wolf–dog hybridization over a long period of time, and even very recently wolves have been used to establish novel 'dog' breeds (e.g. the Czech wolfdog). Thus it is easy to imagine the possibility that dogs originated in Asia and were subsequently transferred to Europe where they hybridized with local wolves.

Second, breeds are often regarded as if they were necessarily associated with a given geographic area. Although this might sometimes be true, there is a need for caution. For example, it has turned out that the Pharaoh hound associated with ancient Egypt is probably a fake 'look-alike' recently created from different types of dogs (see Box 5.4). Thus most recent breeds have a polyphyletic origin, and were created by the use of a divergent and now untraceable sample of dogs. Breeds are not Linnaean entities; they represent a transiently frozen state of a dynamic population that has historically experienced admixture, introgression, and genetic isolation (Neff *et al.* 2004).

Possible location(s) of domestication
In 2002 Savolainen *et al.* developed a model to account for the geographic location of domestication. Researchers compared a 582 bp mtDNA sample from 654 dogs and 38 wolves. Dogs were represented by a wide range of purebreds, and by individuals that belonged to some locally recognized morphological category or were strays. This distinction may be important because in the case of purebred dogs there is reason to assume a relatively closed gene pool, whereas there is no evidence for this in the case of other dogs. Purebreds are far more typical of Europe than of other parts of the world, and it is not clear whether this affected the results (see also Savolainen 2006, Leonard *et al.* 2005).

The phylogenetic analysis revealed six distinct clades of dogs (labelled A–F) which were quite unequal in size. Clade A incorporated more than 71% of the dogs, and nearly 96% of all subjects belonged to three clades (A, B, or C). This indicates that dogs in these three clades represent nearly the whole genetic variation in the mtDNA of recent dogs. Most of these clades also included wolves; however, as indicated above, the presence of wolf mtDNA in a clade should not be regarded as evidence for the origin of these dogs or the clade as a whole (Table 5.1). Instead it was assumed that greater variability (e.g. the presence of unique mtDNA sequences = haplotypes) provides an indication for the location of domestication. Dog samples

were categorized and tabulated according to their origin into seven geographical areas (see Savolainen 2006; see Box 5.3). The frequency of clades A, B, and C is very similar across Europe, east Asia, and south-west Asia. This suggests a common origin of these dogs from the same founding population. However, the genetic variation differs among these three clades, being the greatest in clade A. Although there are many measures of diversity, 68% of the haplotypes found in east Asia are unique to this region, while the same calculation yielded 45% for Europe and only 25% for south-west Asia. Similar results for clade B and further statistical evaluation

Table 5.1 Summary table for various wolf and dog mtDNA sequences reported in the literature. So far six different clades that contain both dog and wolf sequences have been identified, by letters A–F (Savolainen et al. 2002) or by Roman numbers I–VI (Vilá et al. 1997). Estimated date is taken from Savolainen et al. (2002). Breed names are reported only if it appears to be a very specific case.

Clade	Wolves	Dogs	Approx. date
Clade A (I)	Eurasian wolves[c,f] Prehistoric European (c.10 000 BP) wolves (?)[f]	Dog breeds[a] Dog breeds and feral dogs[c] Dingo[a,d]	15 000 BP
		Pre-Columbian dogs[b] Indian feral dogs[e] Prehistoric (c.3000 BP) European dog[f]	
Clade B (II)	European wolves[a]	Dog breeds[a] Dog breeds and feral dogs[c] Indian feral dogs[e]	15 000 BP
Clade C (III)	European wolves[c]	Dog breeds[a] Dog breeds and feral dogs[c] Indian feral dogs[e]	15 000 BP
Clade D (IV)	European wolves[a] Prehistoric European (c.14 000 BP) wolves (?)[f]	Two dog breeds[a] Lapphunds[c]	
		Elkhund[c] Indian feral dogs[e] Prehistoric (c.4000 BP) European dog[f]	
Clade E (V)	South-western Asian wolves[c] European wolves[c]	Korean and Japanese dog[c]	
Clade F (VI)	South-western Asian wolves [c] European wolves[c] Prehistoric (c.10 000 BP) European wolves (?)[f]	Siberian husky and Akita[c]	

[a] Vilá et al. (1997); [b] Leonard et al. (2002); [c] Savolainen et al. (2002); [d] Savolainen et al. (2004); [e] Sharma et al. (2003); [f] Verginelli et al. (2005). (?) uncertain status.

suggested that these dogs most likely originate from somewhere in east Asia.

The east Asian origin of dogs may be further supported by two independent lines of research trying to trace the descent of American and Australian dogs. Early fossils of domestic dogs in the northern part of the American continent raised the possibility of an independent domestication event. In order to find a more definite answer, Leonard and colleagues (2002) successfully isolated mtDNA from 13 specimens recovered at archaeological sites dating back to pre-Columbian times (*c.*1400–800 BP) and 11 Alaskan dogs that lived *c.*420–220 BP, and also included mtDNA sequences of recent dog breeds and wolves. The phylogenetic tree provided by the analysis showed a clear separation between American wolf and ancient dog samples, and all but one of the pre-Columbian and Alaskan dog samples clustered in the clade that was earlier described as having Eurasian origins (Vilá *et al.* 1997). Within this clade researchers also identified an interesting subgroup that showed a very close genetic similarity to extant American dogs that once ranged over a vast region from Mexico to Bolivia. This could be a genetic signal of the early dog population that colonized the New World migrating with humans through the Bering Strait. These results show no evidence for an independent domestication event in the Americas, so all dogs seem to be descendants of Eurasian individuals (for some contrary evidence see Kopp *et al.* 2000).

Although it has been assumed that ancestors of the dingoes were taken to Australia by humans, their exact origin was not clear. Based on morphological similarity (which could be the result either of homology or convergence; Chapter 1, Box 1.3) there have been arguments for African, Indian, or east Asian origins (e.g. Corbett 1995). Savolainen *et al.* (2004) used 230 dingo mtDNA samples, including material from 19 dingoes living before the arrival of Europeans in Australia, in order to find a molecular genetic clue. The results pointed to a very restricted variation in dingoes, in comparison to both dogs and wolves. All dingo mtDNA sequences belong to clade A, supporting an east Asian origin. In addition, more than 50% of all dingo samples have the same haplotype and all

other haplotypes are separated by a few mutational steps in the sequence, which are only present in dingoes. Because a very similar argument can be made for the singing dogs of New Guinea, it seems most likely that the ancestors of dingoes can be traced back to a colonization event by a few individuals immigrating from the east Asian dog population. Japan experienced a similar colonization from east Asia, with the exception that these dogs remained in closer contact with humans and did not evolve a wild population (Kim *et al.* 2001).

Migrating wolf populations and dog breeds without a certain region of origin make these evolutionary models fallible. Moreover, the fragmentation of the present-day wolf population and random extinctions (in both wolves and domesticated dogs) make it likely that the present living animals are not representative of the once established populations. One way to make the predictions of the model stronger is to include very old DNA samples from extant wolves and dogs. This is difficult, because DNA molecules soon decay, but luckily modern laboratory methods can recover ancient DNA samples. Recently a group of researchers succeeded in sequencing mtDNA from five specimens living in the Apennine region of Italy 3000–15 000 years BP (Verginelli *et al.* 2005). According to the archaeologists associated with this research the three oldest finds (dated approximately at 14 000, 10 000, and 10 000 BP respectively) could not be assigned unambiguously as dog or wolf because the bone fragments were too small. They could belong either to a wolf or a wolf-sized dog. The skeletal remains of the other two finds were described as dogs and dated to 4000 and 3000 years BP. Samples from 547 purebred dogs and 341 wolves were included in the phylogenetic analysis of the five prehistoric canids. The analysis found that two of the ancient canid samples, one of them being 10 000 years old and other 4000, were included in clade A which is assumed to have originated in east Asia (Savolainen *et al.* 2002). Even if the older specimen was a wolf this would suggest a very early association between domesticated canids and European wolves. Verginelli *et al.* (2005) suggest that dogs in this clade could be the descendants of two domesticated populations evolving in Europe and in east Asia, contradicting Savolainen's hypothesis. It is likely that our understanding may change with inclusion of additional ancient dog and wolf samples from other parts of the world, but at the moment we cannot judge whether the picture will become clearer or even more blurred.

When were dogs domesticated?

In 1997 a consensus long shared by most archaeozoologists was questioned when a group of researchers suggested that the first event(s) of dog domestication happened much earlier than had been thought. Based on a calculation introduced above (Chapter 5.3.2, p. 110), Vilá *et al.* (1997) suggested that ancient wolf populations might have been domesticated more than 100 000 years ago but at least much earlier than the commonly assumed 15 000 years ago, perhaps between 50 000–100 000 years ago. This time window seemed to fit well with the beginning of colonization in south Asia by *Homo sapiens* (see Figure 5.1). Thus the first encounter between wolves and humans might have been the trigger for domestication (Csányi 2005). The lack of such early fossils was explained by assuming that early dogs were morphologically not distinguishable from wolves, partly because hybridization between wolves and dogs continued for some time before the separation of the wild and domesticated forms. Alternative accounts suggest that this date might refer to the time when the dog population to be domesticated (or its ancestors) split from the ancestors of recent wolves.

Today most researchers would agree that this date is probably an overestimation, although there are still arguments in favour of an earlier date than that suggested by the archaeological record. In line with this, it is also possible to calculate the dates of origin for the clades defined by Savolainen *et al.* (2002, Box 5.3). In the case of clade A the results of this calculation depend on the number of estimated founder wolves (for details see Savolainen 2006). Assuming a single wolf mother as ancestor of all dogs belonging to this clade, the date falls between 40 000 and 120 000 BP. Probably a more realistic approach, based on the involvement of several female wolves, indicates a date around 15 000–20 000 BP. A similar calculation for clades B and C (with single wolves as founders because simpler structure of these clades) results in an estimated date of 13 000–17 000 BP. Although the different clades indicate that domestication events might have happened at different locations (involving different wolf populations) it is less likely that these have been separated by several tens of thousands of years because once domesticated, dogs were likely to spread rapidly among human populations. Thus it is more likely that domestication events took place in a relatively restricted time period, probably around 15 000–20 000 BP.

Similar calculations for the American dog sample suggest that soon after having been domesticated dogs joined migrating human populations on a journey to the New World (Leonard *et al.* 2002). Thus they joined probably not the first but more likely the second wave of humans crossing the Bering Strait around 10 000–12 000 BP. Phylogenetic calculations indicate that dogs arrived in Australia approximately 5000 years ago (Savolainen *et al.* 2004). They are probably representatives of a dog population that was already on the way to domestication, but we have no clues whether or how subsequent selection acted on this isolated population, and some not disadvantageous 'dog-like' traits might have survived in these canids.

Is there a phylogenetic relationship between breeds?

The main question here is whether dog breeds can be classified into a biologically meaningful system based on evolutionary considerations. Kennel clubs apply an arbitrary categorization system which is based on a mixture of physical similarity, traditional working utility (if any), and doubtful information about origins. Just as in the case of 'real' species, where phylogenetic research verified (or sometimes changed) most of the evolutionary relationships put forward by zoologists on the basis of palaeontological and morphological analyses, the systematic comparison of the genetic material present in dog breeds could shed light on their origin and genetic kinship. The problem was attacked from many directions in spite of the general understanding that most (if not all) breeds have a very muddy history and are the products of multiple, poorly documented hybridization events. Breed formation (when the population is reproductively isolated from other dogs) has taken place over an extended period of time. Some breeds were already formed several hundred years ago, while others are just being established (see also Neff *et al.* 2004). In addition, there is an older and extensive tradition of establishing breeds in Europe than in most parts of Asia. In general a considerable part of the European dog gene pool has been isolated from the wolf for a longer time than in most parts of Asia, where novel 'breeds' are now being created from various dog populations.

The comparison of mtDNA haplotype distribution in breeds mirrored this supposed process of hybridization (Vilá *et al.* 1999). A relatively small (but comparative) sample of breeds suggested differences in genetic variability. Some breeds (e.g. Golden retriever or German shepherd) had 4–6 different haplotypes while in others (e.g. Border collies) only one or two mtDNA sequences were detected. However, there was no clear breed-specific pattern.

The lack of breed-specific mtDNA urged others to sequence and compare microsatellite DNA (e.g. Koskinen and Bredbacka 2000, Irion *et al.* 2003). In 2004 a huge effort by a large group of researchers resulted in the genotyping of 96 microsatellite loci for 414 dogs representing 85 breeds. This database proved large enough to carry out a detailed analysis on a pool of dogs that represented most breeds living under human reproductive control (Parker *et al.* 2004). Using multivariate statistical methods they were able to categorize the dogs according to breed, and in parallel they were also able to assign most (99%) of the individuals correctly into the respective breed category based on the DNA sequences. This indicates that there are breed-specific genetic units. It should be noted that on average 4.8 dogs per breed were used for setting up the categories, which might not be truly representative for some of the breeds. The real test of this approach would be to classify dogs that belong to the same breeds but originate from a different geographical location.

Although there was a closer relationship between some 'ancient' breeds and wolves (see below) most other breeds (predominantly of European origin) could not be separated from each other in a tree-like fashion. Thus for most breeds classic rules of evolutionary relationship do not apply. It seems that ecological or economic demands, local possibilities, fashion, or just pleasure in creating a novel kind of dog were constant forces for hybridization among different populations of dogs. A more detailed comparison of the sequences suggests a broad categorization of breeds. Although it may be tempting to name these categories on the basis of one or more breeds which are included, overlaps and other exceptions make any such attempt of doubtful value (Figure 5.3).

A recent comparison of haplotype diversity of mtDNA and Y chromosome microsatellites in dogs and wolves provides an interesting, although not unexpected, twist to the story of breeds. Sundqvist *et al.* (2006) have found that within different breeds of dogs the Y chromosome markers show a lower diversity than the mtDNA. Importantly, no such bias was found in wolves. Thus DNA data seem to verify the well-known practice that in the development and maintenance of dog breeds a smaller number of male dogs are mated to many females (*artificial polygyny*), in contrast to the mating pattern of wolves which breed monogamously.

The problem of 'ancient' dog breeds

The term 'ancient' has usually been associated with a dog breed when there is some hint of evidence for its relatively early presence in human history. Various simple depictions, sketches, or colourful pictures indicate that certain morphological types of dogs could have preceded others, even if there is no direct evidence that the living representatives of a similar breed were direct descendants of these animals.

With the advent of genetic analysis, the term 'ancient' is applied to dog breeds if these dogs show a greater genetic similarity to wolves. Although this agrees with the assumptions of phylogenetic analysis, it is not necessarily true for dogs. In general, reproductive isolation between species ensures that the genetic divergence is a function of the time of this split. However, by crossing dogs with wolves for a few generations, one could end up with an 'ancient' breed that is now more closely related to wolves. Thus, in the case of breeds that are more similar to wolves the separation from the ancestor did not take place, or, more precisely, there was continuous or regular hybridization between the two populations. It follows that these breeds represent transitional forms (and not 'ancient dogs'), and only those breeds that cannot be separated from each other but only from the

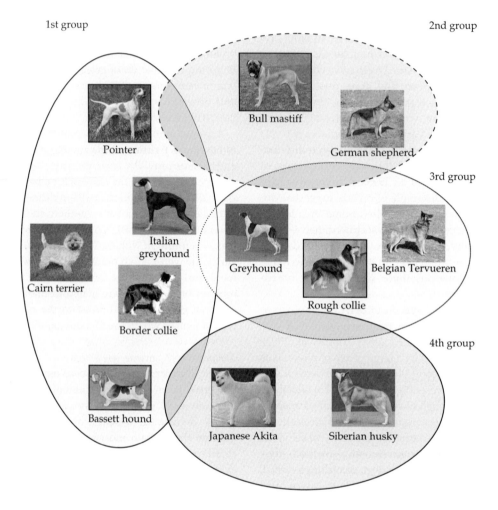

1st group

2nd group

3rd group

4th group

Bull mastiff

German shepherd

Pointer

Italian greyhound

Cairn terrier

Greyhound

Belgian Tervueren

Border collie

Rough collie

Bassett hound

Japanese Akita

Siberian husky

Figure 5.3 Groupings of various breeds based on similarities in DNA sequence. This summary of the genetic similarity among dog breeds provides a strong argument against "real" phylogenetic relationship. The figure (redrawn and modified) from Parker and Ostrander (2005) shows that neither "functional" nor "morphological" similarity explains the presented relationship. In addition to indicating one representative breed of the category like the original authors (in black frame), we show here some "atypical" breeds and others that look similar to one of the representatives of one category but actually belong to a different one.

wolf represent a truly significant step towards speciation in dogs (see below). The finding that Siberian huskies are genetically closer to wolves than are German shepherds (Parker *et al.* 2004) should not be taken as evidence that huskies represent an older breed or are more 'wolf-like'. First, basenjis are even more closely related to wolves although they do not look like them, and second, in order to be useful as sledge dogs huskies need to have dog traits and should not display behavioural similarity

to wolves (Coppinger and Coppinger 2001). Thus even if some hybridization with wolves took place (as is often claimed by the native North Americans), individuals that were behaviourally wolf-like would be selected out rapidly.

A present-day consensus
The sequencing of the dog genome in 2005 brought this species to the forefront of biological and medical interest (Lindblad-Toh *et al.* 2005). Apart from

making detailed comparisons between the human, mouse, and dog genomes, researchers also modelled the possible evolutionary history of the dog. According to their somewhat simplified model, dog evolution can be described by two major steps, separated by two bottlenecks where the descendants have lost a part of the ancient genetic variation. Based on a founding population of 13 000 animals (see also later) they put the first separation event (domestication) at *c.*27 000 years ago, assuming multiple locations. Much later, perhaps only 10 000–1000 years ago, the dog population went through another bottleneck at the time of breed creation. Interestingly, even this transition retained a large part of the previously existing variation, as breeds are not characterized by a dominance of uniform haplotypes. Most breeds still reveal on average four haplotypes, and the average frequency of the most common haplotype is around 55%, although large differences between breeds have been observed.

5.4 Some concepts of evolutionary population biology

Recent theoretical and comparative genetic work allows us to look at the process of dog domestication from a population biological point of view. Although neither wolves nor dogs form ideal populations for such investigations, models developed by such analyses can provide help in organizing our present knowledge and suggest ways of planning the collection of new data. However, we should never feel constrained by these models because they often mirror the assumptions of the researchers, and the real events in dog domestication might actually have been more complex.

5.4.1 The question of founder population(s)

As the genetic variability of any population could be critical for its survival, the number of founders is likely to determine the amount of variation for any selection to act on. Small number of founders might lead to random effects on the phenotype because of genetic drift. Smaller populations are at risk of dying out, especially if selection is too strong. Thus some domestication events have left no or little trace in the present genetic record. On the contrary,

relaxed selection (see below) might increase the chance of survival.

The mtDNA relations among recent dogs revealed by phylogenetic analysis could be parsimoniously explained by the involvement of only a few female wolf-like canids, assuming that dogs in each clade (Vilá *et al.* 1997, Savolainen *et al.* 2002, see p. 114) were descendants of a single mother. It seems more plausible that each female wolf represents a local domestication event in which a large set of individuals participated. This is also supported by recent observations that neighbouring wolf packs are similar to each other genetically and female wolves tend to stay nearer to their original group (Lehman *et al.* 1992, Chapter 4.3.4, p. 82). In addition, the fact that mtDNA has been transmitted from only a few wolf matrilines does not necessarily mean that the founding population was small, because there is a chance that certain family lines have died out (Leonard *et al.* 2005). Diversity can be great: for a given type of gene (*DRB*) within the major histocompatibility complex (MHC, involved in immune functions) 42 different haplotypes were identified (Seddon and Ellegren 2002). Because the chance of novel mutation since domestication was judged to be very small, Vilá *et al.* (2005) assumed that a minimum of 21 dogs would be needed to explain present-day variation if each individual carried 2 unique versions of these alleles. However, such a scenario is unlikely and therefore they ran a computer simulation to estimate the size of the founding population. Assuming no novel mutation, and the decrease of allelic variability by genetic drift, the estimates showed that domestication might have involved a single population of up to 1000 animals, 2–4 populations consisting of 100–200 individuals or even more but smaller founding populations (e.g. 6 populations with 60 wolves). It is conceivable that in early times anthropogenic niches could support only a limited number of wolf-like canids (or packs), and the reproductive separation of large number of dogs from wolves might have been also problematic (Leonard *et al.* 2005). However, the evidence for the limited number of domestication events, and the relatively small number of wolves in any given founding population, represent too small a variation to account for the observed allelic divergence. To explain this discrepancy, Vilá *et al.* (2005)

supposed that the relatively large present-day allelic variation could be the result of regular or occasional hybridization with wolves. Interestingly, other models of dog domestication are based on a much larger number of founders. Lindblad-Toh *et al.* (2005) assumed a starting population of 13 000 individuals which went through different numbers of bottlenecks where some reduction of variability would be expected.

5.4.2 On the nature of selection

In some respects the beginning of dog domestication can be compared to the colonization of an island. The ancestors of dogs choosing this novel, anthropogenic niche, which offered unexploited resources, enjoyed decreased intraspecific and interspecific competition. This could lead to a population expansion because more individuals could produce offspring, many of which would not have had such chances in their former habitat. This process is often described as relaxed selection, when previously handicapped individuals enjoy an increase in their fitness. The result is both an increased population size and also a diversification in phenotypes. The change in genetic diversity has two sources. First, the number of individuals carrying rare alleles increases in the population changing the allele frequency, and second, animals with previously lethal or maladaptive genetic material will also have the opportunity to breed. Although the effect of this latter process is likely to be small, both kinds of events could influence the fate of the emerging population. Hence both mechanisms increase the genetic diversity of the population, and this increased genetic variability provides a wider range of possibilities for novel selecting factors acting subsequently.

Reznick and Ghalambor (2001) argue that the combination of an opportunity for population growth with subsequent directional selection could promote evolutionary changes because in small founding populations selective forces often lead to extinctions. Ancestral dog populations might have undergone rapid reduction of population size because of some selective factors, but founding populations had a better chance of survival. In addition, selection could have acted faster if the number of preferred individuals was greater.

However, even anthropomorphic environments have their limits. Any single human group could provision only a small group of dogs. Therefore selection could have set in locally before it would have been optimal from the viewpoint of diversification. However, if early ancestors of dogs dispersed in human populations that were rapidly colonizing large areas, dogs might have ended up having larger genetic divergence compared to the ancestral wolf population at the centre of the domestication. This might provide the genetic background to the observations that dogs display greater phenotypic variability than their 'wild' ancestors, which emerged slowly and only with some considerable time lag after the start of domestication.

Although it is generally accepted that early environment of dogs was less selective, it is actually quite difficult to find a genetic proof for such a hypothesis. Björnerfeldt *et al.* (2006) assumed that effects of such relaxed selection could be traced in the mtDNA if they compare the rate of synonymous and non-synonymous (functional) mutations in wolves and dogs. Their research revealed that the ratio of non-synonymous and synonymous mutations was on average about twice as great in dogs as in wolves. It is likely that truly disadvantageous mutations have been removed from both populations, and therefore the non-synonymous alterations detected in the mtDNA change the effectiveness of the transcription process only slightly. Thus it can be argued that the environment of dogs is more tolerant for the presence of less-deleterious (non-synonymous) mutation; in other words, the selective constraints of the mtDNA have been relaxed. Extreme relaxation of selection, when modern veterinary medicine enhances the survival of individuals carrying deleterious mutations, can increase the ratio of deleterious mutations in the population, especially when such dogs are not excluded from the breeding population.

It is very likely that ancestor dog populations were affected very early on by directional selection (see Figure 5.2). There are arguments that in early times smaller animals had a greater chance of surviving in the anthropogenic environment. This could be because the constantly available but low-quality food associated with a scavenging lifestyle was more advantageous for smaller animals, so the size of these dogs changed in the direction of other canids with similar habits, such as coyotes or jackals (Coppinger and Coppinger 2001). Alternatively, humans might have preferred to interact with small dogs (e.g. hunting), and they selected them in preference to larger individuals (Clutton-Brock 1984, Crockford 2000). Both selective forces pushed the population in the same direction, which is also supported by the fossil evidence from the beginning of domestication. Archaeological finds from around 5000 BP suggest a modification in the selective environment because larger dogs emerge, some of which are actually larger than some wolves. Importantly, however, small dogs continued to exist. Not only could this be one of the first signs of artificial selection, it might also indicate that by this time (at least with regard to size) the previously more-or-less homogenous population had separated into two or more subgroups. The selection for special forms of dogs could be described as disruptive selection. The preference for large dogs could have originated from the need for companions that provide protection for the house and possessions or for animals and their herders, and are able to move rapidly with the humans across large areas (Coppinger and Coppinger 2001). Similar artificial disruptive selection could have been practised when people selected for dogs that demonstrate certain elements of wolf behaviour (e.g. hunting behaviour, see Box 8.2).

In some texts artificial selection is described as 'destabilizing', by pointing out that it affects the neuroendocrine control of the organism (Belyaev 1979). However, this use of the term is misleading because the effects of selection are measured by the changes in allele frequency and not by the effect that some alleles might have on the phenotype.

5.4.3 Changes in reproductive strategy and effects on generation times

An interesting consequence of dog domestication is the emergence of a diannual oestrus cycle. In contrast to wolves (and with the exception of a few breeds) females of domesticated canids can give birth to two litters per year. Tchernov and Horwitz (1991) argued that this trait could also be an adaptation to the anthropogenic environment, where large amounts of food could be utilized by a greater number of smaller animals, together with earlier maturity. Accordingly this would fit with the predictions of r-selection which assumes that there is a trend for high fecundity, small size, short generation time, and the ability to disperse offspring widely. Although this is plausible, most features of the dog's reproductive behaviour do not fit this picture. Dogs and wolves do not differ in the duration of gestation, relative size of offspring at birth, or lifespan of the adults. Moreover, selection of tameness could bring about most of these changes (Belyaev 1978, Chapter 5.6.3, p. 134) (see Box 5.8).

Regardless of whether this change is a response to environmental challenges or was caused by human factors, it is possible that dogs halved their generation time relatively soon after diverging from wolves. Thus we could suppose that twice as many generations of dogs as of wolves have lived during the last 8000–10 000 years. Even if, as findings suggest, mutation rates (based on synonymous nucleotide changes not affecting the protein) are the same for wolves and dogs (Björnerfeldt *et al.* 2006), shorter generation time could have produced increased variation because dogs had a higher chance of incorporating mutations occurring during the formation of the gametes.

5.5 Emergence of phenotypic novelty

Looking at dogs and watching their behaviour makes one doubt their close genetic relationship with wolves. Superficial judgement suggests a long list of 'novel' traits distinguishing dogs from their ancestors. Here we investigate the emergence of novelty from a proximal perspective; that is, what kind of mechanisms are behind the phenotypic difference between wolf and dog. It turns out that possible changes could have affected different levels of biological organization which are strongly coupled in the process of epigenesis that determines the adult phenotype.

5.5.1 Mutation

The changes in protein structure caused by genetic mutation are often regarded as the most straightforward explanations for the emergence of novel traits during evolution. Intensive research in recent years has found that protein-coding genomic sequences are very complex structures. Genes have segments that regulate gene transcription (*enhancers, promoters*), and DNA sequences for the protein-coding part (*exons*) are interspersed with elements that are not transcribed (*introns*). Thus the effect of mutations in the regions that are translated into proteins depends on their exact location. Some mutations might render a protein totally unable to fulfil its function, whereas others only modify the biochemical character of the protein to some degree. In the former case the outcome may be fatal to the organism, but the latter situation often has less serious consequences.

A recent detailed study provided good evidence of how a potentially deleterious mutation emerged in the dog population and was transmitted and fixed in different breeds (Neff *et al.* 2004). The *mrd* gene produces a protein (P-glycoprotein) which plays an important role in preventing various kinds of (potentially toxic) molecules entering the blood circulation of the brain. It turned out that in different breeds, dogs showing an adverse reaction to these molecules (some of which are veterinary drugs) had a mutant version of the gene. As a consequence of this mutation the gene lacks a four-nucleotide sequence which results in a shorter, truncated protein, which probably cannot fulfil its normal function. After extensive molecular genetic work and the comparison of different breeds for the presence of this mutant allele, it has been suggested that the mutation probably happened in a herding dog living in England in the first half of the nineteenth century. Unfortunately, this dog was among the ancestors of the present-day collies. However, later descendants of collies have also contributed to the establishment of other breeds, so in some cases this mutant allele was passed on, and today it is also present in the long-haired whippets (Neff *et al.* 2004). The tracing of such mutations is truly a kind of detective work, and not many researchers have undertaken it.

Coding regions of many genes are composed of repeated nucleotide sequences of varied length (*variable number tandem repeats,* VNTR). Very often alleles will differ in the number of such repeat sequences that are translated into amino acid chains. The protein products of these alleles, which differ in the number of tandem repeats, differ in their biochemical activity or affinity when interacting with other molecules. It is assumed that mutations changing the number of these tandem repeat sequences will retain the basic function of the protein but slight deviations could affect the resulting phenotype. Fondon and Garner (2004) showed that a contraction in the allele of the *Alx-4* gene could explain the extra dewclaw in Pyrenean mountain dogs (Great Pyrenees) in the homozygous condition. This observation is strengthened by the fact that a similar extra digit develops in mice homozygous for a non-functioning version of the same allele. This finding is potentially interesting because it seems to provide a relatively simple genetic explanation for a marked morphological change which is often taken as evidence for a 'big leap' in evolution. Only dogs seem to have this condition; extant wolves showing this trait are mostly hybrids (Ciucci *et al.* 2003).

In another case a positive correlation was found between the length ratios of two repeats within the VNTR region of the *Runx-2* alleles and *clinorhynchy* (dorsoventral nose bend) in dogs of different breeds (St Bernard, Bull terrier, Newfoundland). This indicates that this protein may play a crucial role in the development of the craniofacial region (Fondon and Garner 2004). As these changes in the VNTR structure proceed by restricted mutational steps, it is likely that phenotypic changes are only possible if lengthened and shortened alleles emerge *de novo*, which determines the progress of selection. However, such correlation does not necessarily mean a causal relation between the genetic change and the phenotypic difference (Box 5.5).

Researchers comparing the human and chimpanzee genomes have suggested that phenotypic changes are more likely to come about if the mutations affect the control of the pattern of expression (location and timing) of the protein and not its structure. So far evidence exists only for the human–ape clade (Rockman *et al.* 2005) but it is conceivable that

Box 5.5 Morphometric differences in wolf and dog

Some features of dogs resemble juvenile wolves, but the concept of general paedomorphism in dogs does not seem to be tenable. More likely selection has decoupled the developmental relationship of some traits while others have remained unchanged. In the case of the head it seems that in the length proportions of the skull (a) which corresponds to relative 'nose length' (palatal length/skull length), there are no differences between (both extant and extinct) dogs and wolves (Wayne 1986b, Morey 1992). The values for dogs fall right on the imaginary line which is indicated by *Canis* species. Such a relationship does not, however, hold for the width and length proportion of the skull (b). Dogs usually have wider skulls than their wild relatives (Wayne 1986b, Morey 1992). Thus the juvenile-type skull form, which would be a case for

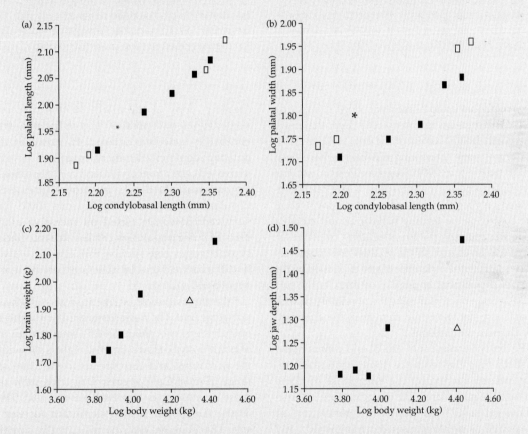

Figure to Box 5.5 Allometric relations for different extant and extinct canids. Data for (a) and (b) are from Morey (1992) and Sablin and Khlopachev (2002); measurements on dingoes were supplied by Justine Philips from specimens in the Melbourne Museum (courtesy of David Pickering and Tara Todd). Dog fossils from Morey (1992) represent North American and European samples from approximately 3000–7000 and 4000–10 000 BP respectively. Data for (c) are from Kruska (1988) and for (d) from Van Valkenburgh *et al.* 2003). (■ extant *Canis* species; ☐ extinct dogs; *, dingo; △ extant dogs).

continues

Box 5.5 *continued*

neoteny, emerges as a combination of (at least) two features of which only one shows a changed developmental pattern. (c) Dogs have about 25–30% smaller brains than canids of the same size (Kruska 2005) and compared to the relative body weight the jaw depth (interdental distance between two molars) is also smaller than expected from a *Canis* species (d) (Van Valkenburgh *et al.* 2003).

Wayne (1986b) assumed that change in allometric proportions might be the indication for artificial selection by humans. Morey (1992) proposed that changes in relation to size might have been the result of two sequential selective steps. First, the size of the wolf-like ancestor

decreased and because of developmental constraints, this was also paralleled by decreasing size of other organs (teeth, brain, etc.). In the second phase selection for larger size took place; however, in the changed anthropogenic environment (relaxed selection) this selection might not have affected all features in the same way. If some kind of decoupling between the traits is assumed then in the absence of morphological constraints, for example, selection for a larger body size (and head size) was not necessary paralleled by longer (larger) teeth because there was no need to eat (or prey on) larger prey. Similar arguments might be made for the decreased relative brain size in dogs.

---

such mutations could be present in dogs. Comparing mRNA expression in three areas (hypothalamus, amygdale, frontal cortex) of the dog, wolf, and coyote brain, some interesting differences have been found (Saetre *et al.* 2004). Dog-specific expression of two neuropeptides (neuropeptide Y and calcitonin-related polypeptide), both of which are involved in the control of the feeding behaviour and metabolism, was found in the hypothalamus. Importantly, there was no control for the possible environmental effects, that is, the difference can be explained by the special experiences of dogs in contrast to the two wild species. (Unfortunately, the possibility of such bias is also not taken into account in the chimpanzee–human comparisons.)

Recently Leonard *et al.* (2005) and others noted that the time elapsed since domestication is simply too short to expect the emergence of many favourable mutations. The mutation rate in functional genes (10^{-5} per gamete per generation) or measured as single nucleotide changes (10^{-7}–10^{-9} per gamete per generation) has probably not offered enough variation for selection in the dog. Thus most of the genetic basis of novel phenotypes in dogs might have been present in the wolf population. Many mutations accumulated during the evolution of *Canis* could have survived in heterozygous animals if the mutations were recessive, that is, the individual had another

'healthy' copy of the allele. In this case only homozygous and possibly less fit animals were constantly selected against. If, however, the anthropogenic environment equalized (or even increased) the chances for survival, then homozygous animals displaying novel (previously disadvantageous) phenotypic traits could have survived. Selection based on recessive alleles present in the population can lead to large phenotypic changes (see section 5.6); one has only to find the carriers and be able to hit on the homozygous individuals.

Take the evolution of size in dogs, which is a polygenic trait. The mean (estimated) wither height of early dogs was about 20–40% smaller than that of extant wolves. However, this height is still within the wolf range and corresponds to the lower size range in most *Canis* species. Dog finds show that these smaller dogs survived for the next 5000–6000 years without further significant decrease in size. This suggests that the reduction in size was based mostly on alleles that were already present in the wolf population, and if all 'appropriate' alleles had been selected no further decrease in size would be expected.

It was the Romans (and possibly also the Chinese) who succeeded in developing an even smaller type of dog. However, this reduction was not proportional for all body parts, but was characterized predominantly by relatively short limb bones. This breakthrough happened when people were able to 'rescue' a (natural) mutation which caused marked phenotypic changes. The propagation of a mutant allele in a population is not a trivial task; breeders need to be able to keep a large population reproductively isolated, and arrange planned matings. The condition of shortened legs is often described as *achondroplasia*, when the bones stop growing early in ontogeny (Young and Bannasch 2006). Crosses between short-legged and long-legged ('normal') breeds most often result in short-legged dogs, and this strongly suggest a (incompletely) dominant mode of inheritance. (The dominant nature of this mutation explains why early dog breeders were able to maintain this allele in the population.) In humans a similar condition is caused by mutations in a growth factor receptor (FGFR3) but so far a similar mutation has not been verified in the affected dog breeds. Importantly, if this mutation occurred in wolves the affected individual had not much chance of survival. In contrast, these small dogs enjoy a clear advantage over large companions in certain human environments (e.g. lapdogs, or dogs used for hunting in burrows, e.g. dachshunds). Thus it is doubtful whether this (dominant) mutant allele could be found in present-day wolves; in contrast, it is probably widely distributed in dogs, because there are other breeds (e.g. German shepherds) where dogs with short legs are born (although this could also be the result of a yet another mutation). Similar arguments could be made for traits such as short hair which is inherited in a dominant fashion over long hair and is possibly not present in wolves.

Thus the large variability of the wolf genome offers some room for directional selection that can lead to large phenotypic changes in dogs without necessarily involving novel mutations. Nevertheless, if mutations occur within this relatively short timescale they can survive in the population if the phenotype has some advantage in certain human environments.

5.5.2 Hybridization

Hybridization between related species (or subspecies) has been often implied as a source of novelty. Descendants of such crosses often retain different fragments of parental characters in unique combination. The greater the phenotypic difference between the parents, the greater is the observed effect (Coppinger and Schneider 1995). Thus the effect of hybridization depends on the time elapsed between the separation event and the hybridization event. However, there is an upper limit for hybridization when phenotypic differences become too large and limit the possibility of hybridization, becoming reproductive barriers. The evolution of wolves suggests that this species was often involved in hybridization events (Chapter 5.4.1, p. 117). During dog domestication various types of hybridization events could have taken place. Early dog-like populations could regularly have hybridized with local wolf populations, and because dogs dispersed very rapidly around the globe some of this mixing might have involved wolf populations which did not contribute to the original gene pool of the dog (see above). There is a long-held view that the genetic material of some local wolf populations could have contributed to the emergence of divergent dog phenotypes (Clutton-Brock 1984). Such assumptions are also supportedly early fossils showing both dog-like and wolf-like traits (Sablin and Khlopachev 2002), and there is also some mtDNA evidence (Verginelli *et al.* 2005).

The problem is that on the basis of the archaeological record it is difficult to discriminate early hybridization events from local domestication events. Molecular data are also very insensitive in this case, and provide only indirect support. For example, mtDNA data will not indicate the effect of male wolves on the dog population (no transfer of mtDNA takes place). Thus the finding that the Indian wolves represent a totally different clade of mtDNA haplotypes does not necessarily exclude the genetic contribution of male Indian wolves to dog evolution (Sharma *et al.* 2003). Similar arguments can be made for the American wolves which lived for many thousands of years along with dogs although no sharing in mtDNA haplotypes has been revealed (Leonard *et al.* 2002). Many think this is unlikely, and assume that wolf-like traits could have been imported by crossing female dogs with male wolves (Kopp *et al.* 2000).

This idea is often supported by historical accounts that in order to 'improve' their dogs some people (e.g. the Inuit in Alaska) regularly hybridize their canid companions with wolves (Clutton-Brock 1984). Although such practices cannot be excluded, other experts dismiss such stories as legends (Coppinger and Coppinger 2001). Nevertheless hybridization with male wolves followed by strong selection and the retention of male dogs only would go unnoticed by current genetic analysis (see also section 5.3.2, p. 114.

Because first-generation wolf–dog hybrids display a set of unwanted behaviours, it is less likely that humans tolerated such individuals when dogs had already been around. Thus the offspring of such matings would be selected out rapidly from the population. Nevertheless hybridization with wolves could have been a simple way of selecting for increased body size. In the case of some present-day breeds there are indications of stronger influence of wolf genetic material (e.g. Norwegian elkhound: Kopp *et al.* 2000), but this might also be the result of a founder effect or genetic drift.

It is also likely that some dog genes found their way into the local wolf population. Thus events indicating local domestication could be feral wolves which had domestic dogs among their ancestors. Recent field investigations on extant populations have also revealed such cases, both in Europe and America (e.g. Randi *et al.* 2000, Ciucci *et al.* 2003). However, the mostly agonistic (or evasive) relationship observed between coexisting dog and wolf populations (Boitani *et al.* 1995), and perhaps the disadvantage of hybrids in associating with either canid species, constrain gene flow to a low level even among coexisting populations.

With the development of breeds the impact of hybridization between dogs and wolves became more limited, with the exception of some recent breeds, in which a few wolves have been utilized in the founding population (e.g. the Czech wolfdog mentioned earlier). In other cases hybridization between different breeds of dogs also yields novel phenotypes, and this technique has also been used in rescuing dog breeds that were in danger of extinction.

5.5.3 Directional trait selection

Anyone who has survived the rearing of a wolf at home could easily put together a list of behavioural traits that would be useful to select for or against. Many authorities on dog domestication have proposed various traits which would be advantageous in an anthropogenic environment, especially if one prefers an affiliative and cooperative companion. According to Clutton-Brock (1984) an ideal dog is small and looks childish with a short nose and large eyes. It is docile and tame and shows a tendency for submission (in parallel also inhibition of attack), is less fond of food and less choosy, and consequently is more ready to share. Making a noise (barking) could also be an advantage.

There are two non-exclusive ways in which to conceptualize the phenotypic changes that took place during domestication. One view assumes that the changes touched mainly upon features of *temperament* (sometimes incorrectly described as 'personality', Chapter 10, p. 221). Temperament refers to a set of behavioural traits which characterize an individual's reactions independently from the actual situation and are not influenced by experience (learning) (Clark and Ehlinger 1987). Accordingly the anthropogenic environment preferred individuals with a certain temperament, the precursor of which was also present in the wolf (e.g. Paxton 2000). Indeed, wolf cubs of the same litter might show large differences in temperament (Fox 1972, Macdonald 1987) which could form the basis of selection. Ideas of selection for 'docility' or 'tameness' can be also considered as a set of traits associated with a special temperament. Frank (1980) describes docile individuals as being less wary of humans and novel stimuli and open to socialization. Thus docility could also be regarded as being bold/fearless/curious and sociable, which are typical 'personality traits' (e.g. Svartberg 2002, Gosling *et al.* 2003). Selection for juvenile character traits would yield a similar list (see below). Thus when investigators invoke selection for docility (or tameness), or for a special form of bold and sociable behavioural type, or for juvenile behaviour, they are actually referring to the same set of phenotypic features. (Box 5.8). Actually, both docility and juvenile behaviour represent special functional organizations of behaviour, so it seems more advantageous to adopt a behavioural model that assumes that certain behavioural types had a selective advantage in the anthropogenic environment. Adopting this view leads to further gains insofar as it can be directly related to wolf natural behaviour without the interpretational difficulties that involve interaction with humans ('tameness') or problems of behavioural development (Chapter 9). In short, certain behavioural patterns in dogs are the result of the response of wolf behaviour patterns to anthropogenic selective factors. A similar view was advanced by Hare and Tomasello (2005), who argued for the selective effects of domestication on emotional/temperamental characters in dogs.

A different approach suggests that in order to be successful in the anthropogenic environment, dogs were selected for human-compatible social behaviour traits (Miklósi *et al.* 2004, Csányi 2005). This view assumes changes in a wider range of social behaviour traits in addition to alterations in temperament (Chapter 8).

It should be stressed that the two views described here are not incompatible, although it is arguable whether selection for behavioural types or sociocognitive abilities took place first. One could argue that the changes in emotional/temperamental characters preceded the emergence of novel sociocognitive traits because the selective factors exerted by humans could only act on a wolf population characterized by genetically altered behaviour. However, even if selective factors could be separated, behavioural types and cognitive performance are tightly coupled at the level of behaviour, thus selection for or against a behavioural trait probably caused correlated changes affecting both systems.

It is more important to note that both concepts assume that there were no special selective forces affecting a narrow (or single) aspect of the behavioural phenotype. Even selection for tameness or docility could potentially involve changes in a broad range of behavioural traits including sociability and aggression. Boldness in approaching strangers, decreased interpersonal distance or flight distance, less specific species-specific recognition system, decreased self-protective tendencies, increased threshold for attack, and increased tolerance for submission could all result in a phenotype that is described as docile or tame. In addition, the actual founding populations (even if only a few might have been involved) could also influence the outcome, and there might also have been differences in the selective factors presented by different human groups. Laboratory experiments found that

even if fruit flies (*Drosophila*) were selected on the basis of the same criteria, behavioural changes that accompanied the process caused marked differences in the different selection lines (Gromko *et al.* 1991), perhaps because in the case of polygenic traits (involving a set of genes with major and minor effects) a different set of (minor) genes might have been touched upon during the selective process.

5.5.4 Selection for plastic phenotypes

The concept of behavioural plasticity has often been raised in relation to wolf–dog comparisons. Frank (1980) argued that the ability of dogs to react to a broad range of arbitrary stimuli and respond with varied action patterns reflects a significant change in behavioural organization. Accordingly, domestication has selected for increased tractability.

The concept of *phenotypic plasticity*, as used here, refers to the difference between genotypes in the degree of responding to environmental challenges. In contrast to the gene × environment interaction, when the effect of a gene on the phenotype depends on the actual environment, phenotypic plasticity means here that a genotype with greater spectrum of reactivity over a range of environments is said to be more plastic (Pigliucci 2005). There are certain evolutionary scenarios when more plastic phenotypes can have a selective advantage, and apparently this also happens in the domestic environment. Continuing Frank's (1980) line of argument, dogs show a more plastic behavioural phenotype because their range of reactions in different environments is larger than that of wolves. Consider the case of attachment behaviour (Chapter 8.2, p. 169). Independently of whether wolves are raised in restricted or enriched human social environments, their pattern of attachment behaviour towards humans has a smaller range than that of dogs exposed to a similar range of environments. Naturally, one way of achieving increased behavioural plasticity is to increase the possibility of environmental control over the genetically determined behavioural programme. As a consequence the trait is more environment-dependent, which increases the role of individual experience and learning in case of behaviour (open programmes, Mayr 1974). However, this change in the mechanism has its costs because such open systems are prone to failure if the environment does not provide the 'expected' stimulation. Such cases may occur rarely in nature, but in a human environment the lack of appropriate stimulation can result in large behavioural differences or malformations (e.g. problem behaviour in dogs). Thus the actual social environment affects behavioural development in dogs to a greater degree than in wolves, and consequently environmental stimulation is expected to have greater effect on dog behaviour in contrast to their wild relatives.

It is therefore a possibility that during domestication dogs with a more plastic phenotype had an advantage; for example, if they were able to react to a broader range of communicative signals (visual and acoustic) emitted by their human companions.

5.5.5 Heterochrony

The evolutionary change in the relative timing of developmental processes (*heterochrony*) has often been implicated as a source of phenotypic novelty (Klingenberg 1998). The idea that the transition from wolf to dog was made possible by such changes has been around for a long time (Bolk 1926, Herre and Röhrs 1990). The morphological and behavioural

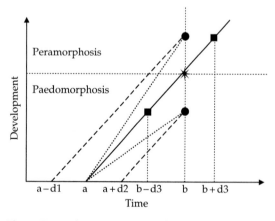

Figure 5.4 A schematic presentation of developmental changes (based on Albrecht *et al.* 1979, Klingenberg 1998). — = development of the ancestor from state a to b (e.g. wolf); ---- = earlier (predisplacement) or later (postdisplacement) w/o change in rate; ····· = slower (neoteny) or faster (acceleration) rate of development; ■ = earlier (progenesis) or later (hypermorphosis) end of development (d1–d3 = arbitrary durations).

comparison of wolves and dogs prompted many to suggest that the latter species has been arrested in a juvenile stage (Box 5.6). The smaller relative size of the dog's head, the shorter nose, many juvenile behavioural characters (e.g. dependent behaviour, playfulness) and the lack of certain patterns of adult predatory behaviour in many dog breeds were used as supporting evidence (see also Coppinger and Schneider 1995, Frank and Frank 1982).

Development occurs in time, so heterochrony is necessarily a relative concept. Usually, the development of a trait between two points in time or during certain developmental stages is compared in the ancestor and the descendant. According to the model proposed by Albrecht *et al.* (1979), phenotypic alterations in comparison to ancestral species due to the heterochrony can be manifested by either changing the time of onset and offset or by changing the rate of development. As a consequence the developing organism passes through fewer (*paedomorphism*) or more (*peramorphism*) developmental stages. The notion that dogs show juvenile wolf characteristics suggests that they do not leave the juvenile stage behind and never pass to the adult (wolf) stage (paedomorphism) (see also Chapter 9.4, p. 210).

Accordingly, the slower growth rate of the dog's head in relation to the rest of the body could explain the observation that a dog will have a smaller head than a wolf of the same body size. Since both wolves and dogs approach maximum size by the end of the first year, and at the same time dogs also become sexually mature, this results in an adult dog having a smaller head/body ratio than the ancestor. This slower rate of development is usually referred to as *neoteny* (Albrecht *et al.* 1979). Note that different variations in initialization time and developmental rate can lead to the same phenotype. For example, later onset but no change in developmental rate (*postdisplacement*) can also lead to the same developmental stage in time as neoteny. Similarly, *progenesis* (earlier cessation of development) also leads to truncated developmental processes, and results in a paedomorphic animal.

So far research has not shown undeniable evidence that the differences between wolf and dog phenotypes are the result of overall paedomorphism. For example, it has turned out that the 'short nose' as a juvenile trait in dogs is an illusion in the

Box 5.6 Heterochrony or developmental recombination in behaviour

Behavioural differences between dogs and wolves have often been explained as a slowing down of development, which results in juvenile traits being retained at the adult age. This theory predicts that in dogs traits emerge later (*postdisplacement*) and develop at a slower rate (*neoteny*) during development. The comparative analysis of various dog breeds does not support this view. Detecting the first emergence of more then 70 behavioural actions in 7 dog breeds, Feddersen-Petersen (2001a) found no evidence for overall neoteny or postdisplacement in dogs in relation to wolves. Although there was a clear variability among breeds, a considerable part of the traits showed even an earlier emergence (*predisplacement*). Note

also that breeds considered to be very similar to the wolf (Siberian husky and German shepherd, Goodwin *et al.* 1997) differ markedly in the timing of developmental events. Apparently, Siberian huskies and Bull terriers show similar amounts of predisplaced traits. This contradicts the idea that morphologically paedomorphic breeds (which also differ from the wolf to the greatest extent) display a slower rate of development. This suggests that either paedomorphism as observed by Goodwin *et al.* (1997) might be related to specific behavioural function (e.g. aggression), or such behavioural variability is secondary and emerges as a result of other physical or behavioural constrains or correlated relationships.

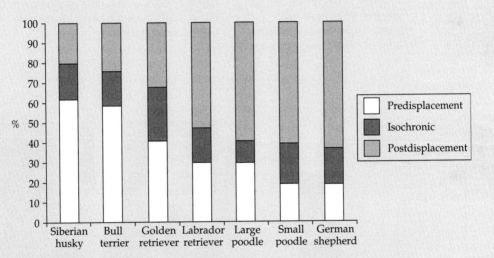

Figure to Box 5.6 The per cent of behaviour traits that emerged (1st day observed) earlier (predisplacement), around the same time (isochronic), or later (postdisplacement) in various dog breeds in comparison to wolf development (based on data in Fedderson-Petersen 2001a).

eyes of the viewer (Coppinger and Coppinger 2001). Both wolves and dogs have the same skull-length proportions, and only the width/length ratios are different, probably because slower relative growth of the face in dogs (Box 5.5). Barking seems to emerge much earlier in many breeds of dogs (around day 9) than in wolves (day 19), whereas

howling has a much later onset (day 1 for the wolf; day 14–36 for the dog) (Feddersen-Petersen 2001b, Chapter 9, p. 201). Thus wolf–dog differences can be partially attributed to changes in the pattern of development but there is no overall pattern that would fit a general trend towards paedomorphism (Box 5.6).

Box 5.7 Correlated changes or phenotypic selection?

The correlative nature of certain phenotypic traits could sometimes make simple problems very complex, because in hindsight it is often difficult to find out which trait was the primary target for selection. For example, looking at the colourful coat of domesticates we might assume that people were selecting for individuals with particular colours, but then it has turned out that selection for tame behaviour leads to changes in coat colour (Belyaev 1979).

McGreevy *et al.* (2004) found that the skull index (skull width/skull length) of dogs correlates with the form of the area for good vision (relatively high number of ganglion cells in the retina: visual streak) in dogs. Dogs having a rounder skull (larger skull index) seem to have a more circular visual streak, whereas long-nosed dogs have a more elongated visual streak, just like wolves. The old finding that dogs have more forward-looking eyes than wolves always used to be taken as evidence for of human preference for

a 'childish' look in dogs. This finding, however, offered an alternative hypothesis. It might be that dogs were selected not for their appearance but actually for their visual abilities, because the more circular visual streak might offer sustained looking ahead (i.e. towards the human). Dogs with such a visual streak might be less distracted by other events going on in the wider visual field. Recently, this idea was tested by comparing the performance of different breeds of dogs in the two-way choice task (see Box 1.2) (Gácsi *et al.* 2009). The results seem to support this idea; breeds with a shorter nose and more forward-looking eyes perform better in this test.

Thus it might be the case that the 'short nose' is the correlated change in the evolution of dogs, because enduring attention has been selected for. This might have enabled the emergence of other skills in dogs which are based on observing humans for longer durations.

Figure to Box 5.7 (a) Short-nosed (brachyocephalic) dogs perform better in using momentary pointing as a cue for hidden food than long-nosed (dolichocephalic). (b) Two representative breeds in the experimental groups: Collie (left); Boston terrier (right). * indicate significantly above-chance performance; & indicates significant difference between the groups. The percentages in the column show the ratio of dogs that choose significantly over chance (binomial test, p < 0.03, at least 15 correct out of 20 trials).

Coppinger and Smith (1990) advocated the view that developmental stages are evolutionary adaptations to particular developmental environments. In line with this, Frank and Frank (1982) suggested a parallel between the developmental environment of a young wolf and an adult dog. Dogs in the anthropogenic environment can rely on a continuous food supply and parental care for an extended duration, and they do not need to defend a territory and fight for dominant status in the group. They argued that such conditions might favour selection for an extension of a developmental stage associated with juvenile traits. Although the idea is appealing, the developmental pattern of several traits contradicts this prediction.

Even if heterochronic changes play a role in the phenotypic evolution of dogs, it might be more fruitful to regard this as one possible feature of *developmental recombination* (West-Eberhard 2003, 2005), which is defined as any novel combination of phenotypic traits expressed during ontogeny. It is very likely that in dogs the relation between some morphological and behavioural traits, which was typical for the *Canis* species, has been changed or decoupled.

5.5.6 The 'mysterious laws' of correlation

Obviously there are some trivial relationships between two or more phenotypic traits, and nobody is surprised to find that animals with longer long bones tend to have longer skulls. A 'mystery' is involved when traits affecting very different aspects of the phenotype seem to be coupled in some way (Box 5.8). Such a correlation between fur colour and behaviour has often been implied and indeed verified for some extent (Clutton-Brock 1984). For example, solid-coloured cocker spaniels show a greater tendency to aggression than particoloured ones (Podberscek and Serpell 1996).

From the relatively small number of genes (estimated to be *c.*19 000 in dogs; Parker and Ostrander 2005) and the much higher number of phenotypic traits, it follows that most genes affect more traits of the phenotype (*pleiotropy*). In parallel, many phenotypic features are determined by a set of genes (*polygeny*). These two kinds of relationships are responsible for correlative changes that depend on the genetic background. If body size is determined by a set of genes that in turn affect a range of other traits, then it is inevitable that if selection for size is paralleled by genetic change, this could alter other phenotypic traits. Selection for 'size' may not always affect the same set of genes because their contribution to the polygenic trait might depend both on the actual genotype and the selective environment. We have to face the fact that there is a very complex relationship between phenotypic traits and the underlying genetic control, which involves not only pleiotropy and polygeny but also complex interaction between genes (e.g. epistatic effects), developmental feedback mechanisms, and the effects of the actual environment.

Very often the basis of correlation between traits is caused by some common underlying role of hormones or neurotransmitters. Most hormones have a very broad range of effects, ranging from influence on morphology (e.g. size), metabolism (e.g. oxygen consumption), to behaviour (e.g. sexual displays). Thus it is conceivable that even a change in hormone levels may influence many aspects of the phenotype. Importantly, such effects can often be witnessed independently, whether these changes are caused by genetic or environmental factors. In addition, observation of one type of effect does not provide necessarily an explanation for the mechanisms. For example, it was assumed that selection for 'tameness' results in reduced adrenal functioning (hypotrophy) (Richter 1959) and this was supported by observations that wild and domestic animals differ in circulating blood hormone levels. However, Clark and Galef (1980) found that environmental differences (sheltered environment) can lead to similar phenotypic differences, because gerbils living without a shelter to hide in (mimicking the domestic environment) were found to show adrenal hypotrophy in comparison to companions that were provided with shelters. Thus the similar phenotype (andrenal hypotrophy) could be the result of the operation of two at least partially different causal chains. The observation that certain environmental changes induce phenotypes resembling the domesticated form in some respects can provide only a limited explanation for the evolutionary factors and the affected genes involved in the domestication process.

Recently, Crockford (2006) has suggested that the changes in thyroid hormone system (thyroxine and triiodothyronine) could explain most phenotypic aspects of domestication, such as a smaller initial body size, piebald coat colour, earlier reproduction, stress tolerance, and tameness. She suggested the following sequence of events: Wolves showing more tolerance towards humans (being 'less stressed') were more successful in invading the anthropogenic environment. Because of the physiological relation between stress and thyroid hormones, such selection could have resulted in wolves with a particular thyroid pattern, which in turn affected a range of phenotypic traits. After many years of selection and breeding for stress tolerance the new canid is characterized by small size, colourful coat, and tame behaviour. The small canid fossil records at the beginning of domestication could provide some support. Crockford's theory is based on three important assumptions: (1) there is a single selective factor involved (stress tolerance), (2) there is a genetic variability in thyroid production which correlates with hormones underlying stress tolerance, and (3) pleiotropic effects of the hormone. Although the environmental stress caused by humans is often cited as a selective factor (e.g. Belyaev 1979), we may suppose that scavenging could have been a recurrent feeding strategy in evolving wolf populations (subspecies) when they cohabited with even larger canids (Chapter 4.3.2, p. 76). A scavenger could evolve various ways to evade direct contact with the food donors. Genetic variability in thyroid synthesis is likely, and there are observations showing differences in dog breeds. Interestingly, the basenji shows a more rapid thyroid metabolism than European breeds (Nunez *et al.* 1970). Unfortunately, very little is known about the genetic relationship between stress hormones and thyroid. Importantly, noting the size differences in Canadian and Alaskan wolves, Jolicoeur (1959) also suggested that the differences in illumination levels could influence growth by affecting hormone balance including levels of thyroid. These north-eastern wolves are not only smaller, but have a shorter snout, and there are also more less-pigmented (pale) individuals in these packs. These later observations support the pleiotropic effects of thyroid.

However it is important to note that Crockford's (2006) theory deals with only one aspect of the complex genetic–hormonal–morphological/behavioural network. One could argue that not stress tolerance but selection for smaller size was the significant factor behind changes in thyroid production. According to Coppinger and Coppinger (2001), the energetic constrains provided by the available food in the anthropogenic environment selected for smaller dogs. This could have affected the thyroid metabolism and there is no need to assume the intervening role of stress-related hormones. Alternatively, individuals were more likely to look for alternative food sources (e.g. human food waste) if expelled from the wolf pack (Csányi 2005). If hormones underlying various forms of sociality (affiliative or aggressive behaviour) have a genetic variability, such lone wolves could be also characterized by a typical pattern of hormone production including androgens, oestrogens, and perhaps even thyroid if smaller wolves are more likely to be losers. Finally, as neither of these assumptions is exclusive, we could assume complex selection factors that acted on (juvenile) wolves leaving their pack and selecting for small and stress-prone characters.

The lesson from all of this is that it might be impossible to isolate a single selective factor, a single trait, and a single causal chain for determining morphological and behavioural changes during dog domestication. Nevertheless these theories might help to determine the direction of research into the strength of particular phenotypic and genotypic correlations which might have been involved in changes observed during domestication.

It seems important to distinguish between two different types of change that are both often described as 'by-products'. In the typical case a by-product is a correlated event that is based on pleiotropic gene effects. One such correlated by-product could the piebald coat that emerges in foxes as a result of selection for certain behavioural traits (Chapter 5.6, p. 132). There are, however, cases in which there is no direct causal relationship between the selected feature and other traits emerging in parallel. Recently, McGreevy *et al.* (2004) found that dogs with a shorter nose (brachiocephalic skull) have more expressed concentration of ganglion cells in the retina. Such an arrangement, which is similar to the focal spot in humans, is assumed to aid in focused vision. Thus selection for short-nosed dogs might have resulted in animals with more enduring powers of watching an object (e.g. a human face) because they have a more defined retinal area and are less distracted by environmental influences. This offers the possibility for better performance in certain cognitive or communicative tasks (see Box 5.7). However, such achievements should not be regarded as correlated by-products of selection for short nose. More correctly, selection for short nose changed the (inner) environment in a way that enabled the utilization of different abilities. Once such dogs are available, selection can act in novel ways on this emerged ability, perhaps resulting in dogs that achieve even higher levels of performance. Recently, Hare and Tomasello (2005) referred to this second meaning of 'by-product' when arguing that the changes in temperament might have allowed selection on other unrelated cognitive abilities. Thus it seems useful to distinguish 'correlated changes' from 'enabling changes' (and perhaps abandon the reference to by-products).

5.6 A case study of domestication: the fox experiment

One of the few long-term experiments in biology started at the end of the 1950s when the Russian geneticist Belyaev set out to replay the evolutionary game of domestication. Being obliged to sort out practical problems of animal management at fox farms, he decided to start a genetic experiment by selecting foxes for special behaviour traits. He argued that people and wild animals (especially dogs, but the idea can be applied also to other domesticated species) could only be part of the same social group if humans have (probably unconsciously) selected for animals showing affiliative behaviour and reduced aggression ('tameness') (Box 5.8). After more than 40 years of continuous selection there is now a population of such selected ('tame') foxes at the Novosibirsk research institute (Trut 1999). Recent interest in the genetic underpinning of domesticated behaviour (Kukekova et al. 2005) initiated various investigations to compare the behaviour of selected and unselected foxes in more detail.

After more than 40 generations, selected foxes display many traits that make them similar to dogs in many respects (Belyaev 1979, Trut 1980, 2001). They show affiliative behaviour, wag their tail, vocalize (whimper) towards approaching humans and lick their hands. These behavioural changes are associated with parallel alterations in morphological traits, such as piebald coat, drooping ears, and curved tail. Further changes affected reproductive behaviour, which became biannual, that is, female foxes were sexually active twice in a year. Although the behavioural traits seem to be stable characters in the selected foxes, the morphological traits were more elusive, and not all animals displayed them in the population. Some traits disappeared during development (drooping ears became erect), and only a minority of the females had a biannual breeding cycle.

Belyaev and his followers stressed the parallels between dog domestication and the fox experiment, and the above-mentioned features leave no doubt that foxes have adopted a range of dog-like traits. However, the differences are equally important. Although the evolutionary relationship between dogs and foxes biases us to a comparison based on homology, it is also clear that foxes represent a different evolutionary clade that separated from Canis 10–12 million years ago (Wang et al. 2004) and has been extremely successful in a different ecological environment (Macdonald 1983). Similarities in ecology and mainly solitary behaviour (Fox 1971, Kleiman and Eisenberg 1973) could provide a base for convergent evolutionary comparison of small species of felids and these selected foxes. It might be the case that at least at the behavioural level selected foxes might be more similar to present-day domestic cats than to dogs (see also Cameron-Beaumont et al. 2002).

5.6.1 The founding foxes and behavioural selection

Fox farming started just before 1900 in several places in Russia because it seemed to be a cheaper way to obtain fur. The foxes used for Belyaev's experiments originated from a farm in Estonia where fox farming had been practised for 50 years. This long separation from the wild population and breeding in captivity

Box 5.8 What is tameness?

There is a widespread belief that during domestication there was a preference for individuals showing a 'tame' phenotype. This is often interpreted as domestic animals being 'tame' by nature, but in fact they become tame only if they are socialized to humans. Importantly, there is no behavioural definition of tameness, which is apparently a complex character that emerges after either being selected for certain kind of behaviour over many generations (Belyaev 1979) or being exposed in early development to the human environment.

In the lack of any ethological definition we could regard an individual as 'tame' if it responds to certain environmental and social stimuli in a similar way to a human. Here is a non-exhaustive list of behavioural features of 'tameness':

- Decreased flight distance (willingness to approach/ not frightened when approached)
- Decreased inter-individual distance
- Decreased agonistic behaviour (both offensive and defensive)
- Decreased activity
- Flexible behaviour pattern
- Rapid acclimatization to novel environments
- No overt reaction to (novel) environmental stimuli
- Little dependence on endogenous stimuli
- Sensitivity to human stimulation (learning) and communicative cues

Note that tameness is a state, but domestication is a complex process. Thus it is misleading to call animals domesticated if they were selected for one or other aspects of 'tame' behaviour.

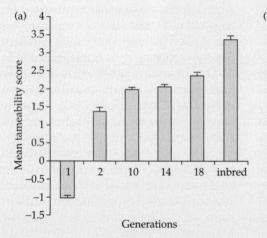

Figure to Box 5.8 (a) The progression of selection for tame behaviour ('tameability') in foxes based on data provided in Trut (1980). Note the rapid change in tame behaviour after just one generation; by the 10th generation most foxes accepted the handling passively. Tail wagging and other affiliative behaviours seem to emerge as population-level behaviour after 18 generations. This two-step process might suggest the involvement of different type of genetic control. The following four-level scoring system was used for selection for 'tameness' in foxes (see Kukekova et al. 2005 for more details): passive avoidance or approach when food is offered (0.5–1); passive behaviour during petting and handling (1.5–2); friendly response to handler, tail wagging and whining (2.5–3); eager to establish contact, licking handler hand, whimpering (3.5–4). (The negative starting value indicates that at the group level foxes showed overall avoidance.) (b) Tame fox-Photo: Elena Jazin.

already made these animals noticeably 'tamer' (Trut 1999) and probably also genetically different (Lindberg *et al.* 2005). When Belyaev started his experiments he described about 30% of the foxes as behaving very aggressively towards humans, 20% as being very fearful, and only 10% could be said to show weak exploratory behaviour ('interest') when approached by the experimenter (the remaining 40% showed ambivalent behaviour, being both aggressive and fearful). The aggressive tendency in the behaviour was a lifelong characteristic of the individuals and seemed to be heritable.

Captive-born fox cubs received very little human contact. At birth they were left with their mother for 2 months; after that they were moved to separate cages in small groups, and finally they were put into individual cages at 3 months of age. The selection process started at the age of 4 weeks and fox cubs were tested monthly until the age of 6–7 months (Trut 1999). To test the fox's reaction towards humans, the experimenter reached a hand into the individual's cage holding a piece of food, and tried to handle and pet the approaching animal. Similar tests were also done in groups of freely moving fox cubs when the animals had the possibility of choosing between approaching the experimenter or remaining in contact with other cage mates. Experimenters were looking for animals that approached the human hand and did not bite when handled or petted. Ten per cent of the females and 3–5% of the males that showed the strongest affiliative tendencies ('tameness') were selected for further breeding, and in parallel an unselected line was also established. Over the years the rules for selection became stricter. At the beginning foxes showed only a marginal interest in the humans; later, however, they not only approached the hand but often vocalized, sniffed, and licked the hand. The behaviour of the selected animals had already changed in the second and third generations, but other correlated changes emerged somewhat later around the eighth or tenth generation.

This rapid change in behaviour shows that the underlying generic variability was already present in the founding foxes (although captive life might have preselected the population) because the occurrence of novel mutation during this short period is unlikely. Selected lines were interbred regularly in order to avoid inbreeding, so the homozygous condition was also not a likely explanation for the altered behaviour. Therefore selected foxes must have harboured a set of specific alleles which affected their behaviour and also other morphological traits (Belyaev 1979, Trut 2001). Selection probably targeted genes that coordinate and regulate gene action at a high level, and thus exert a genome-wide pleiotropic effect.

5.6.2 Changes in early development

Selected and unselected foxes showed marked differences in the emergence of sensory abilities and also in exploratory behaviour in the presence of humans in the captive environment (Belyaev *et al.* 1985). Although all foxes were able to smell, taste, and respond to touch from the day after their birth, selected foxes opened their eyes and reacted to various sounds on average 1–2 days earlier (predisplacement) than unselected ones (reaction to sound 15–16 days; eye opening 18–19 days). Although both unselected and selected foxes spent the same amount of time in walking up to the age of 30 days in the open field test, after 35 days unselected foxes showed reduced activity, and spent more time near the cage walls. As time passed unselected foxes more frequently growled at and threatened the experimenter; in contrast, selected foxes continued to show high levels of activity and interest towards humans. The change in behaviour of the unselected foxes was taken as an indication of the end of the sensitive period. In contrast, in selected foxes the socialization period was extended to about 65 days after birth, approaching the range found in dogs (Scott and Fuller 1965, Chapter 9.4, p. 209).

The relatively extended selection process could have targeted several parts of the affiliative behavioural system, which are very difficult to separate because the foxes might have undergone various developmental changes during the different steps of the selection process. Potentially two different processes are associated with the socialization period. First, at the beginning the cub gathers experience about its own species through a range of sensory channels, which will serve it later in species recognition, and possibly also in recognizing kin or even individuals (Hepper 1994). It is to be expected that the sensory system is biased towards conspecific stimulation; that is, socialization is more rapid when such stimuli are present (Chapter 9.3, p. 207). The testing of 1 month old fox cubs might have selected for those individuals that showed the least preference towards conspecifics and the same time were more attracted to food. Decreased preference for conspecifics can be explained by genetic differences that caused less intensive learning about conspecific cues. If there is individual variation, humans could be found attractive after 35 days by animals having an extended sensitive period. Even minimal contact (e.g. during feeding and cage cleaning) and the earlier testing events could have resulted is some preference towards humans in some cubs that developed a less strong social tie towards their group mates. Thus the selection changed the species-recognition system in foxes by making it less dependent on species-specific cues.

Second, the late selection tests biased for those animals in which fear behaviour emerged later (or never). Although the relationship between learning about companions and the stepping-in of the fear response is not clear, if there were any dependence (e.g. cubs developing a stronger preference earlier became fearful earlier) this was most likely interrupted by the selection.

5.6.3 Changes in the reproductive cycle

In farmed foxes the breeding season starts in mid-January and lasts about 2 months. During selection it was noticed that many individuals, especially females, showed an unusual pattern of sexual activity. The vaginal smears of some females showed sexual activation as early as October–November. A quantitative summary of such extra-seasonal readiness for mating in females showed that these occurred between 10 October and 15 May (Trut 2001). However, such matings rarely resulted in offspring, and only a small number of the females showed a truly biannual (autumn/spring) oestrus cycle. The majority of selected foxes still come into heat in February, although there was a considerable variation ranging from the end of December to the beginning of March.

The investigation of hormonal changes over a whole year pointed to interesting similarities and differences between selected and unselected foxes (Osadchuk 1999) (Figure 5.5a, b). There is no difference in the seasonal pattern for progesterone and oestradiol, although blood levels of the former are usually lower throughout the year in selected foxes. In unselected foxes the mating season is preceded by raised levels of both hormones. Interestingly, the oestradiol reaches higher levels in selected foxes during proestrus, but the progesterone level shows an even more pronounced change by showing a 50% increase during oestrus. It is, however, important to notice that no such changes are present during the autumn (Osadchuk 1992a,b). This could be explained by assuming that only a few foxes in the sample used for these studies showed extra-seasonal sexual activity. (Nevertheless, it would be useful to know the hormonal pattern for those individuals that display unusual mating activity.)

Similarities during pregnancy are also evident. Both types of foxes show a similar tendency for decreasing progesterone concentrations, although selected foxes start from a higher level, and the blood concentration never goes below that measured in unselected foxes. Ostradiol shows fewer marked changes in selected foxes, but it is higher during the preimplantation period and during the last week of the pregnancy.

The annual pattern of testosterone is also remarkably similar in selected and unselected foxes. Both lines reach peak levels of the hormone in January and February (although in some studies testosterone levels are higher in unselected animals; Osadchuk 1992a, 1999) but in unselected males the sharp decrease in concentration is prolonged in March and April. The presence of a sexually active female enhanced testosterone levels in selected males but they usually had a lower base level, and made less frequent mounting attempts. Interestingly, in contrast to what one would expect selected males were generally more aggressive towards females outside the breeding season (see Figure 5.5).

Finally, similar observations were obtained with regard to the hormone cortisol (the main corticosteroid in carnivores). Selection did not seem to change the annual pattern, which was usually lower in the spring and summer, and tended to increase in both sexes in the run-up to the mating season (Trut *et al.* 1972). The main difference was

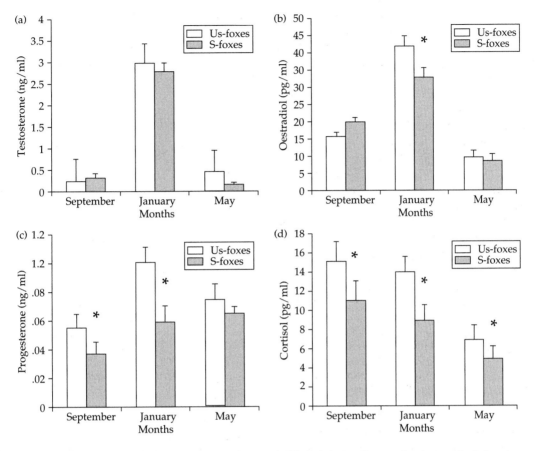

Figure 5.5 The effect of selection for reduced aggressive and increased affiliative behaviour ('tameness') on hormone levels (based on Trut *et al.* 1972 ,Osadchuk 1992a,b, 1999). (a) The only difference in testosterone concentration is in March/April when it decreases more rapidly in selected foxes (not shown). Selected foxes are characterized by lower oestradiol (b) concentration in January, lower progesterone (c) concentration in September and January, and lower cortisol (d) concentration for most time of the year. * indicates significant difference between the two selection lines. Us, unselected foxes; S, selected foxes.

the consistently lower concentration of this hormone in the selected foxes, which was especially apparent in females, sometimes reaching 50% difference from unselected animals.

5.6.4 Have we got domesticated foxes?

Describing the effect of the behavioural selection on the foxes, Belyaev introduced the idea of *destabilized selection* by assuming that the selected foxes experienced some kind of control failure at the level of the genetic machinery. Actually, there might be an alternative account that has already been applied to the dog. One could hypothesize that the major effect of

selection was that in the affected foxes the degree of environmental control over the behaviour is larger. In the case of socialization this has been achieved by making the learning process less specific for conspecifics and also extending the time of sensitivity. Thus selected foxes have more time to learn (or at least habituate to) various living and non-living objects in the environment, which could also result in decreased fear.

In the case of the reproductive system the same effect was achieved by a reduction of hormone levels (progesterone, testosterone, cortisol) but retaining to some extent the sensitivity (*reaction norm*) of the system because both behavioural and hormonal responses to the opposite sex were relatively similar in selected and unselected foxes. Thus external stimuli can still evoke the behavioural response in selected foxes but it can be more graded because there is a wider range between the base and maximum levels of the system. In the case of progesterone this might be true for internal stimulus, when implantation of the embryo results in extremely high hormone levels.

The greater environmental control of behaviour parallels the case of dogs, in the sense of Frank (1980) who referred to this situation as dogs having a better ability to react to 'arbitrary stimuli'. Thus selection in foxes might not simply result in decreased aggressive tendencies in behaviour but in a system that has a larger 'freedom' for showing different levels of aggressive behaviour that is tuned in during the epigenetic process involving experience and learning. Although these foxes have passed some important hurdles on the road towards domestication, it is too early to describe them as being truly domesticated.

5.7 Conclusions for the future

It may be time to give up an oversimplistic approach to dog domestication. Even if we assume that there were special *Canis* populations which formed the basis for the process, this does not explain why it happened only at a few locations. It might be that special environmental/ecological or anthropogenic events initialized the process. These early dogs rapidly found a way into most human communities around the world, where domestication continued at different speeds and extent. At present it seems that neither an evolutionary genetic nor an archaeozoological approach will provide a full picture on its own, and the search for further clues must be based on collaborative investigations that use refined methods for collecting data (e.g. use of DNA from extinct dogs, specific collection of DNA from extant dogs).

The difference between dogs and wolves cannot be attributed to a single genetic or developmental process. Neither hybridization, mutation, or heterogenic changes can explain the phenotypic diversity in this species on their own. Dogs seem to be an example of mosaic evolution (West-Eberhard 2003) where various phenotypic traits have been dissociated and the changes have been controlled by a wide array of genetic and epigenetic mechanisms.

Although present-day wolves are genetically the nearest relatives of dogs, which of the many possible ecological variants of the wolf was the ancestor is still an open question. As the fox selection experiment proves, dogs or dog-like creatures could potentially have been domesticated from any *Canis* species; nevertheless, this does not exclude that a particular wolf variant (perhaps with a scavenger lifestyle) provided easier 'material'.

Further reading

The massive volume on developmental plasticity and evolution by West-Eberhard (2003) provides many alternative evolutionary mechanisms for explaining phenotypic novelty. An up-to-date account of the dog genome is provided by Ostrander and Wayne (2005) and Ostrander *et al.* (2006). A detailed account on a broader perspective of domesticated animals can be found in Herre and Röhrs (1990).

CHAPTER 6

The perceptual world of the dog

6.1 Introduction

Without describing the perceptual world of dogs, we have little chance of fully understanding their behaviour. The capacity of any perceptual system is tightly coupled to the survival of the species in its niche. Thus in comparison to a generalized mammal the sensory organs of dogs could reflect specific adaptive processes as a result of their divergent evolutionary history, different environmental challenges, developmental experience, and genetic and individual variability.

The variability in morphological and behavioural traits can also affect perceptual abilities. For example, larger dogs usually have larger sensory organs. Although it has not been clearly established whether variation in size is also reflected in the number of receptor cells, such a relationship has been often observed in comparisons at the species level. Similarly, different breed-specific skull forms determine the area of binocular vision, and other variations could affect hearing (pricked or hanging ears) or olfactory ability (form and size of the olfactory organ and breathing pattern in short- or long-faced dogs, e.g. bulldog vs pointer).

Individual sensory capabilities might also depend on the actual developmental environment. Environmental stimulation can affect the survival of the neurons (and their connections) which either centrally (in the brain) or in the sensory organ determine the functional aspects of perception. For example, young kittens which were restricted to seeing only vertical black bars on a white background had problems later in navigating in an environment where obstacles were placed horizontally (Hubel and Wiesel 1998). It seems that the lack

of exposure to horizontal shapes prevented the recognition of this visual pattern later in life. Similar effects have also been shown for the olfactory receptors, in which early exposure to different odours modifies odour perception (Mandairon *et al.* 2006). Thus the developmental environment of the dog will significantly influence its later perceptual abilities.

From the practical point of view, sensory organs can be divided into two main parts. The *physical processing unit* prepares the stimuli for neural processing by mainly physical means. This is often an active process that is also under neural control (e.g. pupil dilatation or ear turning). The *receptor unit* is the first step in central neural processing.

6.2 Comparative perspectives

Research on the perceptual abilities of a species is basically a comparative investigation. One most widely used, although somewhat arbitrary, reference species is the human, simply because we have the most understanding about our own abilities. Based on homology the wolf would provide the most useful comparison, but here research is basically non-existent (but see Harrington and Asa 2003). Such comparative work would be particularly interesting because of the assumption that the perceptual abilities of dogs decreased markedly during domestication (Hemmer 1990).

Comparisons with other species (e.g. laboratory rats, Rhesus monkeys) might be problematic because they often fail to account for morphological differences such as the absolute or relative differences in the size of receptive organs (e.g. area of the olfactory epithelium), the number of receptors, or the size of the brain region devoted to perceptual processes. For example, dogs are usually described as *macrosmats* (having a better olfactory ability) on the basis of having a very large olfactory epithelium in contrast to primates (including humans). Although in some cases comparative experiments could not find differences in sensitivity for certain odorous substances in monkeys and dogs, this does not provide evidence that the two olfactory systems are also equal in other abilities (Laska *et al.* 2004). For example, larger relative brain areas allow for larger or more enduring memory capacity.

The comparison of the species on the basis of performance in learning tasks is also problematic. Monkeys can learn a delayed matching task (choose between two stimuli on the basis of a sample stimulus shown earlier) much faster if it is based on visual stimuli than if auditory stimuli are used (Colombo and D'Amato 1986). In dogs, by contrast, the difficulty of acquisition is probably the reverse (see below).

6.2.1 Cognitive aspects of perception

In many textbooks perceptual abilities are portrayed as part of cognitive processes (e.g. Shettleworth 1998). Indeed, perception is an active process controlled by the central nervous system. It involves regular sampling of the environment for significant stimuli (scanning), and is affected by mental representations, which focus the process of information gathering (attention and filtering), which also guide the process of recognition. Such mental representations might have a genetic component; for example, recognition of the so-called sign-stimuli takes place without any prior experience, or in other cases mental representation is established as a consequence of a learning process (e.g. search image).

From the functional point of view, environmental stimuli can be analysed in various ways. Detection means that the perceptual apparatus is able to transform the environmental stimulus into a meaningful neural signal which is capable (at least in principle) of exerting an effect on behaviour. Further neural analysis can quantify the stimulus, and finally it could also relate the percept

to other mental representations to determine its similarity (*discrimination*) or identity (*recognition*).

Although perceptual abilities can be investigated at the level of receptor cells, central neurons (e.g. single-cell recordings) or brain regions (e.g. lesions, brain waves), we restrict our interest to the behaviour of the intact animal (Blough and Blough 1977). Note that not all perceptions that result in neural activity are expressed in behaviour. Thus in order to reveal a behavioural change dogs are often put through a learning procedure which is aimed at building an association between the (presumed) perception of an event and a change in behaviour. However, learning abilities are not independent of the evolved adaptive behaviour; that is, if successful performance is to be expected, learning tasks ought to be ecologically meaningful. Ethologists have long emphasized that stimuli are not equipotent in eliciting behaviour patterns. For example, dogs learn to respond more easily to the locus of a sound when they have to make a choice of behaviour action (go left/right) than if they have to produce or withhold an action (go/no-go). In contrast, this latter task was more efficient for learning differences in sound quality (Lawicka 1969). Similarly, McConnell (1990) found that dogs could be trained much faster to sit by using sustained sounds with a decreasing frequency, and at the same time they learned to come in more rapidly if high-pitched repeated sounds were used.

Such differences also exist between modalities. In recognition tasks dogs are asked to find a match to a sample stimulus presented by the experimenter from a set of two or more stimuli. In this kind of training dogs seem to learn relatively fast if olfactory stimuli are used (see below) (Williams and Johnston 2002), and do quite well with auditory stimulation (Kowalska *et al.* 2001) but, although little formal testing has been done, they usually have problems with visual versions of the task. While no comparable experiments have been done using the same procedure, and slight alterations, such as arrangement of the stimuli or the nature of behavioural response requested, might have a positive effect on the dogs' performance, it can be assumed that species differences could have an evolutionary basis, and preferential types of reactions to certain stimuli might be rooted in behavioural adaptations to the natural environment (Shettleworth 1972).

6.2.2 Experimental approach to study perceptual abilities

In order to precisely establish the limits of perceptual abilities it often seems necessary to put the dog into a somewhat unnatural situation, in which both the task to be learned and the stimulus environment handicap the dog in revealing its true abilities.

First, there are problems with the nature of the stimuli and their mode of presentation. Experimenters often prefer simple stimuli, reaction to which indicates a specific sensory ability. This is in contrast to the natural situation, where events or objects produce complex stimuli affecting various senses. In other cases the special abilities of dogs are not taken into account. In the visual modality dogs seem to be more sensitive to moving stimuli than to stationary ones. Thus visual sensitivity to non-moving stimuli might not represent the maximum performance of their visual system. Similarly, olfactory stimuli of conspecifics or humans are often presented on a cold, unnatural surface, which can also obscure the dog's perceptual ability.

A second problem is to ensure that the dog is presented with the stimulus that we want it to be exposed to. This can be achieved by using special equipment to measure the physical qualities of the stimuli. For example, when testing for colour vision the colours presented should not differ in saturation or brightness. When natural sounds are played back, the experimenter should have evidence that the loudspeaker emits the same range of frequencies that make up the natural sound. To date, most problems relate to the presentation of olfactory stimuli because we have only very limited means of controlling for the quality and quantity of the perceived stimuli (see also below).

Third, it is also important to ensure that the stimuli have really been perceived by the subject. For example, visual stimuli have to be presented at the right distance, or the dog should be allowed or even 'forced' to sample olfactory cues by sniffing. The utilization of one sensory organ rather than another could also depend on the particular circumstances (Szetei et al. 2003).

Finally, as noted above the choice of the appropriate learning task could be decisive in revealing perceptual abilities. Such tasks should be put into a context which is as natural as possible and involve as little training as possible. Only a few animals will succeed in complex and complicated learning tasks with a lot of preconditions, making the results less general, and such protocols are less likely to be reproduced by others (Table 6.1).

6.3 Vision

There are indications that the predatory lifestyle of wolves has left its mark on the vision of dogs. Experts on the visual sensory system describe the dog as a visual generalist, indicating that the dog eye seems to be designed for functioning under a wide range of circumstances (Miller and Murphy 1995). Dogs (and wolves) are active throughout the day, although with peak activity at dawn and dusk. In general the visual system of the dog performs relatively well under low light levels, and is quite sensitive to motion of objects. In contrast, it is less sensitive for detecting details or complex patterned and colourful stimuli.

6.3.1 Physical processing

There seems to be a relationship between body size and overall diameter of the eye (Peichl 1992). McGreevy et al. (2004) measured a variation in eye size between 9.5 and 11.6 mm, which correlated with both skull length and width. This approximately 20% difference seems to be substantial, and knowing that larger eyes are often seen as adaptations for night vision, it would be interesting to know whether dogs with larger eyes see better in dark conditions.

There is also a considerable variation in the angular position of the eyes, which determines the visual field. If the frontal plane of the eyes subtends a small angle the visual field becomes larger, and in parallel the size of the overlapping fields decreases. Generally, shorter (brachycephalic) skulls have more forward-oriented eyes (McGreevy et al. 2004). A smaller overlap restricts binocular vision, which could be disadvantageous for a predator depending on depth perception.

Table 6.1. Comparison of perceptual abilities in dogs and humans revealed by behavioural testing. Unfortunately, the perceptual abilities of dogs and humans have been compared for only a very limited set of parameters. Research has shown that the values obtained on the basis of behavioural performance are very sensitive to the experimental methods and conditions as well as to individual differences. This means that dog and human can be compared directly only if it can be ensured that the observations were done under comparable conditions. Individual variations in olfactory acuity depend not only on the genetic background but also on the actual inner state (hormonal, health, etc. e.g. Walker et al. 2006) in the case of both dogs and humans. Very often only 1–2 dogs were tested which is a problem when the aim is to compare species (dogs versus humans)

Perception	Dogs	Humans	Nature of the difference	Reference
Vision				
Wavelength of cone sensitivity	Dichromatic vision: with maximum sensitivity at 430 nm and 555 nm	Trichromatic vision: with maximum sensitivity at 420 nm, 534 nm and 564 nm	Dogs lack the sensitivity to discriminate between middle to long wavelengths (e.g. yellow vs red)	Jacobs et al. (1993)
Overall visual field	c.250deg	c.180deg	Dogs have a wider visual field	Sherman and Wilson (1975)
Monocular/binocular field	135–150deg/30–60deg	160deg/140deg	Dogs have a more restricted binocular visual field	Sherman and Wilson (1975)
Angle of the field of best vision[a]	5deg	0.5–0.7deg		Heffner et al. (2001) (R)[c]
Visual acuity	6.3–9.5 cycles/degree	67 cycles/degree		Neuhass and Regenfuss (1967)
Temporal resolution (for cones/rods)	60–70 Hz/20 Hz	50–60 Hz/20 Hz	Dogs are more sensitive to rapid movements	Colie et al. 1989
Brightness discrimination (grey shades)	Weber fraction (average) 0.22–0.27	Weber fraction (average) 0.11–0.14	Dogs are less sensitive to different shades of grey	Pretterer et al. 2004
Hearing				
Ears	Mobile ear pinnae	Fixed ear pinnae	Dogs can adjust their ear pinnae to the direction of the sound source	not known
Hearing range	67–44 000 Hz	31–17 600 Hz	Dogs can hear in the 'ultrasonic' sound range	Heffner (1983)
Best frequency	4000 Hz	8000 Hz	Best frequency is lower in dogs	Heffner (1983)
Localization acuity	8 deg	1.3 deg	Less accurate localization of sounds in dogs	
Olfaction				
Threshold (ppb)[b] to carboxylic acids with 3–7 carbon atoms	0.1–10 ppb (lowest concentration)	3.1–31.6 ppb (average)	Dogs appear to be more sensitive	Laska et al. (2004) (R) Walker and Jennings (1992) (R)
n-Amyl acetate	0.0001–0.0002 ppb (lowest concentration)	9.1–167.5 ppb	Dogs appear to be more sensitive	Walker et al. (2003, 2006)

[a] Width of the field of best vision is estimated from retinal ganglion cell densities (Heffner et al. 2001); [b] parts per billion; [c] R, review.

Depending on the shape of the head, the angle of the total visual field varies around 250deg, and the binocular field ranges between 30deg and 60deg (Figure 6.1).

The movement of both the body and the head changes the distance between the stimulus and the retina, and objects get out of focus. By changing the shape of the lens (*accommodation*) the projection of the virtual image can be kept on the retina. This capacity is relatively restricted in dogs because they cannot project an image of an object on to the retina if it is closer then 33–50 cm to their eyes (humans, by contrast, can focus on objects as near as 7–10 cm) (Miller and Murphy 1995). Near- or far-sightedness is the result of improper focusing. A few studies suggest that a significant proportion of the dog population is affected by such problems; elderly dogs in particular may suffer increasingly from such conditions.

A special light-reflecting layer located behind the retina provides further support for the view that dog eyes function well at low light levels. By directing light back to the eyeball, the tapetum lucidum enhances the capacity to see under unfavourable conditions. Thus the minimum threshold of light for vision is lower in dogs than in humans.

6.3.2 Neural processing and visual ability

Colour vision
The dog's retina consists of two types of receptor cells that are non-uniformly distributed. The *rods*, which represent 97% of the receptor cells, are responsible for monochromatic vision in the dark. The maximum peak sensitivity of the visual pigment in the rods (rhodopsin) is at 506–510 nm, also

indicating an adaptation to dim light. The remaining 3% of photoreceptors (*cones*) can be divided into two classes depending on their pigment content (*iopsin*). Cones are responsible for colour vision, and the maximum sensitivity of their iopsins at either 429–435 nm or 555 nm suggests dichromatic vision (human vision is trichromatic and we posses relatively more cones, c.5%). Using human colour vision as a frame of reference, the dog's visual system seems to perceive two hues. Wavelengths in the violet and blue–violet range are probably perceived as 'bluish', wavelengths that would appear to us as 'greenish-yellow' or 'yellow-red' are probably sensed as 'yellowish'. Wavelengths that fall between these frequencies are probably perceived as white or light grey. These assumptions are supported by the observation that dogs have problems in discriminating green-yellow, yellow, orange, or red from each other, and greenish-blue vs grey (Miller and Murphy 1995).

Brightness
Sensitivity to brightness often improves perception of coloured patterns because natural colours often differ in brightness. According to recent results, dogs are less sensitive to differences in grey shades than humans. Their performance was about half as good in a discrimination task based on a choice of the simultaneous stimuli which was also repeated with human subjects (Pretterer *et al.* 2004).

Visual acuity
Visual acuity depends on how many cones are connected to a single ganglion cell. Primates reach the lowest ratio, 1:1. In cats (and dogs are probably

Figure 6.1. The perceptive world of dogs and humans differs to a large extent. (a) Dogs, small and large, as we humans see them. The perspectives of the German shepherd (b) and the cavalier King Charles spaniel (c) as they see us.

similar) the ratio of ganglion cells to cones is 4:1 (Miller and Murphy 1995). The measurement of peripheral or central neural activity, or behaviours suggests that the visual acuity of dogs is about 3–4 times worse than that of humans. This means that dogs can distinguish the details of an object 6 m away that a person could distinguish from 22.5 m. Although this would explain the lack of interest dogs show in visual details, the experiments were based on different methods.

Most cones are located in the central portion of the retina, where their ratio may reach 10–20% of the total number of photoreceptors (Koch and Rubin 1972). In humans this corresponds to a well-defined circular area of high-acuity vision in the retina (the fovea), but such a structure is less obvious in the dog. Nevertheless, higher concentration of cones and ganglion cells can be observed in central areas, but their distribution is more elongated. This so called *visual streak*, which has also been observed in wolves, is thought to provide good vision in a narrow range of the horizontal plane, and it could be advantageous for a predator scanning for prey. Interestingly, a recent study has found that the extension of the visual streak varies with the head shape; brachycephalic skulls with more forward-oriented eyes appear to have a more circular area of high ganglion cell densities, resembling to some extent the human fovea (McGreevy *et al.* 2004; see Box 5.7).

Motion sensitivity
In general, predators should be sensitive to motion. Although experimental data are lacking, there are some suggestions that dogs can discriminate moving objects at a distance of 800–900 m but the range falls to 500–600 m if the objects are stationary. Movement sensitivity of dogs is also supported by data showing that their eyes have a greater temporal resolution than ours; that is, they are able to notice shorter durations between two light flashes produced by the same light source. This could explain why dogs have problems with watching television, in which the refresh rate of the screen is about 50–60 Hz (adjusted to the human eye). For dogs the optimal value would be 70–80 Hz or more (Coile *et al.* 1989), which actually corresponds to that provided by video projectors (see Pongrácz *et al.* 2003). This enhanced sensitivity for motion could be

important for laboratory experiments with dogs, where dogs might sense minute movements that go unnoticed by humans.

6.3.3 Perception of complex visual images

The observation that dogs can learn to discriminate various shapes, such as a circle and an ellipse, goes back to the experiments by Pavlov (1934). Similarly, dogs can be trained to choose between objects that differ in shape, such as a cube or a prism (Milgram *et al.* 2002) but systematic experiments are lacking. (but see Range *et al.* 2008).

Dogs also show attraction to biologically meaningful but static visual images, such as the silhouette of a dog on a screen (Fox 1971), their own mirror images, or video image of dogs, but their interest declines sharply when they are unable to make social contact with the image (and probably because of the absence of odour cues). Dogs also briefly explored a robotic toy dog (AIBO) when seeing it for the first time (Kubinyi *et al.* 2004). The effect is stronger in younger (inexperienced) dogs, but they also showed rapid habituation.

6.4 Hearing

6.4.1 Physical processing

Apparently there is very limited physical processing related to hearing. Upon hearing sound stimuli, dogs aim to bring their hearing apparatus into the optimal position for perception. Sensitivity of hearing is increased by the outer ear which directs the sound waves into the ear canal. In this regard the most striking feature in dogs is the large variability in the size and shape of the outer ear. There are no data on whether surgical changes to the outer ears affect hearing, and how drooping ears modify auditory processing. Anatomical measurements show that the size of the tympanic membrane changes with the overall size of the dog, but this does not seem to have a marked effect on hearing (Heffner 1983).

6.4.2 Neural processing and hearing ability

Changes in air pressure (sound waves) are transmitted by the tympanic membrane and the bones of the ear to the so-called *organ of Corti*, which is a snail-like tubular structure. The final decoding takes place by the auditory neurons sitting in the basal membrane and sensing these pressure changes by means of projecting 'hairs'.

Hearing range
The most critical feature of hearing is the frequency range that can be sensed by the auditory neurons. By emitting pure tones at a given intensity (60 db) the hearing range (*audiogram*) can be determined experimentally (Heffner and Heffner 2003). Audiograms of different species are usually compared by values of lowest and highest frequencies, and the frequency of best hearing. In the lack of any data on wolves, the comparison of dog and human audiograms show similarity at the lower range but dogs hear well above the frequency range of humans (dogs 67–45 000 Hz; humans 64–23 000 Hz) (Heffner 1998). Thus dogs can hear at high frequencies that are imperceptible for us (*ultrasound* in human terms).

Localization
Although hearing can be useful for recognizing and identifying certain individuals or special signals, its primary function in terrestrial vertebrates is probably the localization of a sound-producing source (e.g. prey). It has long been known that animals with small heads (smaller distance between the ears on each side of the head) hear better at high frequencies. One reason for this could be how the brain calculates the position of the sound source relative to the animal, by relying on the difference in arrival times of the sound wave at the two ears (for details see Heffner and Heffner 2003). This creates a selective pressure to extend the hearing range towards higher frequencies (smaller difference in arrival time) in small-headed species, whereas no such need is present in large animals. This relationship would predict a lower maximum hearing frequency in larger breeds (Heffner 1983), but no such effect has been found. Apparently, both a Chihuahua and a Saint Bernard have their highest hearing frequency at 47 000 Hz. Thus it seems that the species-specific hearing ability for high frequencies which is determined at the level of auditory receptors did not change during selective modification of body/head size.

A further interesting relationship has been found between the size of field for best vision (estimated from retinal ganglion cell densities) and sound localization acuity. Comparison of different mammalian species has revealed that animals that have a relatively narrow field for best vision can localize sound sources more precisely (Heffner and Heffner 2003). The difference between humans and dogs fits this picture, because we can tell apart stimuli which are positioned at an angle of 1.3 deg in front, whereas dogs identify them correctly only at angles of 8 deg or more. Unfortunately, a breed comparison has not yet been carried out.

6.4.3 Perception of complex sound forms

There is very limited evidence on perception of complex sounds in dogs. Playback habituation experiments provide some evidence that dogs can sense the difference between different types of barks emitted by the same individual, as well as the same type of bark produced by different dogs (Molnár *et al.* 2008). In a study reported in Heffner (1998), dogs were shown to be able to form two categories of sounds ('dog' vs 'non-dog' sounds) after having been trained on a set of different stimuli. Later, dogs could also successfully categorize sounds to which they were not exposed during the training (e.g. howling).

The dog's ability to discriminate human spoken words was reported by Buytendijk and Fischel (1936). This was based on training a dog to perform an action reliably on hearing a command, which was followed by tests in which the phonemes of the spoken word were changed systematically. They noticed that the beginning of the words is of more significance for the dog, because it was more likely to fulfil the command if the change occurred at the end of the word. The dog probably started to react as soon as it heard the familiar phonemes. Reporting similar observations, Fukuzawa *et al.* (2005) also found that some dogs have problems in recognizing or reacting to commands played back on a tape recorder. In addition, the context of the presentation, including the distance and visibility of the experimenter, also affects the dog's performance.

Certain physical properties of complex sounds can have a more direct influence on the behaviour of dogs. Training experiments showed that dogs could be trained faster to perform a passive action (sit and stay) to a long note with descending fundamental frequency. In contrast, approaching the trainer on command was acquired more rapidly if a sequence of short notes with rising frequency was used as the training stimulus (McConnell 1990, see also McConnell and Baylis 1985).

6.5 Olfaction

In contrast to vision and hearing, dogs have more than one sensory system devoted to olfaction. Apart sensing most odours by receptors in the olfactory cavity, dogs have a vomeronasal organ which also opens into the nasal cavity, has its own layer of receptor cells, and is specialized for the detection of species-specific chemical signals (e.g. sex pheromones). In addition, the trigeminal nerve (innervating the face) also seems to be involved in the process of olfaction. Unfortunately, the general and specific contribution of these systems to the olfactory ability of dogs is not understood, so in what follows no attempt is made to specify the sensor or sensors which mediate the olfactory cue (Table 6.2).

6.5.1 Physical processing

Although it is not always obvious, olfaction is an active process. By sniffing at the odour source the animal can enhance the concentration of the molecules in the nasal cavity and enhance the possibility of contact between the chemical and receptor cells in the olfactory epithelium. Dogs often vary their frequency of sniffing when orienting on olfactory tracks (Thesen *et al.* 1993); more frequent sniffing was also observed when dogs searched in darkness (Gazit and Terkel 2003). The inner surface of the nose is covered with a mucous substance which affects the retention of the chemicals for smelling because it preferentially absorbs hydrophilic odorants rather than hydrophobic molecules. This also could also explain that different molecules are sensed at different concentrations.

6.5.2 Neural processing and olfactory ability

In both absolute and relative terms, dogs have a large olfactory epithelium. Various studies have estimated the size of the dog's olfactory epithelium around at 150–170 cm^2 (German shepherd) in contrast to humans who have only $c.5$ cm^2. The difference in the number of olfactory neurons is correspondingly large (dogs 220 million–2 billion; humans 12–40 million). It is not clear how this quantitative difference supports the superior olfactory ability of dogs, but it may contribute to more sensitive detection or to the detection of complex odours.

The crucial aspect of detection is whether the olfactory neurons sitting in the epithelium have protein receptors on their outer surface which are sensitive for the chemical concerned. Each neuron expresses one type of receptor, and neurons sharing the same type of receptors send their message to the same part of the brain. Based on comparative analysis involving the human genome, researchers have estimated that in dogs about 1300 genes are involved in coding the receptors in the olfactory neurons, which is about 30% more than the number of such genes in humans (Quignon *et al.* 2003). The larger number of olfactory neurons and receptors indicates that in comparison to humans there are more neurons expressing the same type of receptor, and there are also neurons expressing qualitatively different receptors. This could mean that in cases when humans and dogs share the same gene, dogs might be more sensitive for the given chemical because they have more neurons in their epithelium. However, as both dogs and humans also have unique genes, there might be a range of odours for which humans have a better sense of smell (see also Laska *et al.* 2004). Since dogs have a larger pool of receptors it can be assumed that in the case of an arbitrarily chosen odour they are more likely to have a receptor showing some affinity to the chemical. Unfortunately nothing is known about possible genetic variations which could provide the basis for the often assumed, but rarely tested, breed differences.

The olfactory system functions very early in dogs. Recent experiments have demonstrated that they are able to learn *in utero*, because after birth pups displayed preference for food that was fed to their mother during gestation (Wells and Hepper 2006). One implication would be that this ability could be useful for learning about 'safe' food, as in

Table 6.2. Wet nose versus e-nose. The most mysterious aspect of the dog's perceptual ability is olfaction. Many individual dogs have demonstrated high level performance in tasks involving the detection or recognition of odours, but systematic research has only recently started. This parallels efforts to develop mechanized methods of odour detection (electric nose or e-nose), but so far dogs are still somewhat superior (Furton and Myers 2001). However, instead of seeing this as a competition between biological and technical systems, insight gained by such work on dogs could help not only in understanding how olfaction functions but also to develop better equipment. The latter could be especially useful when the work is actually dangerous or unhealthy for dogs (e.g. detection of narcotics). This table presents a non-exhaustive list of recent studies that have tested the performance (reliability) of dogs under (real or simulated) field conditions in various tasks

	Type of work	Odour involved (no. of dogs if applicable)	Results reported	Potential problems limitations	References
Narcotics	Detection	Active drug, decomposition product		Toxicity to dogs	Furton and Myers (2001) (R)[a]
Explosives	Detection	Active chemical, solvents, contaminations	80–90% correct location; 95% detection rate with 5% false positives	Finding the odour signatures to which dogs are sensitive; toxicity to dogs	Furton and Myers (2001), Tripp and Walker (2003)
Explosives	Detection	Explosives (N = 7)	Habituated to path with no explosives but now 1 explosive hidden: (found by 53% of the dogs) Novel path with 1 explosive hidden (found by 96% of the dogs)	Habituation to tracks decreases detection performance	Gazit et al. (2005)
Explosives Humans	Detection Detection	Explosives (N = 7) Live human and/or cadaver scent (N = 11 and 12)	88% in darkness; 94% in light 50–85% correct performance in various simulated scenarios	Training on two different tasks worsens performance	Gazit and Terkel (2003) Lit and Crawfold (2006)
Cancer (melanoma)	Detection	Histocompatibility complex dependent odour (?) Volatile cues from melanoma tissue (?) (N = 2)	Correct signalling with affected patients 6/7 and 3/4 respectively	Not clear yet whether dogs could be used as screens for detection of melanoma	Balseiro and Correia (2006) (R) Pickel et al. (2004)

Diabetes	Detection	Body odour (?) (N = 37)	No special training; dogs become alert (e.g. bark) before the hypoglycaemic episode	Only individuals with the 'right' temperament are suitable (38% of patients with dogs)	Lim et al. (1992)
Epilepsy	Detection	Body odour (?)	No special training; dogs become restless (bark, whine, jump up) before seizure	Only individuals with the 'right' temperament are suitable (c.5–30% of patients with dogs respectively)	Edney (1993) Dalziel et al. (2003)
Scent identification	Matching to sample, two-way choice	Human scent of different body parts (N = 3)	Trained dogs (N = 3) (chance 50%) H^b-hand vs no scent 93.1% H-hand vs S-hand 75.7% H-elbow vs S- hand 58% H-elbow vs H-hand 76.8%	Dogs could have problems in matching different parts of the body by odour (could have been the result of the particularities of the training)	Brisbin and Austad (1991)
	Matching to sample, six-way choice	Hand odours (N = 8)	Trained dogs (chance 16.6%) 31–58% correct	Performance depends on the experimental protocol used	Schoon (1996)
	Matching to sample, six-way choice	Pocket and hand odours (N = 10)	Trained dogs (chance 16.6%) 100% correct with recent samples; 33–75% with older samples	The ageing of the scent (after a couple of days) impairs performance	Schoon (2004)

[a] R, review; [b] H, handler, S, stranger; (?) = supposed odours.

rodents. However, such a functional value could be questioned because in the case of dogs (and wolves) the feeding of the pups is dominated for a long period by milk, then by regurgitated food, and finally by meat brought to them. Thus such early odour learning could be simply the manifestation of a general mammalian trait, but in addition it may also play a role in learning about odours that have a role in social life.

Olfactory acuity

This refers to the lowest concentration of a chemical that can be still sensed. The results of many early studies are difficult to compare because there were marked differences in the experimental methods, the dogs used (breed, age, experience), and the chemicals studied. Recently, Walker *et al.* (2006) have developed a procedure which, if used systematically with different dogs and chemicals, has the potential to make findings comparable. During the training two dogs learned how to obtain the odour sample by pushing the small lid of a box presenting the stimulus in order to get a sniff, and then indicate the presence of the substance by sitting. In the first phase of training a fixed concentration of n-amyl acetate (1 part per billion, ppb, 1 in 10^9) was used; in the final stages the concentration of chemical was decreased to 0.03 ppb. The sensitivity of the dogs for this odour was tested in the range of 6–0.2 parts per trillion (ppt, 1 in 10^{12}). In the testing session the dogs had to indicate the location of the odour by sitting near the appropriate box after sniffing five alternative boxes. The overall performance of the dogs was similar; the threshold concentration was in the range of 1.1–1.9 ppt. This value is approximately 10 000–100 000-fold lower than observed for humans, but it is in the range that was found in mice (Walker *et al.* 2006). The performance of dogs is remarkably good, and these animals detected lower concentrations of n-amyl acetate than found by another study (Krestel *et al.* 1984). The long duration of the training (*c.*6 months) is a disadvantage, but this could be shortened after more practice with the procedure.

Olfactory recognition

Another issue is whether dogs can identify certain objects/stimuli exclusively by their odour. This has

some practical bearing, because it is strongly related to the problem of whether (and how) dogs can identify people by their smell (see below).

In the case of simple odours dogs perform well if they have to match any of the trained odours to a mixed set of trained and non-trained odours (Williams and Johnston 2002). After a simple training procedure to indicate the location of the matching odour by sitting, four dogs were subjected to a sequential learning task. The subjects were trained on 10 different odours, one after another. Dogs moved to the next odour only if they showed high level of matching accuracy with all previously learned odours. The overall accuracy of the dogs was over 85%, but, more interestingly, they needed progressively fewer training trials in order to attain this performance. In the case of the first odour a high level of performance was obtained after 25–30 trials, but by the ninth compound dogs performed above the criterion after 10 trials on average (Williams and Johnston 2002).

6.5.3 Categorization and matching in working situation

A dog faces two types of problem when exploring conspecific odour traces (Bekoff 2001). It could be interested in whether the odour just encountered belongs to a particular class of familiar odours (e.g. females in heat), or whether this odour is the same as another one sniffed at nearby a few seconds ago. The first case could be described as the ability to *identify a category*, while in the second the dog's goal is to find out the identity of the two stimuli (*matching*). In the detection tasks the dog has to indicate the presence of some specific (trained) odour(s) against a background of other neutral odours. During the work the dog has to rely on its memory of the trained odours. In order to make the task easier detection dogs are mostly specialist, being utilized only for a given type of job (Figure 6.2). Thus some dogs search for explosives, others for narcotic drugs, or for combustion accelerants. Apart from the training procedure, the success of these dogs depends mainly on the chemicals used for training. For example, in the case of dogs being trained to detect explosives, the aim is to present the dogs with as many chemicals as possible that could be used at any concentration and combination for making weapons (Furton and Myers 2001). However, the problem is that the subsidiary materials used for making explosives often provide more pronounced olfactory stimuli than the 'active' ingredient. In the case of biologically active substances odour stimuli could be the result of a chemical degradation process, so these compounds should also be incorporated in the training set (Furton and Myers 2001).

Thus the number of actual odours can be quite large, and the training procedure has to be varied in order to establish a wide range of possible 'samples' in the dog's memory. Well-trained dogs can show an explosives detection rate of over 95%, which seems to be an absolute maximum in natural situations. With this performance dogs are still better than 'artificial noses' employed in similar

tasks, which have an error rate of *c*.10% (Tripp and Walker 2003). One potential problem with detection dogs is that they habituate to the search routes if they never find anything. In this case dogs are likely to miss a novel, potentially dangerous, odour source, which could be problematic when dogs are regularly used for monitoring the same area (Gazit *et al.* 2005).

From a cognitive point of view, matching odours presents a more complex process than their detection. It is not enough to train the dog on a series of odours; it must learn that despite all its knowledge of the significance of odours, the most important task is to confirm or deny that the two odours to be compared come from the same source. For many years dogs have been employed by the police of various countries to make such decisions when there is chance for a match between the *corpus*

Figure 6.2. Dogs working for us. (a) Training for searching for explosives; (b) detection of drugs; (c) human scent identification trial; (d) training for following scented trails.

delicti (some evidence found at the crime site) and a sample obtained from a possible human suspect (Schoon 1996). Apart from the juridical problem of how such evidence can or should be used in courts, this task is also very challenging from the point of olfactory perception. In the simplest case, odour samples taken from the same part of the body within a short time should indeed be identical. If trained dogs are tested under such conditions they perform very reliably, reaching 100% correctness (e.g. Schoon 2004). The root of the problem is that we know too little about human body odours, their components, and how they change over time. The individuality of human odour has several sources, some of which have a clear genetic basis (including sex, race, or components of the immune system; see Boehm and Zufall 2006), whereas others have an environmental origin. The later can include diet (as well as smoking or medication), clothing, or the action of bacteria on the surface of the skin (see also Schoon 1997). In a study aimed at separating the genetic and environmental effects of human odour, Hepper (1988) found that trained dogs could correctly match fraternal twins, and also identical twins who were either adults or ate different diets. However, dogs reached the limit of their discrimination ability if they had to choose between identical twin infants eating the same diet.

6.5.4 Perception of natural substances and conspecific odours

Specific odours play a major role in signalling reproductive status in dogs, and dogs of both sexes are able to discriminate among these pheromones which can originate from the urine, faces, vagina, anal sac, and many other organs. One component of these odorous substances was identified as a methyl-*p*-hydroxybenzoate produced by the oestrous female (Goodwin *et al.* 1979), which elicits mounting behaviour in the opposite sex. Male dogs show a clear preference for female vs male odours, but an even greater preference is shown for odours produced by oestrous females; the corresponding preference in females surfaces only if the female is in oestrus (Dunbar 1977). These results are in close correspondence with the behaviour of dogs kept in groups (Le Boeuf 1967).

The source of the odours affects preference. In beagles, oestrous female urine and vaginal secretion was more attractive for males than odour samples from the anal sac (Doty and Dunbar 1974). However, it is not clear whether this effect is due to differences in quality or quantity of the chemical substances. Importantly, the attractiveness of sexual odours depends on various other factors, including the experience and inner state of the perceiver or the producer. A study on six beagles did not find an effect of male sexual experience on the preference for odours collected from oestrous females (Doty and Dunbar 1974), but beagle males show less interest in female odours if the donor was treated with testosterone in adulthood, and an opposite effect is obtained with estradiol (Dunbar *et al.* 1980).

The sebaceous gland located in the intermammary sulcus produces a mixture of fatty acids during the period of suckling (Pageat and Gaultier 2003). Although the effect of this pheromone, known as an *appeasing pheromone*, is not entirely clear, a synthetic analogue was found to have calming effect on dogs in stressful environmental situations which include firework noise (Sheppard and Mills 2003) and waiting in the veterinary consulting room (Mills *et al.* 2006). From the behavioural reaction of many dogs this pheromone seems to have a biological effect, but there is wide individual variation. Until we understand its original biological function during the suckling period, its practical usefulness may be limited.

In dogs, olfactory cues play an important role in kin and individual recognition. Although Mekosh-Rosenbaum *et al.* (1994) reported only slight preference in pups 20–24 days old in contacting home cage bedding over bedding from another litter, and this ability decreased with age (66–72 days), Hepper (1994) found that pups (28–35 days old) were able to discriminate between their own and strange bedding. This discrepancy can be explained by the fact that in the former study all dogs were housed in the same room and fed on the same diet, and these factors were not controlled for in the other work. Hepper (1994) also reported that adult dogs living separated from each other do not retain memories of their siblings; in contrast there was a marked mutual preference in mother–offspring relation, which was maintained over 2 years after separation. Both mothers and their offspring choose to approach the relative in a two-way choice situation. Hepper (1994) argued that the preference for siblings is mediated by familiarity with certain cues, determined partly by common genes signalling kinship, whereas recognition of the mother may be based on a set of individual cues.

There are some indications that other natural odours are not only perceived by the dog but also influence behaviour directly. In order to enrich the environment for shelter dogs, Graham *et al.* (2005) tested the effect of various naturally occurring scenting substances on the overall behaviour. They found that over a period of a few days, similarly to humans, lavender and chamomile exerted a relaxing effect on dogs housed alone by increasing resting time.

Very little is known about the significance of human odours for dogs. Dogs seem to prefer certain areas of the body for olfactory exploration in children. Millot *et al.* (1987) reported that dogs sniffed more at the face and upper limbs of a child, which might indicate that odours produced at distinct parts of the body are either more perceptible or provide specific information.

6.6 Conclusions for the future

Despite their practical usefulness, we know still very little about perceptual abilities of dogs in general. This is unfortunate, not only because such understanding would enhance our possibilities of obtaining dogs that are better at certain working tasks, but also because there are a lot of biologically interesting problems. The large morphological variability offers a very interesting possibility to test for physical (bodily) influences on perceptual ability, in addition to genetic effects and developmental flexibility.

Little is known about whether environmental enrichment or exposure to certain specific stimuli improves perceptual abilities. Early perceptual learning could have a positive effect on dogs, especially when we expect them to rely on their olfactory skills in working scenarios.

Further reading

Lindsay (2001) provides a recent summary on the perceptual abilities of dogs including taste, touch and pain with reference to some neural mechanisms. A similarly useful comparative account with a focus on wolves is given by Harrington and Asa (2003).

CHAPTER 7

Physical–ecological cognition

7.1 Introduction

The distribution and type of food, the need for navigation, and many other factors determine the ecological challenges to be faced by any species. The behavioural solution to these problems depends on the evolutionary history of the species, including its perceptual and mental abilities. From genetic predispositions and developmental experience, individuals obtain some sort of mental representation of their physical environment. Investigating the behaviour of dogs in various types of environments will help us to understand the nature of these mental representations, their constraints, and the interaction between them and behaviour. Dogs' mental representations of the physical aspects of the world differ to a large extent from ours, but currently the planning of many experiments does not suggest that researchers take these issues seriously.

Over the years researchers have adopted two different strategies in looking for the nature of environmental representations in dogs. The *ethological approach* favours the investigation of abilities for which there has been selection in the wolf's natural environment, and might have been retained after the split of the two species (e.g. hunting in groups on live prey, or navigating in space). Researchers favouring a more general *comparative programme* prefer to use tests which have been developed (mainly in monkeys or humans) for revealing some special mental skills, such as reversal learning or matching ability. Perhaps it is best to regard these approaches as complementary, partly because both face problems. First, dogs might have been selected for special skills which might interfere with abilities inherited from the wolf. Second, selection might have been relaxed for some skills because for many generations there was no selection for high levels of performance. Third, some dogs living in an anthropogenic environment lack the necessary experience to show their full range of natural abilities. Fourth, the comparative programme often neglects the natural behavioural skills of dogs and the task setting is often questionable from an ecological point of view.

7.2 Orientation in space

The experience of studying many animal species shows that they have invented a wide array of both behavioural and mental mechanisms in order to navigate successfully in various environments. In dogs, spatial orientation can be based on visual, auditory, and olfactory cues; the last is especially interesting, because this does not form part of our own orientation skills. It seems that dogs prefer to relate environmental information to their own body in space (*egocentric orientation*) but under some conditions they are able to rely on the spatial relationship between two (or more) environmental objects (*allocentric orientation*) (see also Fiset *et al.* 2006). The experimental modelling of navigation is based on the assumption that dogs (like wolves) need to localize moving prey. It shows the anthropomorphism of researchers that most of these experiments involve visual stimuli, and less attention is given to the olfactory stimuli which probably play a comparably important role in dogs.

7.2.1 Path following

Tracking in dogs is based on the natural ability of canids to locate a moving odour source by following the odorous stimulus left behind. Despite much anecdotal evidence and successful training of many working dogs, the mechanism underlying this ability has been given little attention. In one study Wells and Hepper (2003) found that only about half of a sample of trained police dogs were able to find the correct direction of a track under controlled conditions. However, the successful animals demonstrated a very reliable performance. This suggests that tracking is based on a complex set of skills and certain individuals might be more 'gifted' than others. The experimenters excluded Clever Hans effects (the handler did not know the direction of the track) and also provided evidence that dogs relied on olfactory cues present on the track. A subsequent study found that in order to find the correct direction of the track the dogs needed to sample at least 3–5 footsteps; a shorter path did not provide enough information for assessing directionality (Hepper and Wells 2005). Looking at the behaviour of the dog during tracking, three different phases could be distinguished (Thesen *et al.* 1993). In the *search phase* dogs localize the track by rapid exploratory behaviour. In the *deciding phase*, they slow down their movements and move 2–5 footsteps along the track. *After making a decision*, the dogs speed up their movements again and follow the path by taking samples of the airborne scent from above the track. Dogs did not change their sniffing frequency, but the relatively long (3–5 s) decision phase ensures that they have the chance to collect many samples. These experiments suggest that dogs may need to judge the difference in concentration between two points of the track. This could be done by comparing the two end points of the odour gradient between the front and back edges of each footstep, or by comparing the overall amount of odours left behind at each footstep. It is still an open question whether dogs rely on the odour itself, on the decayed odour, or on odours emerging from the disturbed surface. However, whichever stimulus is utilized, dogs must be able to react to small concentration changes which come about over time: only 2 s elapse between the first

and fifth footstep! It is important to note that dogs are unsuccessful in following continuous tracks (Steen and Wilsson 1990), which suggests that they need to be presented with spatially separated, intermittent odour information. Thus tracking could be regarded as a case for the allocentric use of spatial information based on odours.

7.2.2 Beacons

Beacons are proximal spatial cues which directly signal the location of the goal or target (Shettleworth 1998). They could be useful in the final phase of localization, such as the burrow of a concealed rabbit, or a pile of rocks close to a rendezvous site.

In a somewhat arbitrary situation (a modified version of the Wisconsin General Test Apparatus), which allowed the dog to move in space, learning about a beacon was documented (Milgram *et al.* 1999). In this test the dog is given a choice between two potential hiding locations (within a distance of 25 cm) one of which is marked by a small (10 cm tall) rod. Under these conditions most dogs needed about 30–100 trials to achieve the criterion level. In the following experiments the rod was moved away from the food location, which resulted in a marked decrease of overall performance in some dogs. In a different study (Milgram *et al.* 2002) dogs could also learn to rely on a beacon if it was displaced by 10 cm from the hiding location. It follows from the nature of beacons (and possibly the behavioural and cognitive strategy associated with their use) that they signal the proximity of the goal. If the distance between the beacon and the goal is increased, then the subject has to take into account other relational information from space, which might have been difficult in the present case. The Lilliputian setup of the experiment and the lack of other spatial information might have prevented the dogs relying on other orienting mechanisms for locating the place of the food.

7.2.3 Landmarks

Landmarks are physical stimuli in the environment which do not indicate the goal directly. On the basis of at least two landmarks the animal can find the goal if it is able to make complex computations

Box 7.1 Can a dog find its way home?

One of the most highly praised abilities of dogs is finding their way home after getting lost. There are many anecdotal accounts of dogs returning home, which appear in more than one book. Writing about the intelligence of dogs, Menault (1869) reports on a dog, Moffino, who returned home to Milan (Italy) after being lost somewhere in Russia after the Napoleonic wars. Dogs travelling on trains, or traversing huge areas to find their masters, were also among the most favourite anecdotes reported by Romanes (1882a).

Unfortunately, this homing ability of dogs has never been experimentally tested, and it is very likely that there is a bias in the sampling when relying on case studies: the reports tell us only the number of successful dogs, not the number that have never returned home.

There is only one study where homing ability in dogs was tested systematically, but exact data were not reported. Edinger (1915), a very enthusiastic doctor, reports that he deliberately left his dog (a German shepherd) at different areas in Berlin (Germany) to see whether it could find its way home. According to his description the dog did not succeed to begin with, and only

the cooperation of the neighbours and other acquaintances made it possible for the 'experiments' to be continued. With practice, however, the dog improved and later it not only returned home but also went directly to other places at which the doctor was to be expected at given times.

Thus miraculous homecomings based on navigation in an unknown terrain are not to be expected from dogs, but they may show good navigation skills after some practice.

Figure to Box 7.1 A dog on the run. Most lost dogs never find their homes, contrary to common belief.

---

based on the distances between itself, the landmarks, and the goal (Shettleworth 1998). Thus landmarks offer the possibility of finding targets even if they are not visibly marked, and also of navigating on a large scale. Complex representations based on combination of landmarks are often referred to as *cognitive maps* of the environment, but the meaning of this term is still debated (Shettleworth 1998). In any case orientation based on landmarks allows for making short-cuts and/or planning novel routes. Such abilities are by many researchers taken as evidence for the existence of a cognitive map. It is unfortunate that, given the many claims for the homing abilities of dogs, very little research has been done in this area (Box 7.1).

Long-term observation of free-ranging wild wolves suggested that they construct a more or less detailed mental representation of their territory,

which might have the properties of a cognitive map. These assumptions were supported by observations that older wolves are more efficient in organizing their directions of travel, and they often take otherwise unused short-cuts if searching for or chasing prey (Peters 1978). Such orienting abilities are very useful, especially in winter, when visual landmarks may have increased significance for finding directions, and efficient spatial movement saves a lot of energy for the pack.

Chapuis and Varlet (1987) brought dogs to a 3 ha field which was covered by thyme bushes and had generally only a few landmarks available for orientation (see also Fabrigoule 1987). Dogs were shown two hiding places of food during a walk on leash from the same starting point in two different directions. When the dogs were released from the starting point after these visits, most of them went first to the nearest location and then chose a path which led towards the second hiding place (Figure 7.1a). This suggests that during the separate exploratory walks, the dogs collected spatial information (in addition to kinaesthetic information) which was then integrated by computing the spatial relationship of the two locations. The behaviour of the dogs during navigation provided further interesting insights. Relatively often the dog did not run from the first location to the second in a straight line; instead, it oriented the path towards the line between the starting point and the second goal. This tactic seems to be advantageous because there is a greater chance of finding the route to the second target, experienced earlier, than the second target itself. This behaviour became even more prevalent if the dogs were tested in a different field with more landmarks. It seems that if given the option dogs reduce the mental load on navigation and, despite its higher energetic investment, they prefer the safe bet.

7.2.4 Egocentric orientation

Egocentric navigation is useful when the environment is stable and lacks useful cues for orientation. While chasing prey, the predator may pay reduced attention to the surroundings. This can lead to situations when environmental cues are not at its disposal if the prey suddenly disappears. Fiset et al. (2000) have shown that dogs can solve such problems by relying on linear egocentric information which codes the spatial relationship between the dog and the location of the object that has disappeared. In a follow-up study looking for the mechanism of this ability, Fiset et al. (2006) found that dogs are able to use very precise directional cues (less than 5 deg of angular deviation); they prefer to rely on directional information and disregard information on distance.

Dogs can find their way back to a target if they are deprived of any visual and auditory environmental information during the outward journey (Séguinot et al. 1998). It is assumed that the information about the distance travelled and the direction and magnitude of turnings enables the dogs (and other animals) to calculate the direction of the return path as well as the distance to the target (path integration). Dogs performed surprisingly well in such tasks when they were walked along an L-shaped 20–50 m path (without the possibility of seeing or hearing) in a large hall. When released at the end of the journey, dogs made the corresponding turn, pointing their body towards the target, and were also able to correctly judge the distance to be travelled before searching locally for the target (see Figure 7.1b).

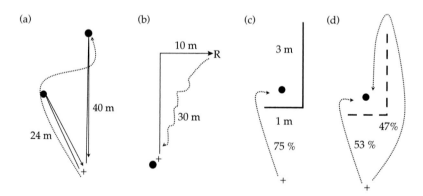

Figure 7.1 The testing of short-cuts in dogs. (a) In a field experiment Chapuis and Varlet (1987) took dogs to visit two baited locations from a starting point. After being released dogs walked first to the nearest location and then took a short-cut towards the furthest place.
(b) Blindfolded and earplugged dogs are taken on an L-shaped route and then released from the end point (R) to find out whether they find their ways back to the baited starting point (Sèquinot et al. 1998). (c) Dogs can perform optimal detouring (choosing the shorter path) when the goal is hidden. In trials with an opaque fence dogs mostly choose the shorter path; however, if they can see the target (food) through the fence, continuous visual contact takes control over the behaviour and acts against the preference for the shorter path (Chapuis et al. 1983).
· · · , the dog's path; —, outward journeg with handler; +, starting position; •, location of reward/target; R, point of release.

7.3 Spatial problem solving

Moving around in space can sometimes be a complex problem involving conflicting information and the tendency to find an optimal solution. Chapuis *et al.* (1983) observed dogs in a series of such experiments when dogs could obtain a reward by navigating around different types of obstacles (see Figure 7.1c, d). The experimenter varied the visibility of the food (using opaque or transparent barriers), the distance to reach the target, and the angular deviation required at the initiation of the route. Based on optimal solution, one would assume that dogs might prefer to walk shorter routes with minimal angular deviations. However, such optimal routes are often distorted by the visibility of the target. In general dogs conformed to the expectation. If the target was hidden behind opaque screens they showed a preference for taking the most optimal routes. However, the visibility of the target modified their orientation such that they tried to maintain a direction which deviated to a lesser degree from the target. Thus the visible goal acted as a 'perceptual anchor' (Chapuis *et al.* 1983) that in some conditions led to inefficient trajectories when the dog had to walk further to reach the goal. There is nothing strange here if the situation is put into an ecological context, because in the case of a ground predator it should be the visible target that controls the behaviour (and transparent obstacles, such as fences, are rarely encountered in nature).

The tendency for direct approach has often been utilized to look for flexibility of spatial problem solving in dogs. Such *detour experiments* have investigated how quickly the dog learns that it has first to move off from the target in order to reach it at the end of the route. Some 6–8 week old pups can solve such a problem without much training (Scott and Fuller 1965), but experience with the barrier

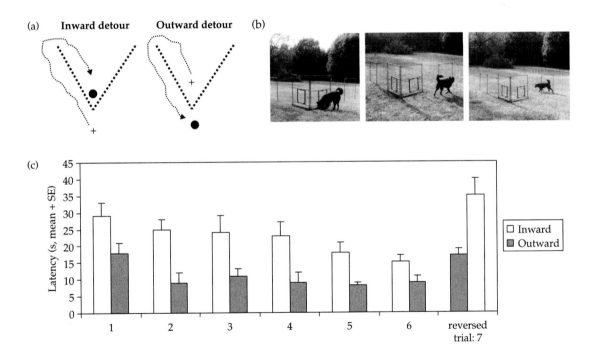

Figure 7.2 (a) Outward and inward detours around a fence represent two different kinds of problem for family dogs. The first is solved rapidly, but the second needs some practice. More importantly, experience with the simple outward task (thus moving around the fence, albeit in different directions) has no effect on solving the inward task faster (Pongrácz *et al.* 2001). (b) Usual sequence of behaviour during solving a detour problem by a naive dog. (c) The latency decreases in outward and inward detour trials.
⋯⋯, the dog's path; +, starting position; •, location of reward/target (redrawn after Pongrácz *et al.* 2001).

facilitates the emergence of correct solutions (Wyrwicka 1958). Relatively inexperienced city dogs learn in *c*.5–6 trials to approach, without delay or hesitation, a target hidden behind a V-shaped transparent fence (Pongrácz *et al.* 2001). Interestingly, we found that it was much easier for dogs to reach the target if they were behind the fence and the target was outside. This may be because the dogs had more experience with getting out from somewhere than getting behind something. However, even repeated experience of getting out from behind the fence did not improve the dogs' skill in finding the target behind the fence in subsequent trials. Thus dogs showed restricted ability to generalize from one type of experience to other solutions of the same task (Figure 7.2).

Recently, the *progressive elimination task* has been used to investigate the pattern of search behaviour in dogs (Dumas and Paré 2006). In these experiments the dog is given the task of collecting hidden food from three locations which are at various distances from its starting position. Dogs showed no preference when the three hiding locations were equidistant, and not surprisingly preferred a target which was closer (*least distance rule*). This was also the case if they had to choose from two equidistant objects and a third one further away. Thus dogs seem to minimize the distance travelled between the locations. Interestingly, the authors argued that this task is analogous to a cooperative hunting situation when the predator is monitoring the movement of both the prey and its companion in the chase. However, hunters do not usually search visually at distant locations. In addition, in the experiment the search was always interrupted after the dog found one food item, and the dog was forced to start the next search from the starting point, which could have brought in problems of memorizing the location which had been depleted earlier. Despite these problems, this task might be useful in finding out the visual–spatial tactics that dogs utilize in a serial search task.

7.4 Knowledge about objects

Perhaps it is worth noting that objects play a more restricted role in a dog's environment than in ours. Most objects in a dog's world are eaten, and only a

few types are used regularly for play. Wolves retain a natural wariness towards novel objects, but in the human environment most dogs become desensitized and are interested mainly in objects that are associated with play. This is not to say that the dog's mind operates without utilizing representations of objects, but these are very likely different from our own. In addition, perceptual, especially tactile, information differs between humans and dogs because of the latter's lack of hands.

One way to show that a species uses object representations is to show that it displays goal-directed search in the absence of visual cues of the target. Dogs (and wolves) have been observed to follow prey even when it is no longer perceived; thus they may control their behaviour by a mental representation of the unseen object. Careful experimental work, which also excluded the role of olfactory cues (Gagnon and Doré 1992), showed that dogs can localize moving objects which disappear behind one of three screens (e.g. Triana and Pasnak 1981, Gagnon and Doré 1993, Watson *et al.* 2001). In this case dogs relied on directly perceived visual information (visible displacement), but in other situations the location of the object was signalled indirectly. For example, the experimenter put an object into a container which moved behind two or three screens. Behind one of the screens the object was removed from the container, which emerged empty from behind the screen. Upon seeing the empty container the witness could deduce that the target was left behind the screen and search accordingly. In such invisible displacements the subject can rely only on indirect information about the location of the object. It is hard to envisage a real-life situation in which a dog would need to employ such an ability. Nevertheless, according to Gagnon and Doré (1993) dogs seem to be able to solve such invisible displacement problems, although at lower levels of performance. There are some suggestions that dogs also deviate from the developmental path of this ability in humans. Before reaching the ability to follow invisible displacement children pass through a developmental stage when they make errors in subsequent hiding trials by searching at the screen which hid the object in the previous trial, despite the fact that they have seen the object disappear behind another screen ('A not B' error).

Box 7.2 Object permanence or rule following?

The reliable performance of dogs searching for targets that disappear behind one of several screens led researchers to conclude that the dog's behaviour is controlled by mental representation even in the absence of the object (e.g. Doré and Goulet 1998). Nevertheless these experiments leave open the possibility that dogs act on the basis of some other search rules. With the participation of an experimenter the test becomes a sort of social game where the human is doing the hiding and the dog is searching. Thus we devised a novel version of the invisible

displacement task (Topál *et al.* 2005b). In this version the training phase is followed by two different types of trials. In the 'No object' trials the target object is never revealed; dogs only see the movement of the container behind the screens, and thus have no clue about the possible location of the target at the end of the trial. In the 'Game' trials the object is visibly given to the owner (who hides it in a pocket) and the empty container is carried around as in other invisible displacement trials. In this trial the dog knows the whereabouts of the object (in the pocket).

(a)

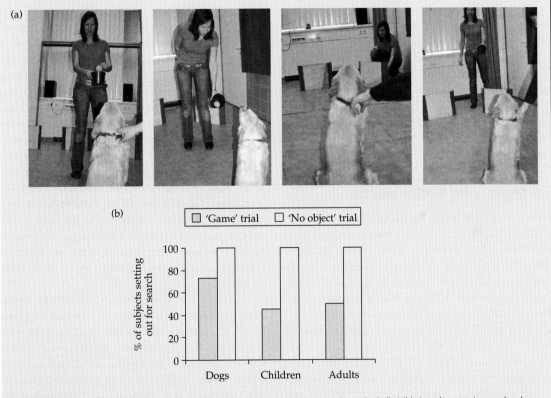

(b) □ 'Game' trial   □ 'No object' trial

Figure to Box 7.2 (a) The hiding sequence in the training: (1) The experimenter places the ball visibly into the container, and makes sure that (2) the dog sees it in the container (3) then experimenter goes behind one screen and hides the ball, (4) at the end the empty container is shown to the dog. (b) In the task for testing invisible displacement a considerable proportion of the subjects also search at the potential hiding location if they know that the object is not there ('Game' trials). Such 'unintelligent' behaviour could be the result of accepting social rules (Topál *et al.* 2005). (In the 'Game' trials the subjects witness that the ball is given to the owner/other person present before the hiding process is carried out, thus no search behind the hiding screens would be expected).

continues

Box 7.2 *continued*

As expected, dogs started to search in 'No object' trials but importantly 50% of the dogs also started to search in the 'Game' trial. However, the search pattern differed between trials as dogs spent more time searching in the 'No object' task.

Such behaviour can be interpreted as a case for social rule-following where dogs recognize that they are players in a hide-and-seek game and the actual place of the target is of less importance. Accordingly, once the hiding is carried out (in whatever manner) the companion 'has no other choice' (in order to avoid social conflict) than to do the search. It is important to note that control experiments (with different dogs) ruled out the possibility that the behaviour of dogs could be explained on the basis of forgetting the location of the ball or other constraints on working memory or object representation. Moreover, behavioural observations also suggested that dogs had some idea where the ball was in spite of setting out to search, because they looked frequently at their owner (who had the ball hidden in a pocket).

Repeating a similar type of experiment with children and adults gave similar results, although the proportion of 'searchers' in the 'Game' trial was smaller in the case of children and adults (Topál *et al.* 2005b).

Interestingly, this 'malfunction' in search behaviour does not emerge in dog pups during development (Gagnon and Doré 1994) but is present in adult dogs (Watson *et al.* 2001). Thus there are arguments that the representational abilities in dogs might rely on different mental mechanisms than those that are in place in 1.5–2 year old children (Doré and Goulet 1998, Watson *et al.* 2001, Gomez 2004). (Box 7.2).

7.5 Memory for hidden objects

If non-mnemonic tactics are excluded, the ability to recall the location of a hidden object is also taken as evidence for the presence of mental object representations. However, the measure of memory is very complicated because it depends on the circumstances under which the experience was obtained, the experience and inner state between memorization and recall, and the inner and external conditions at recall. For example, using the above-mentioned visible displacement procedure dogs could remember the location where the object disappeared for up to 4 min (Fiset *et al.* 2003). After witnessing the disappearance of the target behind one of three screens another screen obscured the view of the screens for various durations, and to reveal their memory dogs had to choose from the same three screens.

One could assume that variations in the procedure (e.g. the nature of the hidden object—a dog's toy in Fiset *et al.* 2003, the number of hiding places, or the distance between the locations—20 cm in Fiset *et al.* 2003) affect the representation of the object and the memory. Something along these lines has been observed by Grzimek (1942) and Heimburger (1962) who tested dogs, wolves, and one jackal in a similar task. The main difference was that the distance between the locations was increased to 3 m and the target was food. Under these conditions the jackal could remember for about an hour, dogs found the food with a delay of 30 min, and wolves located the hidden target only after a 5 min delay. Although the reason for this species difference remains unknown, and might be independent of the task, the main result is that memory duration is sensitive to the task requirements.

Testing a few dogs, Beritashvili (1965) found longer memories when dogs had to find a hidden target in a large room. In this case the dogs also remembered the location of disappearance the next day. By hiding two food items, which had different value for the dog (bread and meat), Beritashvili (1965) showed that dogs can also remember the content of a particular location. After 1–5 min waiting time, in most cases dogs visited the location of the meat first and the location of the bread second. Although these experiments might have been done under better-controlled conditions (e.g. olfactory cues could influence the choice), these pilot results raise the possibility that dogs can

Box 7.3 Logical inference or social cueing?

Erdőhegyi *et al.* (2007) set out to investigate dogs' ability for deductive inference (Call 2004). Their assumption was that in the case of two possible hiding places the dog can infer the location of the target if it is shown the empty location which does not hide the object. Importantly, the human's informing act was explicitly communicative; that is, first she caught the dog's attention, calling it by its name, then she lifted the container to reveal its contents (or that it was empty) for 3 s while alternating her gaze three times between the dog and the manipulated container.

When the human informant revealed the contents of both boxes, or only the baited box, dogs performed correctly. In contrast, when dogs were shown only the contents of the empty box they preferentially chose the empty container.

These results suggested that the dog could not infer the location of the toy object by exclusion (a). Alternatively, one may assume that the dogs' performances reflect a preference for the 'socially marked' container (even if it was obviously empty) rather than their inability to make inferences by exclusion (see also Agnetta *et al.* 2000). To test this idea, in a subsequent experiment (involving a little trick with double boxes) (b), the human informant manipulated both containers in the same communicative way (looking at, tapping, gaze shifts between the dog and the container) but otherwise the situation was the same as above. Now the dogs chose the baited box more frequently than was expected by chance. This suggests that dogs have the ability for simple inference but social cues can easily override their performance. (see also Box 8.7)

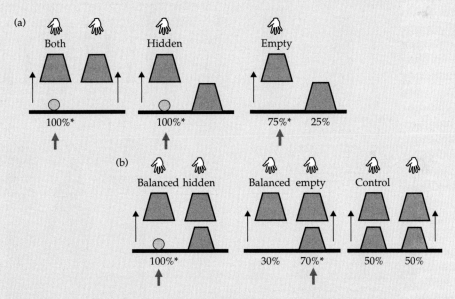

Figure to Box 7.3 Dogs can use simple inferential logic but only if social cues do not bias the situation. (a) Dogs prefer to choose the box that was touched by the human. (b) In the double box experiment, if boxes on both sides were touched the dogs show a preference for the correct hiding place. (Percentage of dogs choosing the ball, * indicates significant difference from chance).

develop complex long-term memories about objects or events. The caching behaviour of wolves (Mech and Peterson 2003) could provide an adequate ecological scenario for which good spatial and object-related memories could be advantageous (Box 7.3).

7.6 Folk physics in dogs?

Recently it has become fashionable to talk about 'folk physics' in animals, assuming that they may utilize some general rules of physics concerning objects and their interactions (Povinelli 2000). As mentioned in the previous section, object perma-

nence—that is, objects continue to exist also if they are not perceived—could also be taken as such a rule. The problem with the concept of folk physics is that it has been derived from human developmental psychology and applied uncritically to comparative evolutionary research. Knowledge about the environment depends not only on the physical

---

Box 7.4 Can dogs count and do they always choose more?

A study by West and Young (2002) showed that dogs might have some sort of numerical competence. (a) The method was based on the so-called *surprise effect* when the outcome of a series of actions violates the expectancy of the observer. In this case dogs witnessed the hiding of two large food items behind a screen. After the screen was removed the dogs saw either two

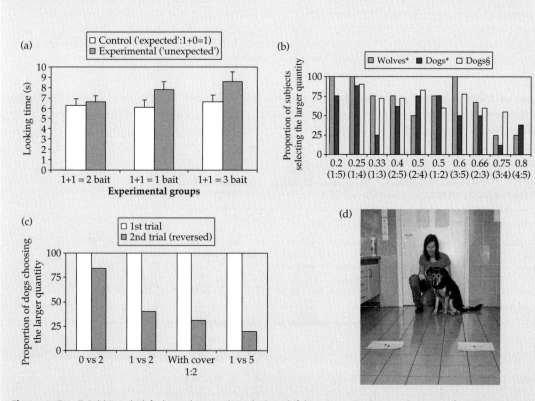

Figure to Box 7.4 (a) Dogs look for longer (mean and standard error) if they witness an unexpected outcome of a hiding test. Each test is preceded by a simple control condition when only one piece is hidden and revealed subsequently. (b) Both dogs and wolves show preference for the larger amount of food in a choice situation, and in general, this decreases as the difference between the quantities becomes smaller, and the absolute amount becomes larger. The discrepant results between the two studies can be explained by differences in the methods. §, Ward and Smuts 2007 (2 trials/session); *, Ujfalussy *et al.* 2007 (10 trials per session); chance performance at 50%. (c) Rapid development of side preference in reversed trials. Only those dogs that chose the larger amount (two vs. one) in the first trial are included in this study. For the second trial the position of the two quantities was reversed. (d) A dog is allowed to watch the two food patches before making a choice

continues

Box 7.4 *continued*

food items ('expected outcome') or one or three items ('unexpected outcome'). Dogs looked for longer at the items if the outcome was unexpected. This difference in looking behaviour led the authors to conclude that dogs show evidence of numerical competence. Importantly, numerical competence is an umbrella term referring to a wide range of abilities including estimation, relative judgements of numerousness, and counting.

Presenting dogs with different quantities of food (e.g. 1 vs 2 items; 2 vs 3, etc.) Ward and Smuts (2007) found that dogs chose the larger quantity if there was a difference of more than one piece between the two amounts offered (at least within the range of maximum 1–5 food items). (b) Thus although the ration of food is the same in cases like 1 vs 2 and 2 vs 4, dogs were

successful only in the latter type of trial. Using a similar method Ujfalussy *et al.* (2007) found similar performance in 10 dogs and 4 wolves.

In the previous experiments the choice situation differed from one trial to the next. Thus it is an intriguing discovery that the performance of dogs drops when the same choice is given repeatedly (Ujfalussy *et al.* 2007). It seems that in two-choice situations dogs rapidly develop a side preference which significantly impairs the performance. (c) Interestingly, dogs also showed this side preference if they were not permitted to eat their choice (food was covered with a transparent box). But they alternated if there was no food on one side (0 vs 2). It seems that the visibility of the food and the similarity to the previous choice trial were enough to constrain the choice behaviour of most dogs.

aspects of these rules but also on the means by which experience is gained (e.g. dogs are not able to lift objects with their paws). Even if some understanding of such rules was demonstrated in very young infants before they had any chance to manipulate objects themselves, it could assumed that such genetic preparedness is stronger in species in which individuals are expected to use objects in a complex way. From the ecological point of view both genetic endowment and individual experience make an individual adapted to the challenges of its environment. Thus the question is not whether dogs are able to act on the basis of rules of human folk physics, but how flexibly they can use their natural skills (Box 7.4). The skills of dogs might vary depending on their physical abilities (e.g. more flexible use of joints in New Guinea singing dogs, Koler-Matznick *et al.* 2003), and their experience. Dogs not exposed to the natural environment or gaining only restricted experience might not be able to show the full range of their capacities (Scott and Fuller 1965).

7.6.1 Means–end connections

Strings and planks do not occur naturally in the environment of dogs (or wolves). In spite of this,

based on observations of how skilfully monkeys (which have hands!) perform such tasks, dogs were set to solve such problems (Köhler 1917/1925). Not surprisingly, the picture was mixed (Sarris 1937, Fischel 1933, Grzimek 1942) but because of the small sample size and uncontrolled factors no clear conclusion was reached. In a recent systematic evaluation of string-pulling skills in dogs, Osthaus *et al.* (2005) found that dogs can learn relatively rapidly to pull a string independent of its orientation if researchers attach a dog treat at the end (Figure 7.3). Next the researchers wanted to find out whether the acquisition of the string-pulling skill also led to the understanding of the 'rule' that the result of the action comes about because the treat is physically connected to the string. To test for this possibility, in a series of experiments dogs were given a choice between two strings of which only one was baited. The overall performance of the subjects was unimpressive, and showed little evidence for favouring the string with the treat. There was a slight tendency to choose the end of the string which was nearer to the bait, but in the case of some clever arrangements this was not the correct solution. Dogs often pawed near the bait even if there was no string to pull. This goal-directed behaviour to reach the

(a) Training tasks (A, B)

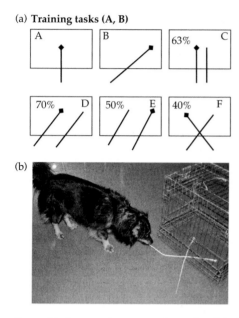

(b)

Figure 7.3 Dogs may show some understanding of simple physical rules. (a) The experimenter offers a choice between two strings, one of which is attached to a piece of meat. Most of the dogs need some training to solve the task (A, B) with a single string, but the complex versions with two strings seem to be beyond their capabilities. Percentages indicate the performance of different groups of dogs in tests C to F. (The value is placed near to the respective string). (b) Dogs are not able to solve the crossed-string problem spontaneously (based on Osthaus *et al.* 2005).

target is also not surprising because it was also observed in the detour tasks. Although these experiments suggest that dogs lack understanding of means–end connections, it should be remembered that this situation is not necessarily natural for the dogs and more variable experience could led to better performance.

7.6.2 'Gravity'

In line with the assumptions of folk physics, another useful rule is that falling objects maintain their trajectory even if they disappear from sight. Comparative experiments have found that infants' and monkeys' (Hood *et al.* 1999) reactions are controlled by this 'gravity rule'. In addition, they also rely on this rule when a connecting opaque tube 'clearly' distorts the trajectory of the object.

In the apparatus used by Hood (1995) the target is dropped into one of three holes, one of which is connected by an opaque tube to one of the goal locations beneath. This arrangement brings into conflict two physical rules: gravity and the physical constraints provided by rigid objects (the tube), which could also be regarded as an understanding of 'connection'. Using the same experimental set-up Osthaus *et al.* (2003) found that at first dogs expect the object to fall vertically even when the connecting tube modifies the trajectory. However, after repeated presentations dogs learned to search in the box that is positioned under the end of the tube. Control experiments revealed that dogs did not come to understand the role of the tube; instead, they invented a simple strategy of searching at the other side of the apparatus. Interestingly, dogs seem to be more flexible in giving up the gravity rule than 1–2 year old human infants, which might be explained by the adult dogs having more experience and/or being more adapted to follow self-propelled objects (e.g. prey) in space (Osthaus *et al.* 2003). This finding also cautions against mechanistic comparison of adult dogs with human infants.

7.7 Conclusions for the future

Despite its practical usefulness in dog training, we know surprisingly little about dogs' understanding of their physical world. In addition, most of our knowledge originates from classic comparative experiments in which dogs were exposed to problems that are based on the ecology of primates and their infants.

The ethological approach emphasizes the ecological validity of the tasks, which in this case should reflect the ecology of the wolf and other *Canis* species. It is very likely that these abilities have not been modified to a large extent by domestication, and therefore dogs (which are easily tractable) could actually provide a first-hand behavioural model for other canids. However, it is important that we can expect the full-blown ability to emerge only if the dog is exposed to right kind of developmental environment.

Comparative work (even including primates) could actually show what kind of alternative tactics are used to solve similar problems. Here the actual mental mechanisms might be of less importance because there are practical constraints in controlling the procedural and experimental variables but the effect of experience (lack, excess, or early exposure) can be investigated in detail.

Further reading

Shettleworth (1998) provides a good overview of issues that relate to cognitive aspects of getting around in the physical environment. Many topics related to physical-ecological cognition have never been investigated in dogs, e.g. timing. See also Healy (1998).

CHAPTER 8

Social cognition

8.1 Introduction

The most striking feature of the social life of dogs is that they spend most of their life in mixed-species groups. This is not to deny that many dogs actually have no relationship with humans or only a very loose one, but if dogs have a choice they seem to prefer to join human groups.

In spite of this obvious phenomenon, the dog–human relationship is most often described by either a *lupomorph* or a *babymorph* model (Chapter 1.6,

p. 16). In the former case the family is visualized as a 'pack' with strongly expressed dominant–subordinate relationships, and the human as the leader. Recent research has shed some doubt on this view of wolf society (Packard 2003; Chapter 4.3.4, p. 82) but many popular books on dogs continue to reinforce it. Sociologists and psychologists have adopted a human perception and 'automatically' utilized a babymorph model (Hart 1995; Box 8.1). These investigations, based on the experience and

---

Box 8.1 Dogs as friends

Interestingly, scientists 'lupomorphizing' or 'babymorphizing' (Chapter 1) about dogs have paid little attention to old folk wisdom about the relationship between dogs and humans when they refer to dogs as *man's best friend.* Recently primatologists have struggled with the definition of the term 'friendship' for primate societies (Silk 2002). Although no definite conclusion has been reached, many important ideas have recently been put forward.

Friendship is clearly more than an affiliative contact and the inclusion of additional criteria seems to be necessary to define any relationship as friendship. Reviewing the literature, Silk (2002) mentions that friendship is characterized as being a form of alliance, providing a social dimension for mutual trade without the need of immediate reciprocation, having a propensity for sharing things and the

possibility of offering social support (and thus enhancing mental and physical health) and engaging in cooperative actions. The largest confounding factor in the case of primates is the often close genetic relationship between 'friends', because in these cases affiliations can be interpreted in terms of kin selection. It is difficult not to notice that the relationship between dogs and humans can also be interpreted in terms of friendship. Obviously, there is no genetic relationship, and ample evidence exist for alliance formation and cooperation, in addition to mutual social support. Thus it might be worthwhile to consider human–dog relationship in terms of a friendship. Naturally this does not exclude asymmetry (dominant or parental) in the relationship in certain contexts, but it includes the possibility of leading and independent life and being an equal collaborative partner.

continues

Box 8.1 *continued*

(a)

(b)

Figure to Box 8.1 Favours that only a friend could do for you. (a) Hunting dogs regularly give up their prey. (b) Guide dogs for the blind not only assist their owner but also disobey if the situation or the safety of the human requires it.

views of dog owners, found that in most families dogs are regarded as members with the rights of a child. Dogs also contribute to the emotional stability of the family (like children) and have a positive educational effect on the children (e.g. Katcher and Beck 1983). The idea that human–dog relationship should be viewed in terms of attachment gained support from questionnaire studies (Serpell 1996, Poresky *et al.* 1987, Templer *et al.* 1981).

Earlier we proposed a third perspective, an *etho-cognitive model* which separates the investigation into two levels. At the functional level the model recognizes that behavioural similarities between dogs and humans (including children) could be the result of convergent evolution, but at the same time, at the level of mechanism the question is, how was the behavioural control system of the wolf affected that led to the observed changes in our dogs?

8.2 The affiliative aspects of social relationships

The affiliative aspects of the dog–human relationship have most often been interpreted as a form of social attachment. Unfortunately, many early

researchers used this term uncritically in relation both to humans and to their dogs. In a recent review on the subject Crawford *et al.* (2006) point out the differences between the framework used for human–human attachment and that applied in companion animal research.

Bowlby (1972) and others referred to attachment as a behavioural system that is based on the interaction between mother and child and has a dedicated function in survival (see also Chapter 9.5, p. 214). Based on this view, Wickler (1976) and others defined a broader version of attachment as a long-lasting attraction to a particular set of stimuli, which manifests in the form of particular behaviours that are directed towards or performed in the presence of these stimuli ('objects of attachment'), in addition to the maintenance of proximity over a period of time. This operational description is in agreement with Bowlby's (1972) assumptions that attachment is a behaviour-controlling system which elicits a particular set of actions in stress situations (e.g. separation from the attachment figure). In practice a functional attachment system can be revealed if the behaviour of the subject fulfils certain criteria (Rajecki *et al.* 1978). The

(a)

(b)

 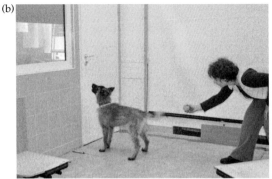

Figure 8.1 Two episodes from the Ainsworth test (see Topál *et al.* 1998). (a) 'Owner, Dog and Stranger' (Episode 2): most dogs play with the stranger in the presence of the owner. (b) 'Dog and Stranger' (Episode 6): many dogs lose interest in playing in the absence of their owner.

individual should display separation stress in the absence of the attachment figure (*caregiver*), seek proximity and contact, and show specific greeting behaviour in the presence of the caregiver, which is at least quantitatively different from similar actions performed towards a 'stranger'. Experimental investigation of the infant–adult attachment is based on the so-called Strange Situation Test (Ainsworth 1969), and attachment in adult humans is measured by semi-structured interviews. Importantly, in both cases human attachment is categorized qualitatively on the basis of form (for details see Crawford *et al.* 2006).

In the literature on companion animals, human attachment to dogs is measured by means of questionnaires which use a continuous scale ranging from 'no attachment' to 'maximum attachment'. This is in contrast to the original model in which the existence of an attachment relationship is a prerequisite and only the form of this relationship is under study. Bowlby's original model does not include a case for 'no attachment', and there is no 'weaker' or 'stronger' attachment. There are only different behavioural patterns which are described as qualitatively different forms of attachment.

The other problem in measuring human–dog attachment is that instruments of different kinds and types are used. Some rely on owners' self-assessment of their overall 'attachment' to the dog (Serpell 1996); others use composite scales based on different set of questions (Pet Attitude Scale, Templer *et al.* 1981; Pet Attachment Scale,

Albert and Bulcroft 1987; Companion Animal Bonding Scale, Poresky *et al.* 1987).

Although direct comparisons with human–human attachment measurements of this kind are not available, the findings suggest important differences. For example, Albert and Bulcroft (1987) report that single, divorced, or widowed people provide higher attachment scores ('stronger attachment') towards their pets than others living in a family. In parallel, adults without children score higher than adults having two or more children in their family. The latter finding in particular makes little sense because one would not expect attachment to change linearly with the number of children. Thus these scales are more likely to measure the emotional bonding of humans to their dogs.

Recent investigations have returned to the original concept as developed by Bowlby for the caregiver–infant relationship, and have used a modified version of the Strange Situation Test (SST), in which the dog is separated from and then reunited with its owner repeatedly, and in parallel it also encounters a stranger repeatedly (Topál *et al.* 1998, Gácsi *et al.* 2001, Prato-Previde *et al.* 2003, Marston *et al.* 2005a) (Figure 8.1). In contrast to the human SST, which aims to assign the relationship to a predetermined category, in the case of dogs the attachment is characterized by means of continuous behavioural variables. This analysis focuses on contrasting the behaviour of dogs towards the owner and the stranger either by comparing various behavioural variables directly (e.g. amount of play, Prato-Previde *et al.* 2003) or by the application of multivariate statistical methods (Topál *et al.* 1998).

In general, dogs displayed specific reactions towards their owners (but not towards strangers) by looking for them in their absence and making rapid and enduring contact upon their return. They also preferred to play with their owner, and decreased play activity in the absence of the owner. A post-hoc factor analysis resulted in three meaningful factors that distinguished three key aspects of the behavioural pattern displayed in the stranger situation. One factor contained behaviours related to the 'stress-evoking' capacity of the situation (*anxiety*), the second consisted of variables describing *attachment* towards the owner, and the third was associated with behaviours related to the *acceptance* of the stranger (Topál *et al.* 1998). Subsequently a post-hoc cluster analysis was applied to categorize dogs in this three-dimensional space using a three-level subdivision for each factor. Follow-up work provided evidence that this pattern of attachment is stable over at least 1 year and is independent of the peculiarities of the testing location (Gácsi *et al.* 2003).

In a replication of the above findings, Prato-Previde *et al.* (2003) questioned whether a dog–human relationship can be characterized as attachment without actually showing evidence for the so-called *secure base effect* (Ainsworth 1969). Accordingly, while exposed to a mildly stressful environment human children use the attachment figure as a safe haven or refuge to which they can return after exploration or when potentially threatening events occur (e.g. the appearance of a stranger). Prato-Previde *et al.* (2003) list three cases in the SST which could reveal the presence of a secure base effect: decreased play and exploration in the presence of the stranger, returning to the owner at threatening events, and playing with the stranger in the presence of the owner. The observations of the dogs' behaviour supported only one of the three conditions, which led the authors to question whether dog–human relationship complies with the features of human attachment. Although Prato-Previde *et al.* (2003) could be right that the present evidence is inconclusive, it is important to point to behavioural differences between dogs and infants. Thus dogs and children might differ fundamentally in their reaction to stress. In the case of infants the SST is usually done in a developmental period when children show a stress response towards strangers, but this is usually not the case with socialized adult dogs. Thus the SST situation might be less stressful for dogs than for children. There are also differences in exploratory and play behaviour. Children show a lower tendency to explore the room as a potential 'territory' than do adult dogs, and in contrast, they show more interest (play) than dogs in novel toys; for dogs, toys are only interesting when manipulated by humans. These differences in behavioural patterns can mask the secure base effect, especially if it is determined on the basis of child behaviour. Thus in the case of dogs novel test designs might be necessary to provide evidence for a secure base effect.

In the case of abandoned shelter dogs, attachment to humans can form rapidly. Gácsi *et al.* (2001) offered adult dogs that had been living in the shelter for at least 2 months a 10-minute period of handling (walk and play) by an unfamiliar experimenter (handler) for 3 successive days. Behavioural observations in the SST test which followed the last handling showed a clear difference between handled dogs and non-handled controls. In comparison to non-handled dogs, handled animals spent more time at the door in the presence of the stranger, spent less time in contact with the stranger, and showed higher scores of contact seeking towards the entering handler. Although the differentiation between handler and stranger was in some instances less pronounced than in pet dogs, these results suggest that a relatively short contact can lead to the reorganization of the attachment system in dogs. In a more recent study using the same methodology, Marston *et al.* (2005a) found that in abandoned shelter dogs physical contact (massage) was more effective than obedience training as a form of handling in evoking patterns of attachment behaviour towards a human handler. These observations suggest that dogs deprived of human contact (shelter dogs) are able and willing to rapidly initiate a novel relationship after a short duration of social contact with an unfamiliar human.

If attachment depends only on the social environment, than adequate socialization to humans should result in dog-like attachment in wolves. In order to test for this possibility we have extensively socialized individual wolf cubs and tested them at 4 months of age in parallel with dog pups that have been raised in the same way. However, results show that in contrast to 4 month old dog pups, wolf cubs of the same age did not fulfil the criteria for attachment (Topál *et al.* 2005a). In the test dogs obtained consistently higher scores for greeting their owner, they spent more time in playing and tried to follow the leaving owner, and stood at the door longer in the absence of the owner (Figure 8.2). In contrast, wolves did not display a preference for the caregivers. Although negative results should be interpreted with care, these observations support a difference in the ability to form an attachment relationship. One might argue that the differences may come about because there is a difference in how dogs and wolves perceive the experimental situation. Wolves may not have been stressed, or may have an altered tendency to express various behaviour patterns. The comparison of dogs' and wolves' overall behaviour in the test situation left little room for such explanations. The only difference found was that wolves explored more at the expense of passive behaviour, but no difference in the amount of play was found. Moreover, if the wolves had not been socialized adequately they should have perceived the entering stranger as more stressful, which should have resulted in enhanced preference for the handler—which was clearly not the case.

These findings also seem to contradict the idea that the behaviour of dogs towards the owner is derived directly from the cub–mother relationship in wolves and has been achieved simply by altering rates of behavioural development. In addition, other observations revealed that by 6–8 weeks of age proximity and contact-seeking behaviour towards their mother gradually decreases in wolf cubs (Mech 1970) and affiliative behaviour is observed mainly towards the pack and not a specific individual (Rabb *et al.* 1967, Beck 1973). At 16 weeks old, wolf cubs were often left alone at a meeting point or rendezvous site where they waited for the return of the hunting group (Packard *et al.* 1992).

Although it has not been specifically tested for, the present evidence provides little support for attachment relationship between dogs (Rajecki *et al.* 1978; Chapter 9.5, p. 214). For dog pups 2 months old, the bitch plays only a minor role in reducing the effect of separation stress (Frederickson 1952, Ross *et al.* 1960, Elliot and Scott 1961) and in choice situations pups do not show preference for their

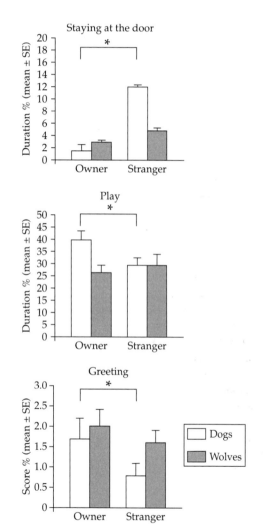

Figure 8.2 Behavioural comparison of socialized dogs and wolves in the Strange Situation Test. Dogs stayed longer at the door in the absence of their owner, played more with the owner, and obtained higher greeting scores with the owner compared to wolves in which no such preferences between owner and stranger were found (For more details, see Topál *et al.* 2005a). (* denotes significant differences between owner and stranger).

mother in comparison with an unfamiliar bitch (Pettijohn *et al.* 1977, but see Hepper 1994). In line with these arguments, Tuber *et al.* (1996) found that in dogs placed into a novel environment the level of stress (measured as cortisol concentration) can be decreased by the presence of a familiar human but not by a familiar dog.

Present results argue that in functional terms there are parallels between the behaviour of dogs and human children in their pattern of attachment (Collis 1995, Serpell 1996). Further support for this convergence comes from the observation that under certain conditions both dogs and children develop similar behavioural malformations that might relate to the attachment relationship. For example, in some cases both dogs and children show abnormal patterns of behaviour (*separation anxiety*) when separated from their caregiver. Overall (2000) argued that the similarities in dogs and children could involve partially common underlying mechanisms, and the dog's condition of separation anxiety could be a good model for the human situation. We know that the lack of a primary caregiver leads to disturbances in human attachment (e.g. Chisholm *et al.* 1995), and in dogs Senay (1966) found that separation from the caregiver at 10 months for a period of 2 months promotes exaggerated behavioural changes that persist long after reunion with the owner.

Some dogs show extreme stress (vocalization, elimination, destructive behaviour) upon separation from the owner, which is interpreted as separation anxiety. So far there is no evidence that this relates to *hyperattachement*, because the affected dogs did not show different pattern of attachment when tested in the SST (Parthasarathy and Crowell-Davis 2006).

The amount of social experience and the mothering style could influence the quality of attachment, which in turn has been implicated in influencing behaviour in other social situations. For example, in human infants attachment seems to predict enthusiasm, persistence, and cooperation at 2 years of age (Matas *et al.* 1978). Based on such findings, Topál *et al.* (1997) discriminated *a priori* two categories of dogs on the basis of the owners' answers to a questionnaire: *dependent relationship*, i.e. dogs living in the flat or house, vs *independent*

relationship, i.e. dogs living in the yard or garden outside the house. They assumed that dogs kept in the house as family members (family dogs) developed a more 'intimate' (emotional) relationship with their owner, whereas dogs living outside the house as a guard or for some other purpose (yard dogs) had a 'looser' relationship with their owner, with little possibility of getting involved in family interactions. In a separation test similar to the one described above, they found that the two groups did not differ in stress-related and exploratory behaviours but family dogs showed more dependent behaviour by spending more time following the owner. In addition, the groups also diverged in a problem-solving task in which they had to obtain a piece of food from under a fence. Yard dogs started to solve the problem on their own, and collected all available food items rapidly. Family dogs behaved in a very 'inhibited' manner; they were reluctant to obtain the food, and frequently displayed communicative behaviours towards their owner (e.g. looking at them), in contrast to the yard dogs. However, their performance in getting the food items rose as soon as the previously passive owner had the chance to encourage them by verbal and gestural communicative behaviour.

These observations suggest that dog–human attachment relationship provides a kind of scaffolding for the emergence of various social behaviours in cooperative and communicative interactions. In contrast to the independent and autonomous problem-solving behaviour of the wolf (see Frank 1980), in dogs the attachment relationship predisposes the dog towards engaging in joint activities with human members of the group (see Chapter 8.9, p. 197).

8.3 The agonistic aspects of social relationships

In contrast to claims by experts in the field (Bradshaw and Nott 1995), modern ethological thought has had relatively little influence on the understanding of aggressive behaviour in dogs. Thus it seems timely to rethink dog aggression in terms of novel ideas that have been introduced by studying other animal species.

There is general agreement among ethologists that the main function of aggression is to divide important but limited resources among group members. When the amount of available resources (e.g. food) decreases, there is an increase in the frequency of aggressive behaviour in wolves (Mech and Boitani 2003), and similarly dogs in groups display enhanced levels of aggression in the presence of food. Thus aggression is an integral part of the behavioural endowment of both wolves and dogs.

Aggressive behaviour in dogs consists mainly of displays that have a signalling function. For the evolutionary biologist the utilization of these signals is problematic, for at least two reasons. First, it might not be advantageous to reveal the next move on the part of the signaller, so it is questionable whether signals evolved for reflecting the inner state or 'intentions'. Second, such signalling systems are not immune to cheating, and individuals could display signals that are not supported by their physical abilities.

The evolution of signals can be put in a different light if we assume that fighting involves not only gains but also costs. Injuries (and also loss of energy) suffered during fights can affect the future chances of the winner, so even favoured contestants should think twice before engaging in fights which could have negative physical consequences. Contests based on mutual signalling could be really advantageous provided that the presentation of the display involves some cost; in other words, the signal provides honest information about the qualities of the signaller. For example, the visual outline of the dog's body, which is emphasized by erect tail and ears, could be such an honest signal, because larger dogs will not only have greater chances of winning a serious contest but there is a genuine relationship between fighting ability and size which cannot be cheated.

In theory one signal could do the job, but in reality dogs have a range of signals that could be utilized during contests. Fox (1970) advanced a hypothesis that the number of signals could relate to the sociality of the species. He argued that the relatively large number of complex displays in wolves reflects the more complex organization of wolf society in comparison to that of foxes.

Elaborate behaviours including greeting ceremonies and the repeated expression of rank relationships evolved a range of signals which are fine-tuned for signalling minute differences in agonistic or submissive tendencies. A wide variety of displays can also be useful for more precise signalling of the individual's fighting potential, which might change over time. Finally, signals that vary in their ability to provide a judgement of fighting ability could also contribute to the settling of contests. According to this view some agonistic displays offer the possibility of assessing the strength or weakness of the opponent before the fight. This process can also ensure honesty in signalling, because cheating would not be useful once the opponent has other means to test the fighting ability. For example, wrestling-type displays could reveal the real strength of the partner without engaging in fighting. Applying this to the case of the dog, we could hypothesize that breeds (individuals) with more constrained signalling abilities may have trouble in living in large social groups because they have problems in communicating their fighting potential. Comparative observations on young poodles and wolves (1–12 months of age) living in groups seem to support this argument (Feddersen-Petersen 2001a), because poodles displayed a higher frequency of agonistic interactions than their wild relatives. These young dogs lunged and bit their opponents apparently without taking notice of the opponent's (submissive) signals. A further study also showed variability in signal utilization in different breeds but no data were presented on how this might have affected the frequency of agonistic behaviour in the conspecific group (Goodwin et al. 1997). It is unfortunate that the comparative investigation of early agonistic interactions involving different breeds was not paralleled by behavioural descriptions (Scott and Fuller 1965).

The chances of winning any contest can be also conceptualized in terms of the *resource-holding potential* of the participants (Parker 1974). The resource-holding potential is determined by fighting ability, information about the disputed resource, and motivation to invest in the contest. For example, hungrier dogs (motivation) and/or territory owners (information about the resource) have a higher resource-holding potential, thus they are more likely to win a dispute. Interestingly, the involvement of many factors in determining the resource-holding potential ensures that two opponents rarely match, which leads to one giving up at the early display phase. It follows that individuals with similar resource-holding potential will contest for longer, and might also risk getting harmed. Thus the manipulation of dogs' resource-holding potential could lead to decreased aggressive tendencies (Sherman *et al.* 1996).

In social animals, such as dogs, winning a contest has both a direct and an indirect outcome. The winner gains control over the disputed monopolizable resource (e.g. territory, food, mate, social partner, object) and at the same time the victory affects the social relationship between the contestants and increases the chance of winning subsequent contests. It also contributes to the privileged status of dominant individuals which can get access to the resources without the need for displaying.

8.3.1 Classification of aggression in dogs

Aggressive behaviour in dogs has been categorized in various ways (Houpt 2006). Although most of these categories are useful from the practical and applied point of view, the theoretical reasoning is often less clear. Ethological reasoning would prefer functional categories which recognize the target of a contest. Thus dogs fight for territory (against non-group members), and resources (e.g. food) or position in the hierarchy (against group members). This distinction is important because it influences the organization of the behaviour pattern, and might also be under different genetic control. For example, we could assume that domestication has had different effects on within-group and between-group aggression in dogs (Chapter 8.3.3, p. 173).

Behaviour actions performed during an aggressive encounter can be categorized either on the basis of their effect on the opponent, or whether they have a signalling or physical role. Accordingly, actions that decrease the distance between the contestants are denoted as *offensive*, and behaviours having the opposite effect are referred to as *defensive* (Feddersen-Petersen 1991). Usually higher-ranking individuals show offensive aggression,

but it can be also witnessed in lower-ranking challengers. In larger groups, dogs or wolves might show both types of aggressive behaviour depending on their opponent. Signals that indicate retreat and aim to terminate offensive aggression are considered as 'submissive signals' or correspond to 'flight behaviour' (Packard 2003).

A different method of categorization considers behaviours having a signalling function and having no potential to cause physical harm as *threats* (e.g. growling). Actions resulting in physical contact or having the potential to inflict pain are described as *inhibited attacks* (e.g. inhibited biting), and finally actions which actually cause physical injury are referred to as *attacking* (e.g. biting) (Feddersen-Petersen 1991).

Using any of these categorization schemes is a valid way to decompose aggressive behaviour in dogs. It is important to see that this categorization does not include any forms of *playful aggression* or *predatory aggression*. It is a common mistake to list these forms of behaviour here, but neither is about division of resources. In the case of playful aggression special behavioural signals (e.g. 'play bow') communicate the non-agonistic inner state of the actors, but this does not exclude playful aggression becoming serious in some cases. In the case of predatory behaviour, it is the primary goal of the initiator to destroy the opponent, which is not the case in a true aggressive contest.

8.3.2 Is there an ethological description of aggressive behaviour in dogs?

The short answer to this question is no. Various authors recognize the similarity between wolf and dog in the units of aggressive behaviour, and some texts provide shorter or longer lists of the behavioural units (Feddersen-Petersen 1991, Packard 2003). Importantly, behavioural analysis has been carried out at different levels of behavioural organization (see also Chapter 2, Box 2.4). For example, Feddersen-Petersen (2001a) argued for seven facial regions (muzzle posture, mouth corner, lips, nose ridge, forehead fur, eyes, ears) which play a role in the expression of aggressive inner state (see also Bolwig 1962). This coding system is based on the mimicking of wolves but can be applied to any dog.

Not surprisingly, Feddersen-Petersen found that dogs have a reduced ability for signalling in comparison to their ancestor. So far, however, there is little direct evidence that the different facial expressions have a functional value, that is, that they reflect differences in the inner state and are recognized by the others as distinct signals. Others suggest the use of a more holistic coding system which is based on overt behavioural units, such as 'avert gaze' or 'chase' (e.g. van den Berg *et al.* 2003, Packard 2003), and finally, Schenkel (1947) uses an intermediate variant by taking into account behavioural details (e.g. visibility of the teeth) and overall body posture (Harrington and Asa 2003).

Qualitative analyses indicate that dog breeds differ in the number of signals used, for example more wolf-like breeds (e.g. German shepherd) have at least nine threat signals in comparison to the two signals in Norfolk terriers (Goodwin *et al.* 1997). But there is little published information on the use of aggressive actions, or their effect on the opponent's behaviour. We do not know whether dogs rely on these signals for assessment, or whether there are qualitative and/or quantitative differences in the aggressive behaviour of different breeds towards either conspecifics or humans. No information is available on the temporal structure of aggressive behaviour in dogs (for a related study on greeting behaviour see Bradshaw and Lea 1993), or whether signalling depends on the rank differences.

8.3.3 Decreased aggression in dogs?

Occasionally experts mention that aggression is reduced in dogs. The problem with this statement is that they usually do not mention what this reduction is relative to. In recent times there has also been a change in our understanding of aggression in wolves, and most observers now report a more peaceful group life in free-living populations than was observed in captive packs (Packard 2003). However, one could still argue that selective changes during adaptation to life with humans have decreased aggression both towards conspecifics and humans. Humans appear even more peaceful than wolves, and dogs have had to show an increased tolerance towards strangers in general

because there is a higher chance of human and dog newcomers joining the group from time to time. Thus there was probably a need to select against aggressive behaviour, because wolves are not tolerant towards strange conspecifics, and only very rarely can a newcomer join the pack.

The rules of agonistic signalling do not apply in the case of interactions with strangers and members of other groups. Attackers pay less attention to submissive signals, so lone wolves are often killed (Mech *et al.* 1998). Wolves with an increased threshold for this type of behaviour would find an easier way into the human community. The higher tendency to share resources in human groups could also facilitate selection for decreased within-group aggression, partly because human and wolf/dog aggressive behavioural pattern is physically incompatible, thus selection could also be aimed directly at reducing the use of these behaviour patterns.

Although there is little experimental evidence, folk knowledge indicates that breed selection achieved separation of within- and between-group aggression in dogs. In addition, selection acted in both directions, lowering or increasing aggressive tendencies. This is probably the case in certain protecting dog breeds that show elevated territorial behaviour towards strangers (dogs, wolves, or humans), whereas in other dog breeds (e.g. hunting dogs) territorial behaviour is much reduced.

Importantly, aggressive tendencies in behaviour can also be modified by changing the sensitivity to behavioural signals. The difference in certain dog breeds' reaction to threatening signals might be rooted in a change in reaction threshold (Vas *et al.* 2005); alternatively, ignorance of submissive signals can also lead to more aggressive behaviour (Fig. 8.3). Finally, in the case of the so-called 'fighting dogs', arguments have been put forward that their extreme and enduring fighting ability may be the result of decreased sensitivity to pain.

8.3.4 Organization of aggressive behaviour and the role of learning

According to Frank (1980), the dog's behavioural actions can be brought under the control of various external stimuli because in comparison to those of the wolf, the motor patterns of dogs are freed from their original motivational background. So far two independent lines of observation seem to provide some support for this idea. Observing predatory behaviour of the wolves and many dog breeds, Coppinger and Coppinger (2001) argued that the fixed sequential pattern of the behaviour has been decomposed into more or less independent units which are displayed at different frequencies in certain dog breeds. An example is 'eyeing', which is usually displayed at the start of the predatory sequence in wolves but seems to be lacking in hounds. In contrast, hunting pointers do not show 'chasing', which is present in the behavioural sequence in most dogs and wolves, and similarly, other hunting breeds should not display 'kill-bite', which is the terminal unit of any hunting predator. To some extent similar arguments were made for aggressive behaviour units (Goodwin *et al.* 1997). In addition, predatory and aggressive behaviour shares some overlapping behavioural units with similarities at the level of execution (e.g. 'bite', 'chase', or 'eye' = 'stare') (Box 8.2).

The importance of these observations for the organization of agonistic behaviour in dogs is further emphasized by the findings that although both wolves and dogs seem to be innately programmed to display most of these actions without much experience, both need to learn the significance of signals displayed by their companions. Ginsburg (1975) describes wolves which had been raised for many months without contact with conspecifics. He observed that these individuals had to spend some time interacting with other wolves in order to learn the 'meaning' of the signals and also how to react to them. Similar conclusions can be drawn from the observations of Fox (1971), who raised single Chihuahuas with cats. When exposed to conspecifics (or their mirror image) for the first time at 16 weeks of age, these dogs were not able to decode the behavioural signals of their conspecific companions but they learned about the signals rapidly during the next 4 weeks of socialization with other dogs. Dogs might also be able to learn about the effects of their signals on the behaviour of the other. Observing wolf cubs, McLeod and Fentress (1997) found that the predictability of the signal decreases with age. They argue that young wolves could learn to

withhold certain signals (e.g. tail raise), and the hiding of 'intentions' could enhance success in contests.

A behavioural system composed of relatively independent behavioural units has the potential for increased behavioural flexibility because its actual performance will depend crucially on environmental feedback (learning). However, what might be an advantageous situation for dog training, might lead to problems if there is a lack of adequate environmental (especially social) feedback. In these cases motor units which might originate from either predatory or aggressive behaviour may become organized into an abnormal behaviour pattern which is detrimental in certain social contexts. For example, the lack of threatening signals in agonistic situations (attacking 'without warning') might reflect a case when the dog is relying on components of its predatory behaviour (which does not incorporate such signals). This might also explain findings that dogs with a history of fighting and biting other dogs are also strongly territorial and tend to show enhanced predatory behaviour (Sherman *et al.* 1996). Territorial aggression and predation share some behavioural units, and in neither case does the actor takes much notice of the attacked party's actions or signals. Thus experience that is lacking or inappropriate could easily result in a behavioural pattern which is elicited in both contexts.

Similar arguments have been put forward in the case of *owner-directed aggression* or *canine dominance aggression*. This type of behaviour, which affects a considerable part of the dog population and seems to be present disproportionately in some breeds, appears to be heritable (Overall 2000). Some imply that abnormal levels of impulsivity could play a role in the development of this condition, but external causes, e.g. lack of appropriate socialization, could also play a role (Overall 2000).

8.3.5 Reaction to human agonistic signals

Over the years ethologists have identified many signals of agonistic behaviour by studying interacting dogs. However, under experimental conditions researchers often elicit aggressive behaviour towards a human stranger who threatens the dog

Box 8.2 Flexibility of the behavioural phenotype

Coppinger and Coppinger (2001) and Goodwin *et al.* (1997) argued that the motor components of predatory and aggressive behaviour show a mosaic pattern by being variably present and absent in certain genetically divergent breeds or breed groups (Table 8.2). Frank (1980) noted that the arbitrary relation between external stimuli and motor components of the behaviour contributes to the behavioural flexibility in dogs which is advantageous in training. Altogether this suggests that the relatively rigid behaviour pattern of adult wolves was decomposed at the genetic level. This also allows for the emergence of an individual-specific flexible behaviour pattern which develops in the course of repeated interactions between the human and dog members of the group. The process leading to such individualistic, habitual patterns of interactive behaviour has been described as *ontogenetic ritualization* (Tomasello and Call 1997).

Such an individualistic pattern of behaviour can emerge in various forms of interactions, and may also include acoustic signalling. Such ritualized behaviour often develops in situations that provide excitement to the participants, such as feeding, going for a walk, or playing (Rooney *et al.* 2001).

Table 8.2 There are some indications that during the evolution of dogs the structure of both the predatory and agonistic behaviour pattern was disrupted. This idea can explain why it is relatively easy to form the motor behaviour of the dog by training. (a) A representation of an idealized predatory behaviour of wolf hunting on prey. Selective breeding enhanced or reduced the tendency of showing some elements of this predatory sequence. For example, in pointers 'eyeing' (orienting towards the prey upon taking notice) is more pronounced ('pointing behaviour') and they will easily learn to refrain from killing (and eating) the game. (b) The comparison of different breeds suggests a fragmentation of threatening behaviour, with some breeds losing major parts of the original motor set. Goodwin *et al.* (1997) argued that the richness of the threatening behaviour correlates with morphological similarity to the wolf.

(a) Idealized wild predatory sequence from left to right (based on Coppinger and Coppinger 2001)

	Orient	Eye	Stalk	Chase	Grab-bite	Kill-bite
Guard dog	F	F	F	F	F	F
Header	H	H	H	H	F	F
Heeler	N	N	N	H	H	F
Hound	H	–	–	H	H	H
Pointer	H	H	F	F	H	F
Retriever	H	N	N	N	H	F

F, faulty behaviour; H, hypertrophied behaviour; N, normal behaviour; –, behaviour absent.

(b) Idealized sequence of threatening behaviour from left to right (modified from Goodwin *et al.* 1997)

	Growl	Stare	Stand erect	Bare teeth	Stand over	Body wrestle	Aggressive gape	Inhibited bite
Siberian husky	X	X	X	X	X	X	X	X
German shepherd	X		X	X	X	X	X	X
Shetland sheepdog	X		X					
Labrador retriever	X		X		X		X	X
Cocker spaniel	X		X		X	X		
French bulldog	X		X					

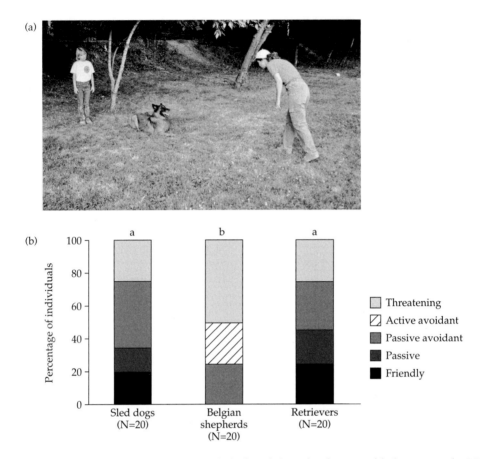

Figure 8.3 (a) The stranger moves slowly and hesitantly towards the dog, which is tethered to a tree while the owner stands c.1.5 m behind. (b) Breed group differences in response to a threateningly approaching stranger. Categories of dog behaviour (for more detail see Vas *et al.* 2005): 'friendly', dog wags tail, tolerates interaction; 'passive', no tail movement, tolerates interaction; 'passive avoidant', averted gaze; 'active avoidant', moves away from the stranger towards the owner, vocalization; 'threatening', sudden movements towards the stranger, vocalization (different letters at the top of the columns indicate significant differences).

(e.g. Svartberg 2002), without actually paying attention to the problem of how dogs recognize human agonistic signals with similar function but often different structure. Vas *et al.* (2005) compared the reaction of dogs to the same person who approached the dog in either a friendly or a threatening manner. They found that the behaviour of many dogs in reaction to the threatening stranger was controlled by the behaviour of the person, and these dogs repeatedly showed the same pattern of behaviour towards a person depending on their manner of approach. This suggests that there are certain aspects (eye contact, body posture, speed of movement, etc.) which determine the signal. At present

there are no experiments investigating the importance of these behavioural features for the effectiveness of the signal. Similarly, we do not know whether dogs decoding the human signal rely on generalized information based on their species-specific signals or whether learning plays a more important role. Knowledge on this topic could be very important because a lot of misunderstanding in social communication is based on the inappropriate signals given by humans (especially by children) (Chapter 3.7.3, p. 59).

Interestingly, there are many assumptions about so-called status behaviours ('privileges') displayed by dominant animals. For example, the dominant is the first to eat, and eats as long as he likes ('Do not feed your dog first!'), it has rights to choose resting places ('You should decide where the dog sleeps and do not share your bedroom with the dog!'), it leads the pack ('Do not allow your dog to cross thresholds first, or lead during walking!'). Although many of these behavioural patterns (and others not listed here) have been observed in dominant wolves, the reliability of these status signals has not been described. It is also uncertain whether dominants rely on such privileges regularly or only under particular circumstances. Thus it is far from proven that humans have to act like a 'dominant wolf' in order to establish an asymmetry in the interspecific relationship. So far questionnaire studies have failed to find relationships between many of these types of interactions and aggressive behaviour in dogs. For example, Podberscek and Serpell (1997) could not detect a significant relationship between being fed earlier or lack of obedience training and aggressive behaviour in English cocker spaniels. Nevertheless it would be useful to know more about the role of these status-related behavioural patterns both in wolves and dogs sharing their life with conspecifics and humans.

8.4 Communication in a mixed-species group

Many ethologists would agree with a definition saying that communicating interactions come about when it is in the interest of the signaller to modify the behaviour of the receiver by using behavioural actions for which there is evidence that they have been selected for such a function. In the long run communicative interactions should benefit the sender, not excluding benefits on the part of the receiver.

Studies looking for the mechanisms of animal communication systems usually focus on the *units* of the signals, the *'aboutness'* of the signals, and *causal aspects* of sending the signal (see also Hauser 1996, 2000). Unfortunately none of these problems is very simple, and our tendency to think in terms of language (the preferred but not unique system used for human communication) makes the situation even worse. For an intuitive introduction to these problems, imagine the behaviour of a dog

just before it attacks, signalling an aggressive inner state like 'anger' (in human terms). The first problem is whether there is a signal unit that corresponds to 'anger'. It is well known that dogs use very different body parts (body, face, tail, and emitting sounds) for signalling, providing the possibility for a wide variation of 'anger' signals. Thus it seems that dogs may have many units for signalling anger, but we still not know whether (1) all possible variations have a separate meaning (or any), and (2) whether there are signals with synonymous meanings. These questions have often been answered by talking about 'graded' signals, implying that there is a continuous change ('increasing signalling activity') in signalling which ranges from 'no anger' to 'great anger', but this does not solve the problem of how this grading of signals is achieved.

Ethologists and psychologists have assumed that the emission of signals depends solely on the inner state of the animal. Thus signalling would automatically parallel changes in inner states, and a previously affirmative animal when attacked by a stronger one would change to signal submission ('the principle of antithesis': Darwin 1872). Thus it was a kind of revelation to find that some signals (e.g. dog barks) can be brought under the external control of neutral cues; that is, dogs can be trained to bark upon a signal (light) in a conditioning paradigm (Salzinger and Waller 1962). However, this was less surprising for ethologists who suggested that animals (e.g. the vervet monkey) in nature use warning calls (about predators appearing in the vicinity of their groups) in a very similar manner to human using words with reference to environmental events ('Leopard!'). Thus our present understanding is that a communicative signal may either indicate the inner state of the sender or refer to events in the environment (Hauser 2000). The problem is how to separate these two different types of signals, because it is difficult to exclude the possibility that the external event (predator) influences the inner state (fear).

Although, according to the definition of communication, signalling is in the interest of the sender, it does not follow that the sender has the intention to signal. Actually, many argue that in the case of agonistic signals it is not always in the interest of the sender to reveal its true intentions, and it is also questionable whether it is in the interest of the observer to signal the presence of a predator to others. This does not, however, exclude the possibility that under certain situations animals signal intentionally.

Despite the claims of many popular textbooks, we know very little about cognitive aspects of communication in dogs. The reason is very simple. Any ethological investigation must describe the units of the signal at the behavioural level, detect the 'meaning' (aboutness) of the signal, and identify the underlying controlling system (whether internal, external, or 'both'). However, this can be done only by making careful observations or designing experiments in which data from two (or more) communicating animals can be collected in a systematic way.

Thus for practical reasons most research has investigated situations when the communicating partner is a human. However, this also limits the information that can be collected and does not solve the problem of intraspecific communication. Apart from investigating what kind of signals are exchanged, and whether the dog's signals have an intentional component, the increased controllability of these investigations has the potential to provide evidence about the nature of the underlying mental processes.

Thinking in the terms of the ethocognitive model, dogs may utilize a representational system which has evolved for dealing with signals of conspecifics and does not differ to a large extent from that of the wolf. Alternatively, their long cohabitation with humans may have selected for special representational abilities in respect to communication with humans, which not only utilize a partially different signal system but use these signals intentionally and in the sense of referring to external events.

Finally, one could think of at least two simple means by which the potential of an animal communication system for sending different messages can be increased (see also arguments on the flexibility of the aggressive behaviour pattern). First, the signalling may be less closely associated with the inner state of the sender, and second, the number of potential signals (units) may be increased. Abler (1997) describes this later notion as the *particulate principle*, and Studdert-Kennedy (1998) argues that the success of human language depended critically on the increase in the number of signal units, which also offered the possibility of a wide range of combinations. Looking at the behavioural potential of dogs, at least theoretically, one could also raise the possibility that the communication system of our companions has shifted in this direction.

8.4.1 Visual communication

Unfortunately, very little is known about visual communication in wolves apart from the agonistic context (see Harrington and Asa 2003) and even in this case little quantitative research has been done. In the case of the dog there have always been indications that dogs can rely on human visual communicative signals, and also that humans understand visual signals given by dogs. However, even in the case of dogs we have little knowledge of how they use visual signals among themselves.

One general point in the case of dog–human interaction is that dogs seem to live in the visual field of the human. This means that the direction that is in focus for the human becomes significant for the dog also. If dogs are deprived of such information (e.g. the human is blindfolded or his head orientation cannot be perceived), they often become hesitant (e.g. Pongrácz *et al.* 2003, Fukuzawa *et al.* 2005).

From the behavioural point of view, the communicative interaction of visual signals can be divided into four stages. First, the sender produces signals for initializing the interaction, next it recognizes that the receiver is in a state to observe the signalling. This state, which often referred to as *attention*, encourages the sender to send further signals, and finally the sender might receive a response from the receiver. In the following we use this simple framework to describe dog–human signalling.

Initialization of communicative interactions
There are some indications that dogs have a strong propensity to initialize communicative interactions with humans by using visual (and sometimes acoustic) signals (looking and gaze alternation)

functionally similar those used by humans. When facing an insoluble problem, dogs often use such attention-getting behaviours. Miklósi *et al.* (2000) first showed a piece of food to dogs, and then hid it at some height out of view in the absence of the owner. When the owner returned to the room, the dogs looked at the owner and displayed gaze alternation between the location of the hidden food and the owner. These actions were more frequent than when no food was hidden or no human returned to the room. A similar phenomenon was observed in a separate experiment (Miklósi *et al.* 2003), in which dogs were trained to pull some food attached to a piece of rope out through the wires of a cage. After having learned how to solve the task, dogs were prevented from getting the food by fastening the rope imperceptibly to the wire of the cage. Characteristically, after a few attempts most dogs stopped trying and looked at their owner who was standing behind them. Importantly, this initialization of communication was not present in socialized wolves that participated in similar experiments. One plausible explanation for this difference is that wolves might be less interested in human communicative signals or getting into communicative interaction with humans. In addition, they might avoid looking at humans (especially at the face and upper body) for an extended period, which could interfere with the possibility of recognizing communicative signals used by humans.

Understanding behavioural cues indicating attention
In the case of visual signals, attention can be recognized by the sensitivity to certain cues which reliably predict gaze direction and visual awareness (body and head orientation, open eyes, etc.). In a series of experiments we have found that dogs are sensitive to behavioural cues signalling attentiveness (Gácsi *et al.* 2004). Dogs were readily able to discriminate face orientation (forwards or backwards) of the human, because they approached the person mostly from the direction of the face when retrieving an object. Importantly, this sensitivity is context-dependent since dogs show no such discrimination in the context of play but only if they are commanded to retrieve an object. Dogs prefer the attentive human when they are given the choice to beg from someone turning either towards them

or away, and there is suggestive evidence that they also discriminate between open vs closed eyes.

Dogs are particularly sensitive to behavioural signals in commanding situations (Virányi *et al.* 2004) when the attention of the experimenter was manipulated systematically. In different trials the experimenter was either looking directly at the dog, standing behind a screen, oriented at another human, or looking into some empty space when commanding the dog to lie down (using the playback stimulus of a pre-recorded verbal command) (Figure 8.4). Depending on the condition, dogs displayed clear variability in their readiness to obey the command. They obeyed the command most often when it was emitted in concordance with facial orientation towards them. They were less likely to obey if the command was seemingly directed at the other person, but they showed a somewhat increased inclination to cooperate when there was nobody in the attentional focus of the experimenter (Table 8.1). Similar outcomes were obtained in experiments where the dog was forbidden to eat a piece of visible food (Call *et al.* 2003, Bräuer *et al.* 2004, Schwab and Huber 2006, see also Box 1.4). The invariable result was that dogs were sensitive to the attentive behaviour of the experimenter. They ate the food when nobody was present and resisted consumption when the human was looking at them. In trials with the experimenter present but signalling inattention (eyes closed or playing a computer game, etc.) the latency for feeding varied but there was an increased tendency for eating. This indicates that dogs use both gestural and behavioural cues to discriminate between attention and inattention.

Visual attention is also important in learning contexts because it seems that dogs use visual cues (eye contact and directed talk) provided by humans to infer whether they are 'addressed' in a particular situation. In the social learning context we have found that dogs learn detouring much better if the human uses eye contact and verbal signals to get the dog's attention (Pongrácz *et al.* 2004). This could also explain why dogs respond better to gestural cues that are preceded or accompanied by visual cues of looking at the object, indicating the goal of the demonstrator or communicator (Agnetta *et al.* 2000).

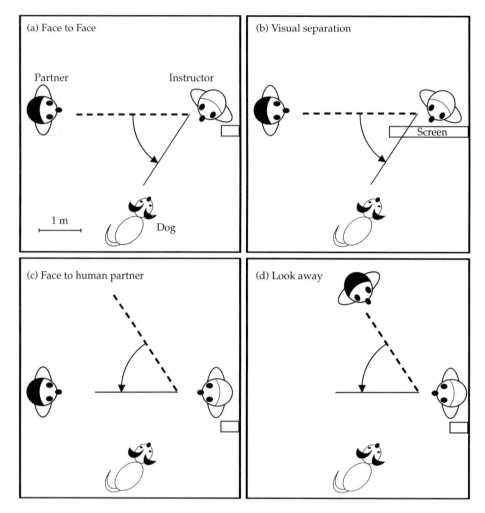

Figure 8.4 View from above of different commanding situations. (a) The command is directed towards the dog. (b) The command is oriented towards the dog but it cannot see the instructor who hides behind the screen. (c) The command is directed towards the other person present. (d) The command is directed at the 'empty space'. The arrow indicates the movement of the head before the command is given. The dashed line indicates the line of sight before the command is given; the unbroken line indicates the line of sight at commanding. (Redrawn after Virányi *et al.* 2004)

Table 8.1 The number of dogs that behaved according to their owner's verbal command (Down!) in the different experimental conditions (see also Figure 8.4). Commands were repeated three times in succession: Down! Down! Down, [dog's name]'! (based on Virányi *et al.* 2004)

Response/Condition	Face to face	Visual separation	Facing human partner	Looking away
Lie down promptly	6	1	0	3
Lie down after the first repeat	11	3	3	3
Lie down when its name called	0	2	2	4
Command ignored	0	11	12	7

Providing information

If the sender has ensured that the receiver is attending to the interaction, the sending of further signals is possible. However, there is a further complicating factor; for optimal communication to take place the sender may need to take into account the actual knowledge of the receiver about the situation. In one study we wanted to see not only whether dogs are able to provide information for humans but whether they also take into account what the human knows or does not know (Virányi *et al.* 2006). (For details of these arguments and their relation to the problem of attributing mental states see also Gomez 1996, 2004.) In the experimental situation the dog is playing with the experimenter when 'suddenly' the toy (a ball) disappears into an unapproachable location. The dog can get the lost toy only with the active involvement of a helper who uses a tool to retrieve the object. According to the experimental protocol the tool is kept in the same place but the toy disappears at different locations. Thus the dog knows the location of both objects, but in the experiment the helper's knowledge is manipulated. In some trials the helper is absent, either when the object disappears or when the tool is placed in a new location by the experimenter. In other trials the helper has no information about the location of either the toy or the tool. Two assumptions can be put forward. First, the dog provides information about the location of both objects (tool and ball), independent of the knowledge of the helper. Second, its signalling reflects the knowledge of the helper, that is, it only signals the location of the object(s) that the helper does not know. The results of this experiment supported the first assumption; dogs preferred to signal the location of the toy but their behaviour was not dependent on the knowledge of the helper. Although this shows that dogs do not seem to take into account what the human can see or has seen (and as a result they obtained some 'knowledge'), the negative outcome might have been the consequence of the complexity of the situation, or else dogs were willing to signal only the place of the motivationally significant object (toy) but not the motivationally neutral one (tool).

Positive evidence was reported by Cooper *et al.* (2003) utilizing the well-known *guesser–knower*

paradigm (Povinelli *et al.* 1990). In this experiment the subject has to choose between two options on the basis of observing the behaviour of two companions. The assumption is that for successful choice dogs might rely on the companion that was observed to have the chance of obtaining the necessary information. Dogs preferred to choose the location of hidden food that was indicated by a 'knower-dog' who was perceptibly witnessing the baiting. Interestingly, this preference showed up only in the first trial and disappeared later, suggesting that the phenomenon is very elusive. In a longitudinal case study with a single dog we have shown that it was able to adjust its communicative behaviour to the state of knowledge of the human partner, and cooperated successfully in the problem-solving task (Topál *et al.* 2006a) (Box 8.3). Nevertheless we are still far from knowing whether dogs are able to recognize some inner mental state of humans or, more likely, if they rely directly on human behavioural cues that are associated with the lack or existence of some knowledge.

Utilization of human visual signals

A very simple method introduced by Anderson *et al.* (1995) provides the possibility of studying the utilization of directed human bodily signals in dogs. In this experiment the dog has to find a piece of hidden food in one of two bowls. To accomplish this task the dog can rely on a cue indicating the correct location, which is given by the human experimenter standing between the two bowls. There are many variations of this procedure (Miklósi and Soproni 2006) but in most cases dogs performed reliably (e.g. Hare *et al.* 1998, Miklósi *et al.* 1998, McKinley and Sambrook 2000). The experiments, in which by now more than 1000 dogs have been tested worldwide, revealed that dogs can rely on the human pointing gesture even if the bowl containing the food is at a considerable distance from the human (Chapter 1, Box 1.2). Dogs were also successful with a few other variations of the pointing gesture, for example, when the human pointed in the opposite direction with the arm across the body (Hare *et al.* 1998, Soproni *et al.* 2002). Dogs were also skilful with gestures that could be considered as relatively novel (not usually

Box 8.3 Representing the other's state of mind

Gomez (2005) describes a method which seemed to be suitable to test for the ability to recognize knowledge or ignorance in others in species without language. Topál et al. (2006a) made only minor modifications to the procedure, which was originally used with an orang-utan, when testing a Belgian Tervueren dog (Philip). The task of the subject is to get a piece of hidden food (or a toy in the case of the dog) by informing the helper human about the whereabouts of either the target object or a tool ('key') which is needed to get out the target object from a holding box. Thus the helper never knew where the target object was hidden (in one of the three identical boxes by an experimenter) but his knowledge about the location of the tool necessary to get the object (i.e. to open the box) was manipulated: He either participated in finding a novel place for the tool ('Relocated condition'), or he was absent during the hiding ('Hidden condition'), or the tool was put in its usual place ('Control condition'). After the dog had learned the rules under the control conditions, it was observed in eight test sessions each of which consisted of three trials (one per condition).

The hypothesis was that if the subject takes into account the knowledge of the helper it communicates only the 'missing' information.

Accordingly, the dog should indicate only the location of the toy in the 'Control' and 'Relocated' conditions, and both objects in the 'Hidden' condition.

The table below shows that the dog indicated (by approach and/or touch) mostly the baited box when the helper knew the location of the tool (Control and Relocated conditions), and there was a suggestive preference for indicating the tool first in the hidden condition.

This result is very similar to that obtained with the orang-utan, suggesting similar mental capacities to solve this task. However, importantly, many researchers would not agree that successful mastering of the task necessarily indicates that the subject recognizes knowledge or ignorance on the part of the helper. Philip's behaviour could also be explained by increased sensitivity to the behaviour of the human (although the experimenters controlled for possible Clever Hans effects), by very rapid learning or reliance on earlier skills (Philip was trained as an assistant dog) or by noting that the indication of the key in the 'hidden condition' might have been caused by the dog being 'more exited' when the key was moved in the absence of the helper (see also Whiten 2000).

Table to Box 8.3

	Approaching/touching				
	Key only	Key then baited box	Baited box then key	Baited box only	Neither key nor baited box
Control condition	0	2	–	6	0
Relocated key condition	0	1	–	7	0
Hidden key condition	0	4	2	1	1

Note: Having been shown the baited box, the dog had no possibility of approaching the key in the 'Control' and 'Relocated' conditions, because the helper picked it up. Therefore the 'baited box, then key' option is irrelevant in these cases.

used by humans), e.g. pointing with the leg (Lakatos et al. 2008), but they failed with gestures in which a pointing finger was the decisive gestural cue. The review of many experiments led us with the hypothesis that dogs are using a simple rule for understanding such directional gestures: Look for some body part extending from the torso of the human body! It seems that in this regard they use a similar rule to 18 month old human infants (Lakatos et al. 2008).

The high level of performance in dogs raised the question about the origin of this ability, because neither wolves nor dogs point like humans. Some have suggested that dogs may rely on the conspecific communication system. Accordingly, dogs (and wolves) do not point with their 'hands', but point with their body when localizing distant prey. (This behaviour was probably selected for in pointers.) Thus in the case of hidden food, the representational system of dogs decodes the human pointing hand in the terms of the directed body of conspecifics. Indeed, dogs are able to use the body orientation of conspecifics for localizing food in a similar situation (Hare and Tomasello 1999). Others stress that this ability is the result of a learning process in which dogs learn to associate the closeness of the human hand (fingers) with the location of food. The problem of this argument is that dogs utilize the signal even when it is absent during the actual choice, and they are much better in relying on these signals than on beacons (section 7.2.2) which are placed a comparable distance from the target. The observation that 2-month-old pups with relatively little human contact (Hare *et al.* 2002, Gácsi *et al.* 2008) can also choose on the basis of pointing gestures makes the exclusive contribution

of learning less likely. Since pointing is perhaps one of the few cross-cultural referential gestures in humans, dogs might have been selected for the utilization of such gestures (Box 8.4).

One way to judge the utility of the above ideas is to compare the performance of dogs and wolves (see Chapter 2, Box 2.2). In one series of experiments we have found that wolves that have been socialized to comparable levels to dogs are inferior to dogs (Miklósi *et al.* 2003) when they have the chance to find hidden food on the basis of pointing signals. Importantly, these wolves could make choices on the basis of other signals (e.g. when the human stands near the bowl with the food, or touches it). In subsequent experiments Virányi *et al.* (2008) reported that socialized 1.5–2-month-old wolf cubs are inferior to dogs in the pointing tests but can reach comparable performance after intensive training. In addition, it seems that if extensive socialization is continued for up to 2 years, wolves reach the dogs' level of performance even without specific training (Gácsi *et al.* 2009).

Other experiments have also revealed that the human hand (or hand action) attains specific significance for dogs. Riedel *et al.* (2006) exposed dogs to a hand action when the experimenter

---

Box 8.4 Referential aspect of communication

There has been a lot of discussion in the literature on whether dogs can also decode the referential aspect of the pointing gesture. In the case of humans there is a general agreement that adults and children of a certain age understand that the pointing is 'about' the object at which the communicator gestures. However, in the case of dogs some have assumed that the comprehension of pointing is possible because the pointing hand (and fingers) has been associated with presentation of food (when feeding the dog) or acts as a beacon for signalling the location. This latter is the more important one because there are results showing that many domesticated animals (e.g. goats) and also foxes selected for tame behaviour comprehend the pointing gesture. (a)

In order to exclude these alternatives we have suggested that the communicative nature of the gesture can be enhanced by making it momentary, that is, the subject does not see it when it makes the choice. Actually, this is also the case with some other visual communicative signals (e.g. play bow) where the partner has to remember the signal. Generalization to other similar but novel gestures could also be a sign for understanding the referential aspect of the signal.

Among domesticated animals (including the selected foxes) only dogs have been shown to comprehend the momentary version of the pointing gesture and choose on the basis of momentary 'leg pointing' in a two-way choice task. (b)

continues

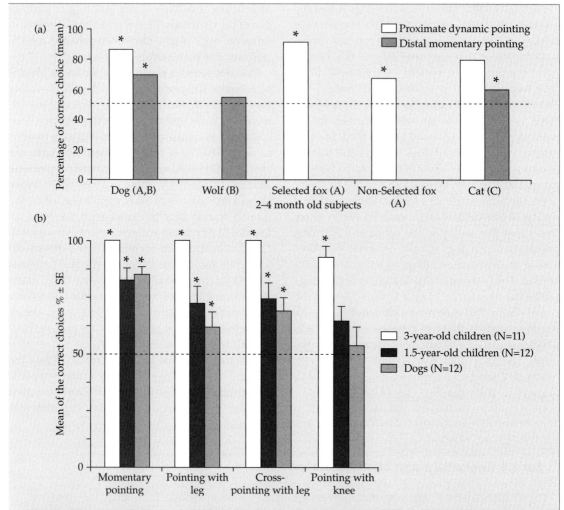

Figure to Box 8.4 (a) Comparison of comprehension of different types of pointing gestures. Note that all 2–4 month-old subjects are successful with the proximate dynamic pointing gesture (they see the pointing hand when making the choice) but only dog puppies show a good performance in the case of the distal momentary pointing gesture (A: Hare *et al.* 2005; B. Virányi *et al.* 2008; C: Kemencei 2007). (b) Adult dogs also comprehend novel pointing gestures but they show a declined performance. Children are shown only for comparison (Lakatos *et al.* 2008). (––– chance level; *significantly over chance level).

placed a wooden object as a marker on top of the correct location. They found that dogs were very skilful in various conditions, for example when they witnessed only the placing of the marker (and could not see the experimenter) or when the experimenter removed the marker after placing it (see also Agnetta *et al.* 2000).

Much less is known about the utilization of other human visual signals. In a similar choice situation dogs can also rely on bowing, nodding, or head

turn, and after some training they are able to choose the correct location on the basis of eye movements alone (Miklósi *et al.* 1998).

In human communication, pointing signals are regarded as referential because they often refer to external events or objects. Whether dogs are able to decode this aspect of the gesture is still debated, but there is some evidence to support it. Soproni *et al.* (2001) repeated an experiment by Povinelli *et al.* (1990) (on chimpanzees and children) in which instead of using pointing gestures the experimenter was either looking into the correct bowl or looking at the ceiling above the correct bowl. The appearance of both gestures is very similar, but from the observer's point of view 'looking into' communicates something about the food, while 'looking above' displays disinterest. In principle both gestures provide discriminative cues for localizing the place of the hidden food, but if the dogs attend to the referential character ('aboutness') of the gesture, they should be correct only in the case when the experimenter is looking at the target (because looking above the food does not *refer* to the location). Interestingly, dogs chose correctly only when the experimenter was looking into the bowl but not when she looked above it (similarly to children and in contrast to chimpanzees, see Povinelli *et al.* 1990). This suggests that children and dogs attended to the referential aspect of the gesture and did not rely simply on the discriminative aspect of the signal, that is, whether the experimenter's head was turned to the left or to the right.

8.4.2 Acoustic communication

Ethologists have spent many years collecting and analysing wolf vocalizations, both in the field and in captivity (e.g. Theberge and Falls 1967, Harrington and Mech 1978, Schassburger 1993, Feddersen-Petersen 2000), but there is little available data on dog vocalizations. Nevertheless there is a general consensus that the two species share most vocalizations (Bleicher 1963, Coren and Fox 1976, Tembrock 1976), with the exception that dogs howl less frequently and are 'noisier' than wolves because of their enhanced propensity to bark in various contexts. For wolves some descriptive data has been collected about the use of most vocalizations in the intraspecific context, and probably most sounds have retained their ancestral function in dogs. Interestingly, however, no detailed investigations have been carried out on dog barking.

Based on Schneirla's theory (1959), Coren and Fox (1976) classified the vocalizations of the *Canis* species according to whether they elicit withdrawal or approach from the receiver. The acoustic patterns of these signals fall into two categories. One type of signal consists of harsh, noisy sounds emitted at low frequencies (e.g. growl, snarl, woof, bark); the other type can be characterized as clear, tonal, and consisting of harmonic sounds at higher frequencies (e.g. whine, yelp, whimper) (Schassburger 1993). Vocalizations belonging to the first category elicit withdrawal in the receiver but, more importantly from the sender's point of view, these are associated with agonistic inner states. Sounds in the other category usually signal friendly or submissive (*appeasing*) tendencies. Comparing several bird and mammalian species, Morton (1977) concluded that this categorization of vocalizations could provide a general rule for the relationship between inner state and the acoustic features of the sound (motivation-structural rules). As we shall see, this idea is not only valid for all *Canis* species but also seems to apply to the only vocalization that has changed during domestication: barking.

The role of barking in communication

Researchers have often noted that in contrast to wolves, in which barking has been described as a signal for warning or protesting (Schassburger 1993), dogs invariably seem to bark in a wide range of contexts. Accordingly the barking of dogs is considered as a hypertrophied by-product of the domestication process (Coren and Fox 1976) having no particular function in either species-specific or cross-species communication.

When Feddersen-Petersen (2000) recorded barks from different breeds in different contexts, she noted that barks vary both in frequency and in the relative amount of harmonics. In comparison to wolf barks, dogs emitted barks at much wider range of frequencies, and barks could be dominated by either harmonic or noisy sounds. Thus it seemed that as well as using barks more frequently, dogs also utilize different acoustic forms (Figure 8.5).

Yin (2002) argued that if Morton's rule is valid then the differences in the acoustic structure of barks should also reflect differences in the inner state. In support of this idea she found that the acoustic parameters of the dog barks depended on the recording context; for example, dogs barking in isolation produced higher-pitched sounds than when the dog was disturbed by a sudden noise (a ringing doorbell).

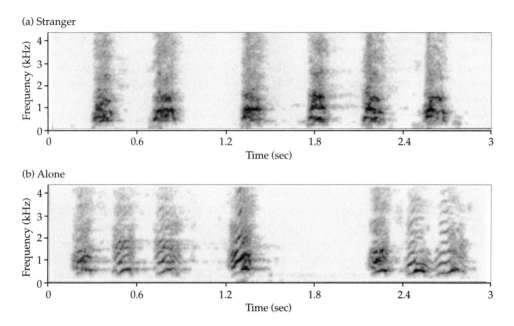

Figure 8.5 Sonograms of Mudi (Hungarian sheep dog breed) barks emitted towards a stranger (a) are clearly different from barks given when the dog is left alone (b). In the former case most sounds are emitted at a lower frequency and faster rate. The contrasting bands visible at a series of frequency ranges indicate a more 'tonal' sound in the 'left alone' barks.

There is, however, another problem to be clarified, which concerns the audience of the signal. Barking has been often observed in dogs living with humans and is relatively rare in stray and feral dogs (Boitani and Ciucci 1995). Thus some researchers have assumed that dogs use barking as a means for communicating with humans, and a few have even gone as far as postulating that barking is an imitation of human speech. In any case we might suppose that if dogs use barking as a signal for humans, we should at least show some skills in decoding it. In order to test this idea Pongrácz *et al.* (2005) recorded the barking of the Hungarian Mudi (a barking, medium-sized herding sheepdog breed) in seven different behavioural situations and played it back to humans who either owned no dogs, owned a dog of another breed, or kept Mudis at home. The listeners had two tasks. First, upon hearing a barking sequence they had to note on five independent five-item scales how aggressive, desperate, happy, playful, or fearful it felt, and they had to assign the same vocalization into one of seven contexts offered by the experimenter ('dog attacks', 'dog is left alone', 'dog is

playing', 'dog is about to go for a walk', 'dog watches his ball', 'dog participates in defence training'). Surprisingly, the experience of owning any dog or being the owner of a Mudi made no difference; all adult humans showed similar performance. In general humans put the barks into the appropriate categories more often than expected by chance, and they also associated the correct emotion with the situation; that is, barks which were recorded from an attacking dog were also described as aggressive. The analysis of the acoustic structure also provided further evidence for the operation of Morton's rule. Barks recorded during attack were noisier and had lower frequencies than barks from a dog that was left alone in the field. Interestingly, another parameter was also involved: the rate of the barking sequence. Listeners found the barking more aggressive if the rate was rapid (shorter time between two barks). It should be noted that the rate of barking offers a possibility of digital coding (number of similar-sized signals during a given time duration) which is especially useful if communicating over longer distance (see also Schleidt 1973) (Box 8.5).

Box 8.5 On the possible 'meaning' of barking

Human listeners were able to allocate barks correctly (significantly above the expected chance level) to categories of different contexts provided by the experimenter (a). Humans also judged the possible emotional content of the bark accurately (Pongrácz *et al.* 2005) (b). It is likely that for both kinds of judgements humans relied (among other acoustic features) on the frequency of barking, because barks with lower frequency were usually regarded as more aggressive and barks with higher frequency were described as being more fearful (c).

In a different study we analysed the possible context-specific and individual-specific features of dog barks using a new computerized learning algorithm (Molnár *et al.* 2008). A database containing more than 7400 barks (from the Mudi breed, see Figure 8.5) which were recorded in six communicative situations were used as the sound sample. The task of the algorithm was to learn which acoustic features of the barks, which were recorded in different contexts and from different individuals, can be distinguished from each other. The program carried out this task by analysing

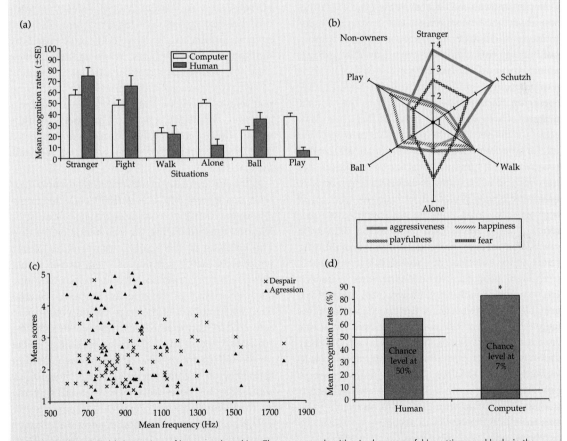

Figure to Box 8.5 (a) Comparison of human and machine. The computer algorithm is also successful in putting novel barks in the correct category (chance level at 17%). (b) Non-dog-owners have no problem in assigning an emotional state to dog barks recorded in different contexts. Note the higher scores for the key emotions on the respective axis. (c) The relationship between barking frequency and emotional scores. (d) Humans seem to have difficulty in matching barks emitted by the same dogs. After practice, the computer algorithm can solve the problem. (*, significant difference from chance).

continues

Box 8.5 *continued*

barks emitted in previously identified contexts by identified dogs. After the training phase the computer was provided with unfamiliar barks recorded in the same situations. The recognition rates found were high above chance level: the algorithm could categorize the barks according to their recording situations and the barking

individuals. Interestingly, the algorithm performed much better than humans. The program was successful both in categorizing the barks according to the predetermined situations (a) and also in matching different barks emitted by the same dog (d). The latter task was impossible for humans (Molnár *et al.* 2006).

Thus it seems that dogs can vary at least three parameters of their bark (frequency, tonality= noise/harmony, rate), all of which are related to the inner state of the sender.

Importantly, in the vocal system of the wolf other sounds are clearly separable, and have a very distinctive overall acoustic structure which is important because this way the receiver can discriminate among them unambiguously. In the case of barking a single form of vocalization is used to signal a wide spectrum of inner states by modifying key acoustic features. This gives an the signaller increased flexibility (in the sense of the particulate principle) but on the other hand it assumes an acoustically skilled receiver like a human. It cannot be excluded that the ability to use barking to express a wide range of inner states was preferred by humans who could rely on this signal even from a considerable distance. Thus it seems that cohabitation with such vocal mammals as humans has had a facilitating effect on the evolution of vocal abilities in dogs.

Humans need relative little experience to decode the meaning of barking. Children from the age of 6 are able to report correctly the two basic emotions (aggressive vs fearful) involved in some situations (attacking vs left alone). People who had lost their vision before birth performed at comparable levels to sighted people (Molnár 2008). In many respects human non-linguistic signalling is also in accord with the motivation-structural rules, so we might be able to rely on this ability in decoding vocalizations of other species, including dogs (see Box 8.5).

Utilization of human acoustic signals
In humans talk is a dominant way of establishing social contact, so it is not surprising to find that humans also talk to dogs, and from the reactions of

their companion many believe that dogs understand what is told to them. It is interesting, however, that humans often use a modified type of speech for verbal communication with the dog. Hirsch-Pasek and Treiman (1981) described this dog-directed speech as 'doggerel' and observed several similarities to the 'baby talk' used by mothers talking to infants. When talking to their children, mothers (or fathers) use the speech register at higher frequencies, talk more slowly and in simpler sentences, rely on a smaller vocabulary, express affection, and also talk from the perspective of the infant. Most of these observations were supported in a detailed comparison of doggerel and baby talk (Mitchell 2001).

Although there are no observations on how dogs react to doggerel, McConnell (1990) collected cross-cultural evidence that humans use specific acoustic features for influencing the behaviour of the dog. The analysis of the acoustic features of human whistles showed that dog trainers prefer to use short, rapid, repeated broadband sounds to stimulate the activity of the dog. In contrast, whistles used for inhibition of the dogs' activity were characterized by continuous narrowband vocalization. An experimental study provided further evidence for the utilization of these whistles. Dogs could be trained faster to come (facilitation of activity) when short, repeated notes were used as training stimulus (McConnell 1990).

The only field observation on dog and human communication during herding work revealed that sheepdogs have been trained to make at least six actions on verbal or different types of whistle commands while moving the sheep (McConnell and Baylis 1985). In absolute terms this number is not exceptionally high, because dog owners report that their dog 'understands' 32 verbal commands on average (Pongrácz *et al.* 2001). The first study investigating 'verbal understanding' systematically tested the capacities of a German shepherd dog that had been trained for acting in films (Warden and Warner 1928). Previous observations revealed that the dog executed two types of actions. Some actions related to changes in body position ('Sit!') or were aimed in general terms to some specific aspect of the environment ('Jump up high!' = the dog jumps up to the object or person nearby). Other actions had a specific goal; for example, the dog had to retrieve a specific object. In general the dog could perform most actions of the first type even when the owner was behind a screen (to reduce the effect of other than verbal cues). In contrast, the dog had difficulties in fulfilling the commands if they related to specific objects ('Go and get my keys!'), probably because in this case the dog could not rely on the orientation or some other bodily signals provided by the owner. When testing specifically for understanding names of objects, the dog was just above chance level (but not significantly) in retrieving the commanded object when it was placed together with two other objects. Nevertheless, dogs can be trained to retrieve objects by name (Young 1991).

The problem of whether dogs also learn the names of objects spontaneously during social interaction with humans (just as happens with human children) has recently attracted a lot of attention after one dog (a Border collie) was found to retrieve more than 200 objects by name (Kaminski *et al.* 2004). This dog also showed some evidence of rapid parsing of a novel utterance with a novel object. In these experiments, when commanded to retrieve an object the dog chose a novel item out of three familiar objects if the command referred to an unfamiliar object name (Kaminski *et al.* 2004). Although there is some disagreement on the interpretation of the underlying cognitive processes (see Bloom 2004), this dog was able to recognize that a novel utterance is 'indicating the name' of the novel object in the presence of familiar objects with known 'names'.

Folk belief assumes that dogs are able to eavesdrop, that is, learn the meaning of certain human utterances by listening to verbal interactions between people. Undoubtedly human infants have such ability, because it has been shown that 18-month-old children preferentially associated verbal utterances to those objects that were in the visual focus of the adult(s) during the emission of the sound (Baldwin and Baird 2001). A similar effect was also found in African grey parrots when they were trained by the model–rival method which was devised on the basis of human–human interactions (Pepperberg 1991, 1992). The above idea in relation to the dogs became plausible when McKinley and Young (2003) presented evidence that dogs can also learn the name of an object when they observe two humans repeatedly naming the novel object during conversation. Dogs verified their knowledge by being able to retrieve the commanded objects (out of three) significantly more often than expected by chance. Very likely dogs are relying on the same visual cues utilized by children, which include cues indicating the attention of the human and the manipulation of the object. A more recent demonstration shows, however, that acoustic cues might play only a small role in this phenomenon, and the increased visual interest of humans towards the object might be enough to increase its salience for the dog (Cracknell *et al.* 2008).

8.5 Play

Although complex social play is one of the most striking phenomena of mammalian behavioural development, its adaptive function is still largely a mystery. Thus Coppinger and Smith (1990) developed theories suggesting that play could have been originated from the need to reorganize the behaviour of the mammalian neonate into the adult pattern. Most researchers, however, maintain that the costs involved in play indicate some adaptive function, which could be different according to species and ecology. In social mammals with complex behavioural patterns play could facilitate the establishment of behavioural routines, provide physical and/or mental exercise, and strengthen individual relations (e.g. Bekoff and Byers 1981).

Specific functional considerations gained some support from the finding that in canids the amount of play correlates with the sociality of the species.

In jackals and coyotes, which are considered to be less social, play occurs less frequently than in wolves and dogs (Fox 1975, Bekoff 1974, Feddersen-Petersen 1991). In addition, in coyotes and to some extent in jackals hierarchical relationships develop before the increased playing activity, which suggest that play has only a small role in the establishment of social relationships. In dogs and wolves intensive playing precedes the establishment of social hierarchy, which offers the possibility of developing social ties independent of the subsequent social relationship. However, there are also differences between the two species. First, although adults of both species demonstrate play behaviour, this activity is more pronounced in dogs, and is not only evident in relation to humans but remains a characteristic behaviour in adult dogs. It should also be noted that whether dogs or wolves play more 'in general' depends on the breed used for comparison. For example, Bekoff (1974) reported increased play frequency in beagles compared to wolves, whereas poodles played less than wolves of the same age (Feddersen-Petersen 1991). Second, there are differences in the pattern of play behaviour both in the type of play routines utilized and also in the use of signalling behaviour used to elicit play. Unfortunately there is no comparative study, but wolves and dogs might differ in 'projects' used during play (e.g. in wolves: keep-away, tag, wrestling, king-of-the mountain (Packard 2003); in dogs: chase object, compete for object, object-keep-away, tug-of-war (and more; see Mitchell and Thompson 1991)). Beagles also incorporated sexual behaviour patterns (e.g. mounting, clasping) in play sequences, which was not observed in wolves (Bekoff 1974). In addition, there is some variability in the signals used during play. Fedderson-Petersen (1991) reported that wolves show expressive facial signals, which she defines as 'mimic-play' and which seems to be absent in poodles. In contrast, the beagles studied by Bekoff (1974) used a somewhat wider range of signals for initiating play and were also more successful in eliciting a response from their companion than wolves. Both studies also note that dogs often use barks as play signals, which was not observed in the case of wolves.

Studying the signalling pattern of play, Bekoff (1977) emphasized that some play signals are able to modify the effect ('meaning') of preceding or subsequent actions (*metacommunication*). Observing playing dogs and wolves, Bekoff (1995a) noticed that play bows do not occur at random but are displayed after or before actions (bites) which have the potential to be misinterpreted by the partner.

The fact that dogs play both with humans and with conspecifics offers an interesting possibility of investigating how they decode human behaviour signals (reportedly dogs also play with monkeys, without needing much experience; Bolwig 1962). Rooney et al. (2001) systematically tested the reaction of dogs to various play signals (play bow, lunge, and both actions presented with inviting verbal utterance). Each signal (which was derived from a previous study observing a large number of dog–human games) was effective in inducing play in the dogs. It is interesting to see a parallel here; vocalization on the part of the human had a facilitating effect on play, just as it does in conspecific dog-interactions. This study also provided further evidence that dogs have the ability to rely on a very diverse set of play signals. This seems to be a manifestation of ontogenetic ritualization (Tomasello and Call 1997) when a behavioural action becomes a part of a communicative signal set through the habitual interactions of two individuals. This might also explain why some dogs use barking as a play signal. At early stage of play development barking might just be one expressive behaviour resulting from the excited state of the dog. But later, after repeated playful interactions, the players might learn mutually to use it as a signal. The possibility of ontogenetic ritualization also makes it difficult to investigate whether the visual (bodily) similarity of the play signal in humans and dogs contributes to its effectiveness. (Note the close relationship between ontogenetic ritualization and the particulate principle, p. 178).

It is a recurring assumption in the literature that 'winning' games affects the hierarchical relationship between humans and their dogs (e.g. McBride 1995). Apart from the fact that there are no data supporting this idea (Rooney and Bradshaw 2003), it also goes against the logic of play because, according to what has been noted above, in dogs play signals help to ensure that any harmful action is/should not be taken seriously. In addition, play is characterized by alternation of roles played, and animals avoid interacting with players that are not willing to engage in role changes. However, it is not rare for some playful interactions to turn into serious fights which can affect the relationship. Thus from the point of view of the participants it seems to be more important to keep on signalling playful intent, which lessens the negative influence of these interactions on the relationship. However, there might be differences in dog breeds as they might be restricted in their ability to display play signals.

Unsatisfied with the simplistic description of complex activities during play, Mitchell and Thompson (1991) developed novel behavioural models. Accordingly, play partners usually have two tasks to accomplish during any kind of social play. They have a goal to participate in the interaction by utilizing a specific pattern of behaviour ('project'), but they also aim to contribute to a common goal in order to maintain play activity. Interacting dogs might have an individual preference for engaging in certain play projects, which might be or might not be compatible with the actual project played by the partners. Thus the task of the players is both to indicate preferred projects and also to respect indications by the other for other projects. Play interactions can be extended if players initiate ('suggest') compatible projects (e.g. dog runs, human chases) but each should also be ready either to give up their own project or entice the other in order to engage in its project (Mitchell and Thompson 1991). Observations of dog–human play found that both partners performed enticements or provocation by refusing to continue participation, or self-handicapping, but only humans performed truly manipulative actions (for a developmental aspect see Koda 2001). Thus it seems that both partners recognize not only the common goal of playing but also that either their own goal may be changed or they have to make the other change its goal. Mitchell and Thompson (1991) suggested that play activities of dogs might be described in terms of intentions, which include having a goal/intention to engage in a given project and also recognizing similar goals/intention on the part of the partner. In similar vein Bekoff and Allen (1998) argued that playing offers a natural behavioural

system in which problems regarding intentionality can be investigated. In agonistic situations it would be disadvantageous to reveal future intentions, but collaborative interactions might have selected for ability in representing the other in terms of intentions. Thus playing between dogs, and especially playing with humans, might increase a dog's skills in attending to the behaviour of the other, and even representing it in terms of intentions.

Rooney *et al.* (2000) compared dog–dog and dog–human object play and found that the same dogs were less competitive and more interactive with humans (in contrast to playing among themselves). Dogs offered an object more often to humans and also gave up possession of an object sooner. These differences led the authors to argue that dog–dog play is under different behavioural control from dog–human play. As support for this idea Rooney *et al.* (2000) refer to Biben (1982) who found that social hunters are less competitive during object play. This suggests that the observed difference could be explained by the lack of cooperative hunting among dogs and the possibility of selecting dogs for cooperative hunting with humans (see also Chapter 8.8, p. 196). Although this model fails to account for cooperative hunting abilities in wolves, it seems to indicate that dogs use different mental representations for framing play with conspecifics and humans. This is also underlined by the finding that dog–human play might influence the relationship between the partners (Rooney and Bradshaw 2003).

8.6 Social learning in dogs

Social learning is an efficient method for obtaining information by observing conspecifics. In these cases the key experimental evidence is that naive individuals gain some advantage if they have the chance to observe skilled performers (*demonstrators*) in comparison to others that do not have this experience. Most researchers agree that, depending on the ecology of the species, social learning can offer an advantage over individual learning (see Zentall 2001, Laland 2004). In contrast, there is much disagreement on the underlying cognitive mechanism that controls the process (see Whiten and Ham 1992).

Despite the high number of social species among canids, practically no experimental research has been done on social learning in wild-living species (Ney 1999). For wolves in captivity there is anecdotal evidence for social learning. For example, Frank (1980) suggested that wolves learned to open the door of their cage by observing humans. Interestingly, this author also suggested that domestication might have acted against this capacity in dogs, and this view was reinforced by some earlier negative findings that dogs did not learn aversively conditioned leg flexion by observation faster in comparison to naive conspecifics (Brodgen 1942). However, in contrast one might assume that familiarity with humans (who show complex social learning skills) facilitated this ability in dogs. Spending most of their time in or near human groups could have provided an advantageous scenario for dogs to use human behaviour as a source of information. The issue is also interesting from a representational point of view because information can be obtained by observing either knowledgeable conspecifics or humans.

There is some evidence for social learning from conspecifics which leads to improved performance. Slabbert and Rasa (1997), for instance, demonstrated that young police dog pups left with their mother until 3 months old, and provided with the opportunity to observe the bitch searching for narcotics, displayed a superior performance when learning the same task later in comparison to control pups. In addition pups can also learn from each other: observing a littermate pulling a small cart on a string facilitates the emergence of the same behaviour later (Adler and Adler 1977).

From other experiments we already knew that 'average' companion dogs are not especially good at making detours. Only after 5–6 trials were dogs able to make the shortest (outward) detour around a V-shaped fence in order to get to the target object (food or toy) (Chapter 7.3). In contrast, dogs improved their performance after watching a detouring human demonstrator: observers mastered the task after 2–3 trials (Pongrácz *et al.* 2001). Importantly, dogs were able to rely on this information when they had earlier contrary experience. In one experiment, dogs were allowed to get to the food/toy through an opening in the fence near the tip of the V. If they were

subsequently prevented from choosing this direct route (by closing the opening), the performance of most dogs deteriorated and fell below that showed by naive individuals without any experience. This indicated that the previous experience of a direct route had a strong inhibitory effect on the dog's ability to devise alternative solutions. However, if after such an experience the dogs had the opportunity to observe a demonstrating human they could overcome this bias and rapidly developed the habit of detouring (Pongrácz *et al.* 2003).

Similarly, dogs learn to manipulate a handle mounted on a box more rapidly if they are exposed to a human demonstrator (Kubinyi *et al.* 2003b). The pushing of the handle to the left or to the right released a ball on the opposite site of the box. Dogs observing a human manipulating the handle were more likely to push the handle (with their nose) in comparison to dogs that either witnessed the experimenter touching the top of the box or played with the experimenter in the vicinity of the box. Importantly, dogs also showed a preference for touching the handle if the demonstration did not cause the ball to roll out (and consequently no play followed). This suggests that dogs have a predominant tendency to follow human actions even if the outcome is not clear, which might explain their tendency to develop curious habits (see also below).

There are two further interesting aspects of social learning. If dogs have no experience with the situation they are in, their behaviour is affected by a greater degree by the actions of the demonstrator. Observer dogs that had some experience with walking around the fence learned the detour rapidly, but they did not follow the actual direction (going from the left or the right) of the detour as demonstrated by the experimenter. In contrast, naive dogs without any experience of this task preferentially chose the same direction as walked by the human (Pongrácz *et al.* 2003). Such a preference for copying human actions was also observed in other circumstances. Dogs were inclined to manipulate the handle of a box after human demonstration even if there was no external incentive to do so (Kubinyi *et al.* 2003b).

A recent experiment has provided some evidence that human behaviour action can actually be used by dogs as a cue for selecting functionally similar behaviour on their part. To show such an ability in dogs Topál *et al.* (2006b) adapted the 'Do as I do!' procedure which had previously been used to show imitative abilities in apes (Custance *et al.* 1995, Call 2001) and dolphins (Herman 2002). We trained a skilful assistant dog (who was trained to assist his disabled owner) to perform on command ('Do it!') an action that matched to some extent the action demonstrated by the experimenter (Topál *et al.* 2006b). These matching pairs of actions had been predetermined by the experimenter, and included turning around the body axis, barking, jumping up, jumping over a horizontal rod, putting an object into a container, carrying an object to the owner/parent, and pushing a rod to the floor. Importantly, the context of the demonstrations was always the same and all means for performing any of the actions were possible. Correct performance therefore depended only on the observer's ability to perform the matching behaviour to the demonstrator's action. After 1 month of training the dog was very skilful in performing the matching action, even in test situations (without reward) with another human (to exclude Clever Hans effects). In some follow-up investigations the dog was also shown novel actions that were not part of the test or the training but were to some extent part of the assistant dog's (trained) repertoire (e.g. opening a door). The results of these tests showed that the dog was able to use the rule he learned during the training, and displayed actions which closely matched the demonstrated ones (Figure 8.6).

This type of experimental procedure seems to be a particularly useful tool for finding out what kind of behavioural model is used by dogs when they observe others in mixed-species groups. Because of the anatomical differences there is only a partial overlap between the organization of actions in dogs and humans. For example, for reaching we use our hands, but dogs use either paw or mouth. When watching conspecifics dogs can rely on the species-specific bodily representation which is only partially useful in the case of observing humans. The comparison of the performance to human or dog demonstration could provide some clues in this regard. (Box 8.6).

The tendency to use the behaviour of others as a model for one's own behaviour can be also shown under more subtle conditions which might help to explain the development of habits in dogs. In one experiment Kubinyi *et al.* (2003a) tested whether dogs spontaneously adopt a novel, arbitrary (actually pointless) behaviour. They requested dog owners to change their route after arriving back from walking the dog. Instead of approaching the door of their house by the shortest direct route they were asked to take the dog off leash and make a short circuit, leading away from the door. At the beginning most dogs chose either to follow the owner or wait at the door until the owner returned, but after a period of 180 walks (3–6 months) half of the dogs not only escorted the owner but also overtook them and finished the circuit earlier. One dog, which was observed after these detours had been terminated, maintained the habit of running ahead over a period of 2–3 months. These results indicate that after a certain amount of experience dogs form expectations about human behaviour. This ability to adopt a virtually useless habit and to anticipate the action of the other also contributes to the manifestation of synchronized behavioural interaction between humans and dogs (Kubinyi *et al.* 2003a). At the level of social interaction such anticipation can also be interpreted as a mechanism for reducing conflict between two parties and contributing to effective cooperation.

8.7 Social influence

Recent theories on mimetic processes prefer to discriminate social influence from social learning (Whiten and Ham 1992), which was not the case in the older literature. Social influence usually does not involve any learning, and the similarity in behaviour is the result of other processes such as changes in motivation when being in a group. For example, Compton and Scott (1971) found that young dogs give more distress calls and eat more if they are together, and increased food consumption in a social situation can also be found after satiation (Ross and Ross 1949). To describe similar parallel action in social animals, Scott (1945) introduced the term *allelomimetic behaviour* which in many respects seems to be synonymous with social influence. However, Scott also formulated a more specific hypothesis by suggesting that allelomimetic behaviour can result in mutual adjustment of

(a)

(b)

(c)

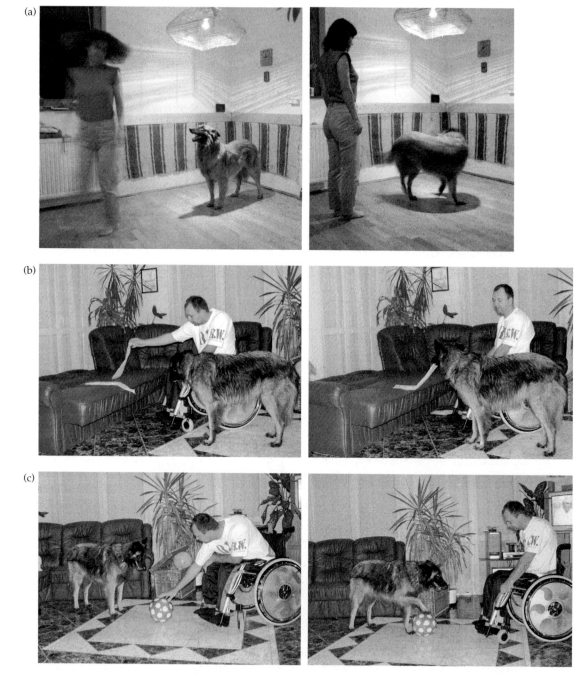

Figure 8.6 Dogs can learn to execute a functionally similar action on the basis of the action shown by a human demonstrator. First, the owner/experimenter shows an action followed by the command 'Do as I do!'. (a) Dino executes the 'Turn around' action that was also part of his training. (b–c) Philip is shown novel actions: 'Pull the sock from the couch!' and 'Put your foot on the ball!' (see also Topál *et al.* 2006b)

Box 8.6 Social learning: what is learned?

One of the most intriguing questions of social learning is what kind of novel information is obtained by an observer. In the past, it was thought that animals learn components of their behaviour by 'imitation' but it has turned out that learning about motor aspects of behaviour is rare, and most often the observer learns about a certain relationship between a behavioural pattern and the environment (Whiten and Ham 1992). Importantly, the effect of observation depends also on the actual experience of the observer.

So far dogs have been tested mainly with human demonstrators, but they can also learn by observing skilled conspecifics (Pongrácz *et al.* 2003). The human demonstration provides an interesting problem because many behavioural actions of the two species are executed differently. Odendaal (1996) notes that a digging human (gardener) might facilitate digging behaviour in the dog observer. Although it might sound trivial, it is not clear what aspect of the human behaviour releases similar action on the part of the dog. The standing human who moves the soil with a tool (spade) provides few visual features resembling a digging dog. Thus digging behaviour in dogs could be facilitated by the smell of the fresh soil, or simply by observing the soil

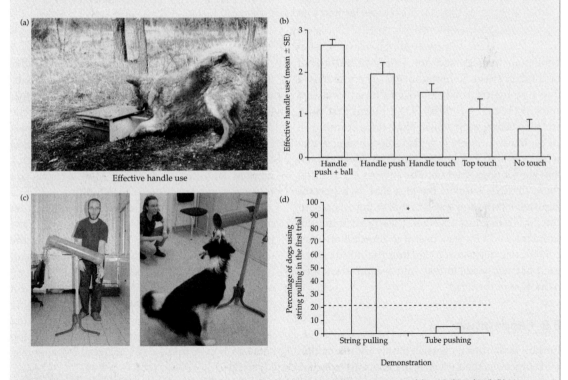

Figures to Box 8.6 (a) The dog is about to maniplate the handle after demonstration by an experimenter (photo: Enikö Kubinyi). (b) Mean effective handle-pushing action during the three trials. A dog received a score of 1 if it got the ball by using the handle in a trial (max. score = 3). The box released the ball in the 'handle push + ball' condition (experimenter pushes handle and ball comes out of box), followed by short play. In all other conditions (excluding the control = no touch) the experimenter touched the box, but no ball was released (for more details see Kubinyi *et al.* 2003b). (c) The string is pulled down on one side to release the ball from the tube (left). Next the dog must try to retrieve the ball. (d) Without a demonstration, only 22% of dogs used string pulling (horizontal line = chance level). After a demonstration of string pulling, 50% of dogs pulled the string. However, if dogs were shown an alternate action (i.e. pushing one end of a tube down by hand) fewer dogs pulled the string (8%). Both values differ from the baseline performance and from each other (* = significant difference) (based on Bánhegyi 2005).

continues

Box 8.6 *continued*

being moved. Finally, dogs could recognize the spade as an extension of the human arm and relate the contact between the soil and the spade to a corresponding action of their foreleg, paw, and the soil. Note that social learning theorists would invoke different learning mechanisms to account for each of these possibilities.

Similar problems can be raised in the case when dogs observe a human using a hand to manipulate a handle (Kubinyi *et al.* 2003b). The action clearly draws the dogs' attention to the handle, but they move the handle using their muzzle not their paw (which would be their anatomically corresponding body part). This is either because dogs have a general preference for using their muzzle in such situations (for pushing small objects), or they have habitually learned that

human hand actions are best copied by muzzle actions on their part (a, b).

A recent experiment (Bánhegyi 2005) provided evidence that the behaviour of the demonstrator affects which kind of behavioural action is selected by the observer dog. Dogs could witness the experimenter releasing a ball from a opaque tube by either pushing it down at the end or by pulling a string (c) Control observations established that without demonstration dogs prefer to push the tube down by standing on their hind legs. Not surprisingly, dogs witnessing the pushing action mostly chose a similar action in order to release the hidden ball, but the rate of this behaviour decreased (and the frequency of pulling increased) if the human demonstrated pulling (d).

behaviour in dogs and not simply in an overall facilitatory effect. To reveal such an effect he staged dog running trials where dogs received food as a reward (Vogel *et al.* 1950). They found that dogs running alone were slower than dogs running in pairs; however, there was also some evidence that when running in pairs faster dogs slowed down and slower dogs sped up. The authors argued that these findings support the idea that each partner adjusted its running speed to that of the other, with the aim of running together. Clearly such mutual mimicry could be very useful in hunting or other cooperative actions (e.g. leading the blind) when each partner needs to take into account the speed of motion of the other.

8.8 Cooperation

Certain goals can be achieved only by interaction with others in the group. Some goals are specific, such as hunting for large prey which would not be possible on an individual basis. At other times goals are more general, such as when a dog 'wants' to play (see above). In both cases the interacting animals can reach the goal only if they pay some attention to the behaviour of the other and take this into account when choosing their own actions. In this sense collaborative activity can be said to lead to the construction of a joint actions.

Although there is limited evidence, the popular literature often assumes complex cooperative hunting ability in wolves. A recent short review on this topic has raised some doubts about this assumption. Asking wolf experts about various forms of complex cooperative hunting pattern in wolves, Peterson and Ciucci (2003) present an ambiguous picture with a marked division of opinion. Most experts are inclined to interpret cooperative hunting in wolves as simple group chases. This does not deny that more complex interactions can occasionally take place, but these could be also explained by the special circumstances and might not be the result of some sort of joint planning (Peters 1978). There are arguments that in most cases wolves do not spend enough time together in the pack practising and learning cooperative actions. This might not apply to founding parents, which could develop such skills over many years of being together (e.g. Mech 1995).

Earlier arguments based on wolf behaviour assumed that the ability to engage in complex cooperative actions was one of the key features that was utilized after domestication. Thus the ability of dogs to hunt with humans or help in herding sheep is based on the cooperative behaviour evolved in wolves.

It is interesting that although dogs have been routinely used in cooperative tasks with humans for many thousands of years, we know very little about this ability. Dogs have often been utilized effectively in roles which were partially incompatible with human behaviour. This includes 'jobs' such as herding, fighting and protecting, or pulling and transporting. More recently dogs have been used as guides for blind people or to assist people with disabilities. Finally, dogs also cooperate in tasks which are designed to provide fun and physical exercise for humans (e.g. agility competitions, dog dancing). The performance in these tasks depends crucially on some form of training, and many regard the dogs' cooperative achievements as being nothing more than enhanced learning performance explicable by simple associative rules (but see Johnston 1997). Such views are also supported by the lack of observations of cooperative behaviour in feral dogs (Boitani and Ciucci 1995). Without denying the role of learning, one may suppose that selection in a human community (where cooperative behaviours play an important role) could also have contributed to an enhanced ability for cooperation which is revealed only when dogs have the chance to interact with humans.

The interaction between blind people and their guide dogs has been used as a behavioural model for inter-specific cooperative interactions (Naderi *et al.* 2001). Experienced dyads were observed when negotiating a novel obstacle course, and the goal was to determine the ratio of actions initialized by either the dog or the human. Although there was wide variation among the dyads, on average dogs and humans each initialized about half of the actions but at an individual level the role of the initiator changed continuously. In most cases neither party initiated more than 2–3 actions in a row, and it was most common to relinquish the initialization after one action. This suggests that there is flexibility in taking the leader's role during a cooperative action. Importantly, the partners have different roles in the task, because the blind person might know the direction of their walk but the dog has the visual information about the actual environment. Thus the leader's role changes over because there is a need to perform a different kind of action. However, each partner has to make his own decision on whether to take the lead or allow the other to do so. It seems that in contrast to most cases

when cooperative animals execute similar actions (*parallel cooperation*), dogs participate in a cooperative interaction that is based on complementation. Reynolds (1993) argued that the complementary nature of cooperative interactions is the hallmark of joint actions in humans, and played an important role in our behavioural evolution.

8.9 Social competence

In recent years there has been much debate about the cognitive aspects of mental processes that may underlie complex social behaviour. The problem in this discussion was that critics of the anthropomorphic approach (e.g. Heyes 1993) successfully forced their opponents to produce mechanistic explanations of behaviour and this distracted the research community from realizing how little we know about the behavioural skills used in social interactions. A few recent research papers seem to reflect this feeling by arguing for a more detailed understanding of social behavioural interactions before embarking on the mechanistic details of cognitive underpinning. For example, Barrett and Henzi (2005) argue that instead of focusing on a narrow aspect of social behaviour (e.g. deception) researchers should take a wider perspective and search for many alternative ways (behavioural tactics) by means of which animals are able to navigate in the social network. Based on the assumption that the nature of social behaviour is under selection, species should vary in their ability to display various forms of social skills (see also Johnston 1997).

In their review Barrett and Henzi (2005) introduced the term *social expedience* which is defined as the ability to 'select whatever tactic is necessary to solve an immediate problem'. Pursuing this idea, we might utilize the term *social competence* to refer to the complete set of social skills which characterize a species. At the behavioural level we recognize different behavioural actions that can be organized in a functional complex during interaction with group mates. In this sense such functional units corresponding to social competence can be regarded as a specific 'tool set' (see also Emery and Clayton 2004) (Box 8.7).

Anyone can experience the difference between social competence in wolves and dogs by sharing

Box 8.7 One case for social competence: the pedagogy hypothesis

Cognitive psychologists describe teacher–learner interactions in humans as pedagogical knowledge transfer which can be defined as the explicit manifestation of generalizable knowledge by an individual (the teacher) and interpretation of this manifestation in terms of knowledge content by another individual (the learner) (Gergely and Csibra 2006). Teaching is often described as human-specific behaviour and is regarded as a primary, independent, and evolutionarily earlier adaptation than many of our other complex cognitive skills (e.g. language). The human pedagogy model accounts for the effective transfer of complex information by parents that often surpasses the cognitive skills of the babies (Gergely and Csibra 2006).

We have hypothesized that this model might be useful to account for similar interactions between dogs and humans. The model has three important components:

• *Ostension* can be defined as communicating about future communicative action. Dogs show preference for establishing eye contact with humans and recognize the cues showing the human's communicative intent (e.g. using eye contact and directed talk provided by humans to infer whether they are 'addressed' in a particular situation; Pongrácz *et al.* 2004).

• *Reference* can be defined as the willingness to follow another's directing cues (pointing gesture, gazing direction) and to identify these gestures as referential cues (referring to objects/subjects). Dogs are proficient in utilizing gaze-shifts (Soproni *et al.* 2001, Virányi *et al.* 2004) in communicative interactions (when they try to identify the addressee of the communicative actions).

• *Relevance* refers to the 'expectancy' of the learner that the information provided is relevant (and novel) and there is no need for further understanding. Dogs often rely on human behaviour 'blindly', which seems to support this notion. For example, in a two-way choice test dogs repeatedly chose an unbaited location on the basis of human cueing (Szetei *et al.* 2003).

It seems that in the learning context, establishing eye contact, addressing the learner by name (ostensive communicative cues), then shifting eye-gaze or pointing to the object to be manipulated (referential cues) accompanied by verbal attention-getting ('Look! I'll show you something . . .') triggers and facilitates behaviour on the part of the dog which corresponds to the demonstrated action. Dogs react in such situations as if a human demonstrator (a tutor) imposed an expected behaviour action on them.

(a)   (b)

Figure to Box 8.7 Eye contact and motor synchronization form the basis of joint attention which is essential in complex cooperative tasks such as dog dancing.

their life with a member of either species. But more seriously, the basic idea is that the dog's version of social competence is more similar to ours, resulting in smoother coexistence. This means that dogs evolved a social tool set that only partially overlaps with that present in wolves, and shares many features with ours. Differences in social competence might come about by the 'invention' of novel forms of behaviours (and abilities) or more likely by the application of behaviours in different ways in different social contexts. The use of barking or the ability to display preference towards objects used or indicated by interacting humans (see above) might provide examples for the former case. Playing might present a case for the latter.

Three examples underline how little we know about dog social competence, and that assumed functional compatibility does not necessarily reflect similarity in the behavioural control. Dogs are often described as being 'jealous' when they try to interfere with an owner who is showing friendly intentions towards another dog. Such dogs often display protest behaviour (e.g. barking), try to redirect the interaction (initialize interaction with the owner), or behave agonistically (attack the other dog). Thus 'jealousy' is an important behavioural tool in maintaining contact with preferred companions. Although this behavioural pattern has not been studied in dogs, we have observed it a lot in wolves during interactions with humans. At the moment it is only a bold assumption that 'jealousy', as a social tool, in humans and dogs (and wolves) is controlled by the same mechanisms.

Similarly, little is known about 'guilty behaviour' but here the situation is probably different. In humans guilt reflects an understanding that one has violated some social rules. Dogs are often observed acting 'guiltily' after doing something wrong (e.g. Lorenz 1954) but it is not clear whether this reflects only fear or expectation of some sort of punishment, or whether the dog is able to comprehend that some rule has been violated. De Waal (1996) supports the former interpretation because one study found that a dog (husky) behaved guiltily independently of whether he or his owner made a mess in the living room (Vollmer 1977). However, this insensitivity to the doer does not provide decisive evidence that the dog is unable to

distinguish between the two situations. The ability of dogs to adopt social rules offers the background for the emergence of guilty behaviour if they recognize the discrepancy between their actual action and the internalized social rule. Nevertheless the jury is still out on whether guilty behaviour in dogs shares features with that in humans.

Another interesting aspect of human behaviour is the notion of 'expertise', which is often described as deliberate practice to improve performance. Helton (2005) argues that dogs meet all the criteria for expertise, and this is indicated not only by the fact that dogs participate in training but also that many individuals refuse to do so. In addition, researchers see a relationship between playing and practice (Bekoff and Byers 1981); that is, play may be an evolutionary invention for practising certain behavioural skills. Thus increased affinity for play in dogs and frequent play between dogs and humans may provide the participants with development of expertise. Again it is not clear whether such expertise in dogs is self-motivated (as in humans), or if it is only enforced by humans using various means for motivation.

Most social situations are very complex. Thus for an organism showing human-like social competence we would expect fine-tuned ('sensitive') behaviour and not an automated set of responses. Thus sensitivity to changes in the social situation can be revealing. For example, imagine a communicative interaction between a human and a dog in which the animal is commanded to retrieve an object ('Get the ball!'). Now imagine the same situation in which the human is replaced by a loudspeaker. What should one expect from a dog (or even a child) with social competence? Is the 'correct' response to retrieve or not to retrieve? People often assume inherently that the dog should fulfil the command because they assume people would do the same. But actually this might not be the case, because from the dogs' (or child's) point of view the situation is quite different. The vocalization sounds strange (the loudspeaker does not reproduce the human vocalization faithfully), it does not come from the same source (the loudspeaker is on the floor), and also there is nobody present to whom the ball could be carried. Given these (and other) differences, the 'normal' response would be not to obey!

When dogs get such commands in natural situations humans usually direct the utterance to them (facial cues signal the look at the dog), the action is usually preceded by attention-getting signals (e.g. the name of the dog), and the human is oriented to the visual space in which the action should be carried out. If the dog seems to be hesitating, humans often repeat the command or give other encouraging vocalizations. Thus it is no wonder that dogs, which are used to interacting with humans in such a behaviourally rich social situation, do not perform well if all these stimuli are suddenly removed. In contrast, 'over-trained' dogs could act blindly (just as humans might) and give the impression of acting automatically.

Recently some experiments have obtained data that show the sensitivity of dogs to such changes in the communicative situation. For example, Pongrácz *et al.* (2003) tested the response of dogs to human visual and verbal signals by projecting the interacting person on a screen. The performance of dogs was reliably good in tasks in which mainly visual signals (pointing) were used. However, if the owner gave the commands verbally, dogs obeyed them to a lesser extent when the human was not present in the room but rather as an image projected on the wall (owners could watch their dogs through a video link). Even this modest performance diminished when no projected image was available, and the commands were only provided through a loudspeaker. Fukuzawa *et al.* (2005) also found that the performance of dogs declined when humans wore sunglasses, or sat, or when the distance between dog and human was increased. Such experiments are very important because they point to key aspects of social interactions which are attended to by the dogs and which might be different from

those that play a role between interacting humans. However, even this might not be necessary, because so far we do not know how infants or young children would behave in comparable situations.

8.10 Conclusions for the future

In comparison to the wolf, domestication led to marked changes in the social behaviour of dogs. Most of these social abilities come to light when living with humans, but there is also evidence that environmental effects alone cannot explain differences in dog behaviour. The contrast to social behaviour in stray and feral dogs suggests that the dog has a very plastic social phenotype which fits into different social environments. Thus dog behaviour as we observe it in dogs with whom we share our lives is dependent both on genetic changes that took place in the course of domestication and also on the social environment in which these changes are manifested.

Until more details of dog–human interaction are known the idea of human-compatible social competence remains a hypothesis. Nevertheless the existence of many potential elements of this tool set has come to light, including attachment and sensitivity to behaviour cues in communicative interactions and learning situations.

Further reading

Tomasello and Call (1997) and Gomez (2004) provide extensive reviews of social cognition in primates which in many respects could be a model for similar endeavours in the dog. Cheney and Seyfarth (1990) give a similar account for monkeys and apes.

CHAPTER 9

Development of behaviour

9.1 Introduction

Dog breeding and training would have been impossible if shepherds and hunters had not acquired some understanding of dog behavioural development. In order to produce a skilful four-legged companion to assist in their work, people worked out methods and procedures, some of which were grounded in the biology of development whereas others were (and are) possibly mythical.

One of the first large-scale studies on dog development was published by Menzel (1936), who reported on behavioural observations collected over a period of 16 years on more than 1000 pups. Although this study did not provide quantitative analysis, it raised most of the main questions on dog behavioural development which have subsequently kept researchers busy. Menzel (1936) recognized that dog development can be divided into periods or stages, and there is close agreement between these subdivisions of dog development and those described later by Scott and Fuller (1965). Interestingly, both publications suggest parallels between periods of dog and human development, although these now seem somewhat far-fetched. Menzel (1936) also stressed the importance of the environment in the development of the offspring. He presented a detailed description of the emerging attraction of dog pups towards humans, and he also noted that with increasing age young dogs became more wary of strangers. Without presenting much evidence he argued that the behaviour of an adult dog can be predicted on the basis of early observation of the puppy. The validity of this idea has become one of the most problematic questions of dog behaviour (see Box 9.5).

These descriptions of developmental periods and other early writings of some ethologists gave the impression that their authors believed in a relatively strong genetic determination of juvenile behaviour. Not surprisingly, such suggestions led to heated debates. Bateson (1981) criticized Scott, whose theory of 'critical periods' relies exclusively on endogenous rules (cited from Scott 1992). Although careful reading of the original papers by Scott and his colleagues shows clearly that this is a misinterpretation of their work, the graphical portrayal of behavioural development provided in the original texts (e.g. Scott and Fuller 1965) is certainly open to such interpretations. Indeed, the popular and dog-breeding literature was quick to interpret Scott's results in the Batesonian way, and this has determined up to now how pups are socialized (and separated from the litter).

9.2 What are developmental 'periods'?

The development of an individual is often described as being made up of a sequence of events from the fertilization of the egg until adulthood. This idea probably has its origin in anatomy, as the developmental stages of the embryo can be associated with changes in morphological features. Although even these developmental changes are not independent of environmental influences (e.g. temperature), they seem to be under relatively well-coordinated genetic control. In contrast, in the behavioural literature development is always portrayed as an interplay between genetic components and environmental influences (*epigenesis*) (e.g. Caro and Bateson 1986). Thus the concept of fixed developmental periods is problematic if it is applied to more complex systems such as behaviour. This is especially true in the case of the dog, where there is a long tradition of using the developmental stage as a reference system for explaining earlier and later behaviour.

Developmental periods are discussed within either a functional or a mechanistic framework. The *functional approach* recognizes that the developing animal is a form of life adapted to its current environment (Coppinger and Smith 1990). Thus dog development can be interpreted in terms of changes in the physical, ecological, and social environment. This environment can be especially complex in social species, where the offspring can interact with group members other than the mother, as is the case in *Canis*. Investigating the behaviour of the offspring in relation to its developmental environment can reveal important aspects of adaptation. However, this approach may be problematic if there is little knowledge of the relevant developmental environment and the main ecological variables involved. As we will see, the developmental environment of the wolf is often used to 'explain' the supposed periods in dog development. However, without much data in this regard, such explanations are no more than possible narratives.

The second possibility is to describe development as a process during which the capabilities of the animal's perceptual and behavioural systems increase over time. This *mechanistic approach* investigates how perceptual abilities emerge and improve, or when and at what rate physiological and behavioural mechanisms (including both vegetative and neural mechanisms) converge to adult functioning. Such investigations often assume that early abilities somehow predict later performance.

The emergence of novel features of the organism is most often used to indicate certain developmental changes or periods. However, it is a mistake to view development simply as a sequence of events. It should be realized that development involves many parallel changes at several different organizational levels in the young animal, and many of these changes are not only sequential, but more importantly, conditional, and they often occur at different levels of biological organization. It is often the case that events at the behavioural level presuppose the completion of preceding events at a different (e.g. neural) level. Thus Fox (1965) argued that neural developmental periods precede related behavioural periods, suggesting that certain behavioural abilities emerge only if the developing neural system reaches a certain point of maturity (Box 9.1). This is most obvious in the relationship between perceptual and behavioural abilities that emerge during development. Although pups' eyes open at around 10–14 days after birth, it takes a long time (several weeks) before they approach the

Box 9.1 Parallel stages in development

Development consists of a number of parallel processes that are realized at different levels of biological organization. If the processes are arranged along an absolute scale, it can be observed that the start and end of a period do not correspond across these levels. Changes at one level depend on preceding changes at another level. For example, neural maturation in the form of emerging and disappearing reflexes seems to precede changes in the animal's overt behavioural abilities (Fox 1965). Important changes occur in brain development when its

growth slows down around 1 month of age (Fox 1965, Arant and Gooch 1982).

It may be useful to fit dog development into the general framework of wolf development in order to identify the targets of selection, even if at present there is a lack of such data. It is possible that wolves have a sensitive period for olfactory learning of conspecific stimuli during the neonatal period, which may not be present in dogs. The extension of the (second) sensitive period in dogs is probably the result of selective processes.

continues

Box 9.1 *continued*

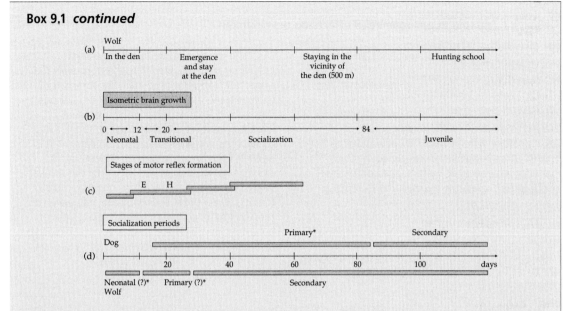

Figure to Box 9.1 (a) Changes in the physical and social environment of the developing wolf (based on Mech 1970 and Packard 2003). (b) The four-stage model of Scott and Fuller (1965). (c) Changes in neural development in the dog puppy. Fox (1965) recognized four distinct stages in the development of motor coordination based on the emergence and discontinuance of reflexive motor reflexes. (d) Described and hypothetical sensitive periods in wolves and dogs. Domestication has probably changed the structure and relationship among these periods in the dog (days are given only for orientation) (E = eye opening; H = ear canal opening; * = sensitive period).

visual abilities of adults. This happens not only because the neural system is not ready for processing visual information but because, in order to function well, a large amount of visual experience is needed which can be gathered only by extensive exposure to the environment.

Although there was a conceptual advantage in dividing development into periods, the complex nature of interaction between endogenous and external events provoked the development of other models. Chalmers (1987) formulated behavioural development in terms of directing and stopping rules which can be induced either endogenously or externally. *Directing rules* describe how the behaviour emerges and increases in frequency, and *stopping rules* refer to the termination of certain behaviour periods in development. This framework offers a possibility of investigating whether the emergence of behaviour patterns such as sucking, play-fighting, or playful mounting is under endogenous and/or external control and whether

their presence or absence in later behaviour depends on endogenous (e.g. maturation) or external effects (e.g. behaviour of the mother). Caro and Bateson (1986) found it useful to provide a summary of the types of events that influence behavioural development. This view differentiates between canalizing, facilitating, maintaining, enabling, and initializing effects (Box 9.2).

These models also help to draw our attention to other, often neglected, problems in behavioural development. First, if there is a conditional relationship between two developmental events it seems logical to suppose that if the first event is late, this will also delay the subsequent event that is dependent on it. Thus, if the eye opening of some dogs occurs later, it might be plausible to suppose that their visual abilities will also be delayed. Second, if such differences in timing seem to have a genetic background (e.g. in the case of breeds) then this should be also taken into account when one compares breeds (Box 9.3). Thus such

Box 9.2 The role of environmental effects on development

Environmental effects can influence the development of different patterns of behaviour in different ways. In providing a general overview, Caro and Bateson (1986) presented a simple schema which seems to be a useful framework to apply in the case of the dog.

Canalizing effects result in decreasing the differences between individuals; such effects, usually referred to as buffers, ensure that the individual obtains the necessary skills under a wide range of environmental conditions. MacDonald and Ginsburg (1981) found that young wolves are able to develop typical social behaviour even if they are isolated from conspecifics for various periods. *Facilitating* events

(A) result in earlier emergence of certain behaviour patterns. Providing the pups with the opportunity to hunt or to learn by observation facilitates the emergence of such skills, although these could also be obtained by individual learning (Slabbert and Odendaal 1999). In some cases, constant environmental stimulation (A) is needed to *maintain* behaviour. A dog-specific example here is the genital licking by the mother which is necessary to stimulate urination during the first 3–4 weeks. Environmental influence can *predispose* the animal to take a certain path of behaviour development. Thus, neonatal exposure to humans (A) enables wolves to form intensive social relationships (B) with humans.

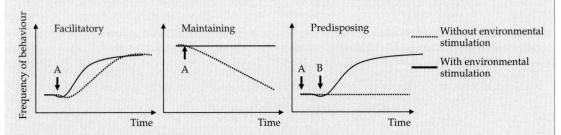

Figure to Box 9.2 A schematic framework for discussion of environmental effect on the developing organism (redrawn after Caro and Bateson 1986).

Box 9.3 Comparative development in dogs

Very few studies have compared behavioural development in dogs between breeds or with the wolf. Feddersen-Petersen (2001a) observed the development of different dog breeds (husky, German shepherd, Labrador retriever, Giant poodle) and the wolf during the first 12 weeks of their life. She observed the first emergence of certain behaviours which belonged to different classes of action (e.g. orientation, comfort, locomotion, etc.). Although her observations are based on a small sample size, they can be used the make some general remarks and hypotheses (see also box 5.6):

• There seems to be some variation in the emergence of behaviour. Although the analysis does not provide data on individual variation within a breed, even so it is striking that in many cases there is a 1 week difference between the early and late developing breeds.
• In most cases the order of the breeds and the wolf is similar across different patterns of behaviour. Huskies seem to develop the fastest and Giant poodles and Labrador retrievers the slowest. However, there is also some variation in the case of some behaviour patterns. German shepherds develop at a very similar rate to wolves

continues

Box 9.3 *continued*

apart from when they show an avoidance response. It seems that morphological similarity to the wolf does not always determine behavioural similarity.

• Feddersen-Petersen's data provide little support for the idea that there is a general pattern of neoteny (slower development) in dogs, because if this were so wolves would show the fastest pace of development. Interestingly, huskies show many behaviour patterns much earlier than wolves (*predisplacement*). She argues that this may reflect artificial selection for superior running abilities. This hypothesis would be supported if other 'running' dogs

(e.g. hounds) also showed a similar facilitated development.

• Breed-specific variation in development cautions against concluding that there is a general time-independent pattern of dog development. If points of behavioural development could be compared between breeds, this would offer a possibility for examining their relative timing. Similarly, these findings suggest that the usual timing of puppy tests at 6 and/or 8 weeks of age may not apply for all breeds. Thus puppy tests should be adapted to the breed in question (Box 9.5).

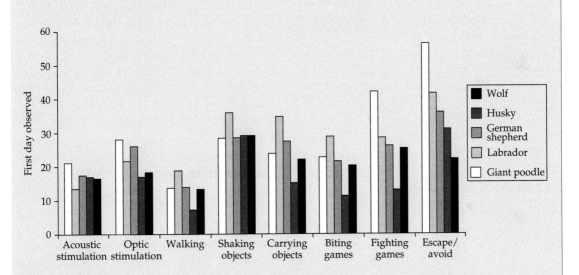

Figure to Box 9.3 The first day of emergence of various behavioural patterns in different wolf-sized breeds. Lower columns in comparison to the wolf values indicate earlier emergence (predisplacement), higher columns refer to later emergence (postdisplacement) (based on data from Feddersen-Petersen 2001a).

comparisons should not necessarily be made along the same absolute time scale but perhaps in relation to specific endogenous or environmental events. Third, the determination of the starting points of developmental periods is often more easily defined than the end points. One explanation for this could be that the termination of the period could be more dependent on the particular environment, and also there are often alternative or

supplemental developmental mechanisms which widen the time window on an individual basis.

9.3 Rethinking developmental periods in dogs

The views of Scott and Fuller (1965) on dog development have had an important influence on how dogs are raised or bred around the world.

Their four-stage model provides the basis for all texts published to date on this aspect of dog behaviour. Even researchers studying the wolf refer to the developmental stages of the dog (e.g. Mech 1970, Packard 2003). This is especially interesting because it is well known that selection has affected dog development to a great extent. Thus there is little reason to suppose that the behavioural development of the dog corresponds to that observed in the wolf. On the contrary, we should assume that dog development has been modified from that displayed by their ancestors. This is true not only for the development of the puppy but also for the developmental environment in which it grows.

In this section we reproduce Scott and Fuller's original framework and identify possible functional explanations based on the developmental environment (ecological and social) of wolves. However, it should be pointed out that the lack of research in this area makes any such parallels questionable. The dates used in the Scott and Fuller model are mainly based on the development of the beagle, and one might doubt whether this breed is a good representative of dogs or provides a useful comparison with the wolf.

9.3.1 Neonatal period (days 0–12)

Wolf cubs spend this period in the darkness of the den dug by their mother a few weeks before their birth (Packard 2003). The nest cavity is usually 2–3 m from the surface and provides a more-or-less stable physical environment. This makes it possible that delay in the development of proper thermoregulation of the cubs does not have an adverse effect. The cubs' perception of their environment is restricted to tactile and olfactory stimuli. During this time only the mother and the siblings are the source of physical interactions and these include stimulation of the tactile receptors around the mouth (suckling) and on the body (cubs 'wrestling' for position in the nest or at the nipple; mother licking for cleaning and elimination), and olfactory stimulation by learning about species-specific odours. Importantly, the mother only rarely leaves the den because she is fed by the male.

Human selection interfered with this system at two points. Dogs are usually provided with an artificial den. This altered the developmental environment of the pups because these 'dens' are usually well lit and open, so that there is a chance that pups are exposed to other (social) stimuli during the neonatal period. Human provisioning of the mother also made the usual contribution of the male in raising his young unnecessary. The lack of selection for 'good' provisioning by males resulted in a markedly decreased paternal behaviour in dogs. Behavioural observations of feral dogs show that human interference destroyed many aspects of species-specific reproductive behaviour in dogs. Feral dog females rear their young apart from the pack (Daniels and Bekoff 1989) and, according to Boitani and Ciucci (1995), one main reason for high infant mortality in feral dogs is that the mother is not able to choose an appropriate nest site for her offspring and leaves them alone too often when looking for food (because the male dog does not provide food). Interestingly, even in human care dog mothers decrease the amount of time they spend with the pups at a very early stage (Malm and Jensen 1997).

9.3.2 Transition period (days 13–21)

Wolf cubs spend this period in the den too, mainly with their siblings and mother. Their motor abilities develop slowly and their exploratory behaviour is restricted to the immediate area underground. This period is characterized by increasing perceptual abilities. It starts with the opening of the eyes and ends with the opening of the ear canals.

Interestingly, there is a large variation in the timing of both eye and ear opening which, at least at the level of the breed, seem to be independent. According to Scott and Fuller (1965), Cocker spaniels open their eyes by day 14 but only 11% same-aged Fox terriers have their eyes open. In hearing, by contrast, the opposite pattern emerges. Here, spaniels seem to be a bit behind as, at this stage, only 61% of pups showed a startle response to sudden sound (the first indication of some hearing function), whereas nearly all the terriers respond to a startle in the same test. Thus, using eye and ear function as indicator dates, there is a difference in the duration of this period between some dog breeds. It lasts for only a few days in Fox terriers and much longer than a week in Cocker spaniels. The delay in Cocker spaniels might be attributed to their drooping ears, because they might need more time to 'learn' hearing.

Direct stimulation between mother and pups decreases in parallel with a decrease in the neonatal behaviour patterns, and motor skills for certain communicative signalling, such as tail wagging, emerge. Puppies slowly gain the capacity to change their behaviour following repeated experience to positive or negative aspects of the environment. Scott and Fuller (1965) mention evidence that in dog pups an operant response to food emerges around day 15, and a few days later they will show similar motor learning to aversive stimulation.

9.3.3 Socialization period (days 22–84)

According to observation of wild wolves, the cubs emerge from the den around the age of 3 weeks (Mech 1970, Packard 2003). This is a major change in their developmental environment, because the cubs are now exposed to novel perceptual stimulation of various sorts, including visual and auditory, and they have the chance to improve their motor skills, partly through interaction with members of their social group including both same-aged siblings and older juveniles from previous years. The socialization period corresponds to the sensitive period (see below) for learning about the social environment.

The role of learning about their companions and the ways of social interactions was investigated by isolation experiments (Fox 1971, Ginsburg 1975, MacDonald and Ginsburg 1981). This work revealed that the perceptual and referential systems are under a different type of control from the action systems. In the case of the latter some sort of genetic components play a major role, so that dogs and wolves are able to display complex motor acts (e.g. communicative signals) without much experience (McLeod 1996, McLeod and Fentress 1997). In contrast, the recognition of the partner as a social companion and response to behavioural signals shown by the partner is experience-dependent. Dogs raised with cats (Fox 1971) needed some

experience to accept other dogs (or their own mirror image) as conspecifics, and had to undergo a period during which they learned to recognize the motivation behind certain social signals, and select the appropriate behavioural action on their part (see also Ginsburg 1975). For example, such experience is needed in order to show the normal pattern of submissive behaviour. Although the exact mechanisms underlying this phenomenon are not clear (and have not been studied), it is conceivable that by this period pups have developed effective perceptual systems that are able to process complex visual and auditory cues. The social interactions also help them to learn how to control motor behaviour. For example, pups acquire the motor skill for biting behaviour flexibly and less painfully (*bite inhibition*) in a social context.

Wolf cubs spend this time around or near the den, but in some cases they will be moved to other dens and later they spend most of their time at rendezvous sites, which means that they get used to a changing physical environment, that is, however, buffered by the stability of their constant social environment. During the first 3 weeks the den provided a physically fixed location as the centre of their world, but now the pups learn to centre their activities on a more dynamic point represented by their family.

Wolf cubs are weaned at around 8–10 weeks of age (Mech 1970, Packard *et al.* 1992), after which they become increasingly dependent on food provided by the parents and older siblings. Importantly, especially early on, cubs have to obtain their share of food by actively begging from the others, as licking at the mouth corner automatically elicits the regurgitation of food by the adult. Cubs rapidly learn to use the food carried home in the stomach of the others, and provocation of regurgitation diminishes only after successful hunts when older wolves carry home uneaten meat. This provides experience of food sharing and competition for food; situations where social hierarchical relationships emerge. Packard *et al.* (1992) found relatively little food-related aggression between adults and cubs in wolves, probably because mothers were able to direct the interest of the offspring from milk to other alternative sources of food (regurgitated food or food carried back after a hunt). Importantly, these interactions provide a social milieu in which the cubs learn.

The situation in dogs is more complicated. In the case of feral dogs the absence of the father and other helpers increases the burden on the female, which might lead to more competition among the pups. In dogs living with humans, the latter might provide additional sources of food when the female decreases lactation frequency (and may not regurgitate), but this situation lacks most of the original social components. Regular interaction of human carers with pups at these times could prove to be important in the process of socialization.

Researchers discriminate between primary and secondary socialization periods, but the exact meaning of these terms is often not clear. Scott and Fuller (1965) made this distinction on the basis of differences in the mechanisms involved (see also Freedman et al. 1961). They argued that primary socialization takes place during an 'imprinting-like' sensitive phase (see below), when the animal learns very rapidly during a short exposure, and that the learning process depends only in part on external incentives (e.g. food). Although never stated explicitly, in their sense secondary socialization refers to processes that are based on various forms of associative learning. This secondary socialization is analogous to taming, when 'wild' animals are familiarized with humans and undergo various forms of learning. Thus, in the framework of Scott and Fuller (1965), both conspecifics (dogs) and humans can bring about primary socialization if dogs are exposed to them during the socialization period. In contrast, Lindsay (2001) distinguishes between primary and secondary socialization on the basis of whether the subject is conspecific or human, which seems to be problematic. According to this view, primary socialization takes place during weeks 3–5 in the native social group, and this is then followed by secondary socialization to the human after weaning, when dogs are separated from other family members.

It should be noted that during this period pups can be socialized to various species including monkeys, cats, or rabbits (Cairns and Werboff 1967, Fox 1971). This capacity is also made use of in raising

dogs to protect livestock, when they are given extensive social experience with the members of those domestic species which they will guard (Coppinger and Coppinger 2001). All this supports the idea that there is little if any genetic component in recognition of the species, but important experiments to provide a more conclusive answer to this problem are lacking.

Another important change in development is the gradual emergence of hierarchical relationships among dog pups. Scott and Fuller (1965) reported that by 11 weeks of age the percentage of complete dominance relationships had increased sharply, although there was a marked difference between breeds. Unfortunately, there are few data on wolves and dogs relating to how early social relationships and experience influence later behaviour concerned with role and rank in the social group. Fox (1972, 1975) suggested that in wolves both genetic components ('temperament') and social experience determine later positions in the dominance order. Relatively stable social positions were observed in wolves (MacDonald 1987) and dogs (Wright 1980) during the socialization period, but this held true only where animals were simply categorized as dominants ('mostly' winning) or subordinates ('mostly' losing). Even in this case there were individuals that switched between categories.

9.3.4 Juvenile period (12 weeks to 6 months or later)

This is the longest and most variable period of development, yet it has been given the least attention in the study of dog development. For simplicity most authors implicitly assume that it extends until sexual maturity (though Scott and Fuller refer to it ending at the age of 6 months, for a reason that is not clear).

Wolf cubs start to follow the pack on hunting trips after 16 weeks of age, and the ensuing period is referred to as time spent in 'hunting school' where both perceptual and motor skills are improved (Packard 2003). These excursions provide an opportunity for the cub to improve its hunting skills, and it also practises mutual interaction and coordination of movement with its companions during group hunts. From the behaviour point of view this juvenile period might be best viewed as ending when the wolf leaves its natal group; this can take place at different dates between 9 months and 3 years of age (Gese and Mech 1991). As most wolves disperse before the age of 2 years, before reaching sexual maturity, in order to be successful these animals need to adopt a range of novel behaviours if they are to establish their own pack (or occasionally be successful in joining another one). Thus wolves retain some of their capacity to develop novel social relationships after the primary socialization period. It is likely that this provides the biological foundation for the development of secondary socialization in dogs that are separated from their families.

The juvenile period is also usually omitted from discussions of dog development, perhaps because it is difficult to give a general account. However, it is important to note that, while juvenile wolves have the opportunity to enrich their social experience at this time, this is often not the case in dogs as they spend most of their time alone after being separated from their siblings and mother. This partial social isolation could have a very critical effect on later life. This underlines the importance of puppy or juvenile dog classes in the case of 'city dogs'. Probably as a result of selection, dogs mature sexually earlier than wolves, usually between 9 and 18 months of age depending on the breed. It seems that in dogs the onset of sexual maturity is independent of behavioural maturation. Thus many breeds of dog do not display fully adult-like behaviour until 2 years of age, which corresponds to the time of sexual maturity in wolves, although they are ready to mate much earlier.

9.4 Sensitive periods in development

It should be remembered that the study of dog development has been strongly influenced by parallel investigations in other species, and the establishment of concepts such as 'critical period' or 'imprinting' has had an affect on how researchers interpreted their observations on dogs. Accordingly, Scott and Fuller (1965) described the 'critical period' of socialization as a well-defined period during development during which experience with the future objects of socialization is essential. The lack of such experience will have a marked detrimental effect on natural behaviour. Based on a series of experiments, Scott and Fuller argued that the 'critical period' in dogs lasts from approximately week 3 to week 12, that is, it corresponds to the socialization period. Importantly, since the original introduction of the 'critical period' concept by Lorenz (1981) and others, and the work of Scott and Fuller, our understanding of it has changed somewhat. First, it has been suggested that *sensitive period* might be a more appropriate term because the time boundaries involved are more varied than originally assumed. Second, in many cases studied so far (e.g. song learning, filial imprinting) organisms express a preference for species-specific stimuli (indicating the presence of a predisposition) which is revealed by rapid learning during a short exposure to such stimuli. Experience with such preferred stimuli (or the natural object itself) has the potential to 'overwrite' earlier exposure to an artificial stimulus (Bolhuis 1991). In line with these findings in other species we shall now try to summarize present knowledge about the start and end of sensitive periods and the specificity of the learning process in wolves and dogs.

Available evidence suggests that the sensitive period for socialization is much shorter in wolves. Observations reported by Zimen (1987) show a close relationship between early human contact (up to 3 weeks of age) and the effectiveness of socialization. Humans could develop a close relationship with the wolves only if the socialization started before this age. Wolves could also be socialized after this age, but they then develop relatively early distancing behaviours when interacting with humans. Similar results were obtained when wolves were raised with both humans and siblings (Frank and Frank 1982, but see also Fentress 1967). Experiments with wild-type (unselected) foxes seem to support these observations, as they displayed fearful behaviour when placed in an unfamiliar environment at 5 weeks of age (Belyaev *et al.* 1985), which is usually regarded as the end of the sensitive period (see below and Chapter 5.6, p. 132).

Interestingly, there is some indirect evidence in wolves that there is also a neonatal sensitive period (or a single sensitive period that starts right after birth). This is based solely on olfactory stimulation, which might be the reason for separating the two phases because in the subsequent sensitive period the puppy obtains mainly (or in addition) visual and acoustic information. So far there is no definite evidence for its existence, but it is widely known that wolves can be socialized to humans only if they are separated from all conspecifics before eye opening and exposed to intensive human contact (Klinghammer and Goodman 1987). Although the role of this neonatal period is not clear, it may be that during this time the animal learns to identify the stimuli which later become the central targets for the processes of socialization. Thus in wolves an early experience seems to control later developmental events.

Importantly, learning in this early period seems to rely on some sort of predisposition and shows features of stimulus specificity. Even if wolves are exposed to humans very early (but after days 11–12) they show a strong preference for conspecifics or dogs (Frank and Frank 1982). Early exposure to humans can to some extent counteract this but, even in this case, it cannot be reversed. Wolf cubs socialized to humans from days 4–6 showed no preference for their caregiver in the presence of a dog (Gácsi et al. 2005). It is important to note that dogs socialized in a similar way show a preference for the human if they are given the chance to choose a dog instead (Figure 9.1).

Thus early exposure to humans enables the development of a wolf–human social relationship, but there seems to be a competitive relation between conspecific and heterospecific stimulation. Stimuli from humans are effective only if they are exclusive, and exposure to conspecific stimulation has the potential to override this effect.

Extensive work in Scott's laboratory provided evidence for an extended period for socialization in the dog in comparison to wolves. Accordingly dogs could be socialized as late as 8–14 weeks (Freedman et al. 1961) (Box 9.4). This elongation of the period is probably the result of selective processes because a similar effect was observed in foxes selected for 'tame' behaviour (see also Chapter 5.6, p. 132). In this species approximately 40 years of selection has produced a sensitive period of double the duration, if one takes it that the emergence of

fear at around 10 weeks of age indicates the end (Belyaev et al. 1985).

Although we have no evidence for the wolf, a further interesting aspect of socialization is that learning can occur after a relatively short exposure. There is experimental evidence that in the dog socialization to humans can develop on the basis of a few minutes of social contact per day, or even when a passive experimenter makes gaze contact with the puppy for a few minutes over a couple of days (Scott and Fuller 1965). Unfortunately, there is very little evidence for the specificity of this process and the role that any genetic influences play. Experiments are lacking to show whether exposure to a similar 'amount' of human or dog stimulation leads dogs to show particular preference for one or other species. It would not be surprising to find some modification of their genetic make-up in favour of a preference for humans.

As a result of behavioural interaction with others the pup not only becomes a member of the pack but develops individualized relationships with others in it. This means that the young will regard the group not just as grouping of familiar animals but also as a social unit composed of certain individuals. The development of a hierarchy in the group presupposes some kind of categorization ability. Interestingly, 5-week old wolf cubs preferred their caregiver to a stranger (Gácsi et al. 2005), but no such preference was observed in puppies (see Figure 9.1), although this does not necessarily mean that they were unable to discriminate between humans at this age.

Usually it is assumed that the increase of avoidance ('fear') of novel stimuli signals the end of the sensitive period because, in practical terms, the animal is restricted in obtaining novel experience. Functionally we could argue that this change in behaviour plays an important role in keeping the puppy within the group. However, mechanistic models of sensitive period termination leave many questions open. In the Scott and Fuller model the emerging avoidance is discussed as a result of determined internal processes (maturation). This is based on the finding that in general dogs show very marked avoidance of humans if they have not had any human experience by 14 weeks of age. However, the avoidance of novel stimuli can be the

(a)

(b)

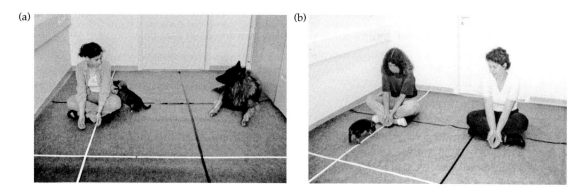

(c)

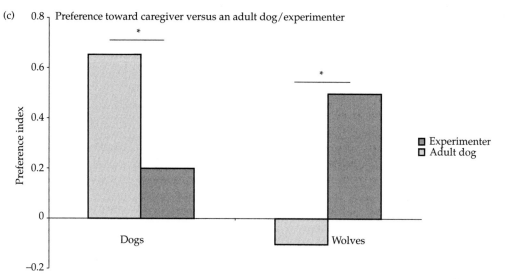

Figure 9.1 Preference for humans in socialized wolves and dogs. At 5 weeks of age socialized wolf cubs and dog pups were tested in a social preference test. In these experiments the subjects could choose between their caregiver and a dog or between the caregiver and another human. (a) Dogs usually prefer a human to a dog. (b) Wolves prefer the owner to the other human. (c) The larger the preference index the larger the attraction toward the caregiver. Dogs spent more time with their caregiver than with the adult dog, but preference vanished when the experimenter played the role of the competing social stimulus. In the case of wolves the results were the opposite (Gácsi *et al.* 2005). The index was calculated as: (relative duration of time spent with caregiver − relative duration of time spent with other stimulus)/ (relative duration of time spent with caregiver + relative duration of time spent with other stimulus). Significant differences are indicated by asterisks (*, $p < 0.05$).

result of learning processes. We must not forget that pups learn about what they are exposed to. Thus dogs lacking human experience learn that humans are not part of their social environment and develop no representations (referential structures; see Chapter 1.8, p. 22) for dealing with humans. These ideas fit with models of the sensitive period that assume termination comes about when initial referential structures are established and the system has gathered the maximum amount

of information that can be stored. This enables the puppy to discriminate clearly between known and unknown and to prefer the former over the later. In this model avoidance would be an indirect result of the lack of experience during the sensitive period.

In the view of these arguments, the lengthening of the sensitive period by selection raises interesting questions. The most parsimonious assumption would be that selection resulted in delayed

Box 9.4 Is there an 'optimal' period for socialization to humans in dog development?

A much-quoted study on 'imprinting' in dogs (Freedman *et al.* 1961, Scott and Fuller 1965) claimed that they have a sensitive period in development in weeks 3–12. For this study Cocker spaniels ($n = 18$) and beagles ($n = 16$) were isolated from humans and exposed at various times to human socialization for 1 week (see table below). After being reintroduced to their companions, all dogs were exposed to humans again in weeks 14–16. Two types of measures were taken at two different times. Dogs were observed during their interactions with humans at the beginning and the end of each of the two socialization periods. Attraction and avoidance were measured by scoring the behaviour of the pups in the presence of the human.

• *Attraction and avoidance at first encounter with humans* (Socialization I-see (a)): The authors note that early behaviour (weeks 3–4) is difficult to assess because of the limited motor ability of the pups. Thus the increased attraction to the handler might simply reflect increasing ability to walk. At week 5, pups show high levels of attraction and little avoidance when they encounter a human for the first time in their lives. Dogs tested in week 7 or 9 display less attraction. In parallel, avoidance changes in the reverse direction (B). If the same scoring system is used, attraction seems to decrease more rapidly than avoidance increases (control dogs are those not given any socialization experience).

• *Attraction and avoidance at the end of the socialization (Socialization I):* One week of socialization seemed to be enough for all dogs: they all reached a low level of avoidance (E). Unfortunately attraction scores were not reported, but generally high levels would be expected.

• *Isolation from humans:* After the socialization period, dogs were put back with their companions. It should be noted that for each group the time between the socialization experience and the final test differed: pups socialized at week 5 spent 8 weeks with companions, but dogs socialized at age of 9 weeks were isolated from humans for only 4 weeks.

• *Attraction to humans at week 14 and 16 (Socialization II-(b)):* Dogs socialized very early (week 2), and control dogs without any human experience, showed little attraction. All other groups showed similarly high levels of attraction. In these groups attraction scores did not improve to a large degree, but dogs receiving socialization with humans for two additional weeks recovered almost completely. This suggests an important and special role of very early stimulation in dogs. However, control dogs never showed much attraction to humans. A randomly chosen dog from this group could not be socialized to acceptable levels even after a further period of 3 months.

Table to Box 9.4 The outline of the 'wild dog' experiment (based on Scott and Fuller (1965); Freedman *et al.* (1961)).

Time of separation from litter (week)	Socialization I (week)	Living with the litter after Socialization I (week)	Socialization II (for all between 14th & 16th weeks
2nd	3rd	11	2
3rd	4th	10	2
5th	6th	8	2
7th	8th	6	2
9th	10th	4	2
14th (control)	—	—	2

continues

Box 9.4 *continued*

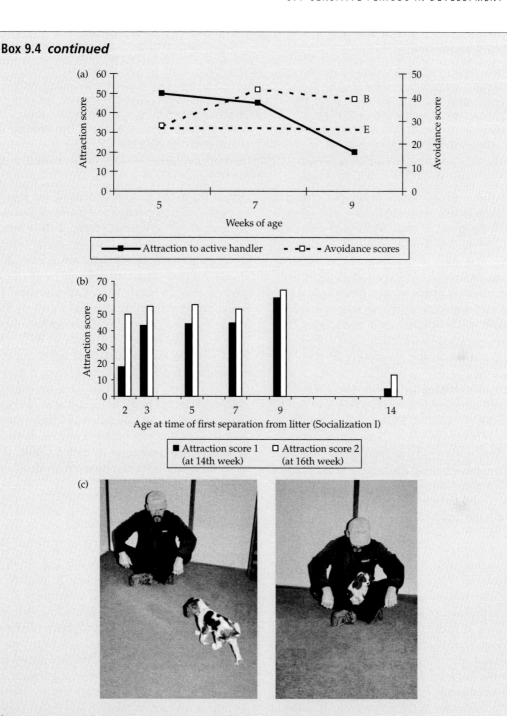

Figure to Box 9.4. (a) Attraction and avoidance scores of 5–9 weeks old puppies obtained from the first socialization phase (Socialization I) (based on Scott and Fuller 1965); (B = beginning avoidance scores; E = end avoidance scores). (b) Attraction scores at start and end of the second socialization phase (Socialization II) between 14–16 weeks of age (based on Freedman *et al.* 1961). (c) Starting at 2–3 weeks of age dogs are generally attracted to a passive human.

continues

Box 9.4 *continued*

These results show that if dogs receive no human stimulation before the age of 9–14 weeks, they cannot be socialized. However, there are data showing that even a short exposure to humans can counteract this, and dogs generalize early social experience to other humans.

Thus there might be a relatively long sensitive period for developing social relationships with humans. In addition, we do not know how the choice of breeds influenced the results, as in some breeds even the duration of the sensitive period might be different.

---

maturation, with avoidance behaviour emerging later. Alternatively, we could assume that selection affected the speed of establishment of referential structures which would maintain exploratory activity. This delay in finalizing referential structures would represent an increased plasticity of the system, which is useful because it has to deal with increased complexity in the social environment, composed as it is by members of two species. Only further comparative work in dogs and wolves, or in foxes, could reveal the plausibility of these two ideas.

It should be stressed that the emergence of extreme avoidance is present only in pups that have had no experience of humans. Exposure to a minimum amount of human experience is enough to reduce the levels of such wariness in dogs (Stanley and Elliot 1962), and these animals retain their ability to develop and maintain social contacts with strangers after the 'official' end of the sensitive period.

Mainly for practical reasons, Scott and Fuller (1965) also introduced the concept of the 'optimal period' for socialization. Accordingly, 'best' results can be achieved if dogs are socialized between 6–8 weeks or 1–2 weeks thereafter (Scott 1986). This advice takes into account that the puppy should get social experience of both conspecifics and heterospecifics in order to develop 'normal' social behaviour. They argued that the puppy should be introduced into its novel human environment before the end (or even better around the middle) of the socialization period, but that it should also spend enough time in the native group in order to gain experience of conspecifics. Although Scott and Fuller were cautious enough to point out many times that developmental periods are subject to variation because of both genetic and environmental causes, their efforts to determine an 'optimal'

period led to the general understanding that dog pups should be separated at 8 weeks or even earlier. Such an indiscriminate practice is, however, not advantageous in the case of many slower-developing breeds. In addition, there is no evidence that at this time socialization would be specific to a particular individual. Most findings show that socialization with one human has a general effect, that is, if during the socialization dogs have experience with some humans there is every chance that most of them can be socialized to other people without much difficulty later. However, it is advisable for the breeder to provide the pups with variable experience of humans, including children, and perhaps people who look different. Purely from the adoption point of view, there is no need to rush to separate the puppy from its native family, especially if the new owners cannot offer a socially rich home environment.

9.5 Attraction and attachment

It is unfortunate that in the old literature these two terms were used interchangeably (e.g. Scott and Fuller 1965, Scott 1992), because in present use attraction (or affiliation) and attachment do not refer to the same aspect of behaviour. *Attraction* could be defined as any form of preference for one class of stimuli over another. *Attachment*, on the other hand, is a feature of the organization of behaviour at a functional level, a property that emerges under special circumstances and involves a complex interaction between perceptual, referential, and action structures (Chapter 1.8, p. 24). In addition, attachment usually describes a particular individualized relationship.

Bowlby (1972) defined attachment in humans as 'seeking and maintaining proximity to another individual'. The functional argument is that remaining close to the object of attachment (e.g. the mother) contributes to the survival of the young because this offers various forms of support (e.g. food, protection from predators, etc.) (see Gubernick 1981, Bowlby 1972). Thus the role of attachment becomes especially important when there is the danger that a mobile offspring distances itself from the parent. For this reason attachment is more important in precocial species, and also rises in importance with increasing motor skills of the off-spring. Accordingly the idea of attachment is based on a special form of social relationships that develops between two individuals. Observational data usually provide little evidence for an attach-ment relationship because the appropriate circumstances occur only rarely. Thus in most cases attachment is demonstrated in a laboratory setting if the animal's behaviour fulfils certain behavioural criteria which are based on the assumption that the subject is mobile and has the motor skills necessary to approach or avoid stim-uli/objects that occur in its environment (Chapter 8.2). Thus a relationship can be described as attach-ment if the subject is able to recognize the object of attachment (*individual discrimination*), shows a pref-erence for regarding it as the centre of its social environment (the *secure base effect*) during explor-ation and when experiencing danger, and displays specific behavioural changes upon encountering the object of attachment after stressful separation (*greeting* and *behavioural relaxation*). In order to test for these criteria experimentally most researchers use some sort of control object (*stranger*) which belongs to the same category as the presumed object of attachment.

A re-evaluation of the older literature on social relationships shows that many of the phenomena described as such do not really reflect attachment but are cases of attraction or affiliation based on genetically influenced preferences and/or the effects of learning. Taking a closer look at the devel-opmental work on social affiliation between human handler and dog pups, it becomes clear that up to the end of the socialization period pups do not develop attachment relationships with humans, and similarly no individualized social relationship emerges towards other dogs (e.g. the mother).

Although no specific experiments have been reported, Pettijohn *et al.* (1977) found that humans were more effective in reducing stress in dog pups than their own mother even when dogs had very little heterospecific experience (e.g. only with peo-ple cleaning their cage). Thus it seems that dogs at this age are attracted to some categories of social objects, without having established a strictly indi-vidualistic relationship. This does not indicate that dogs are not able to recognize their mother or sib-lings, for which there is experimental evidence (see Hepper 1994; Chapter 6.5.4). It shows only that young puppies are indiscriminate in the individu-als to which they direct their behaviour when it comes to surviving dangerous and stressful situ-ations. Perhaps in the case of young puppies any group member or the group as a whole could pro-vide protection and there is no need for the help of a particular individual.

Recently we have obtained experimental evidence that 4-month-old puppies have developed an attachment relationship with their owners (Topál *et al.* 2005a) but wolves at the same age did not (Figure 8.2). From the functional point of view wolf cubs might receive the same protection from all members of the pack, thus there might be no need for individualistic attachment. These differences suggest that such early attachment in dogs is the result of selective processes.

Scott (1962) also noted that social attachments may form throughout life, but at that time he was probably referring to affiliative behaviour. However, the statement is probably true in the case of dogs, which retain the ability to develop attachment rela-tionships later in life and to many people (Gácsi *et al.* 2001). Such flexibility could be advantageous in allowing the dog to join a different group of humans even later in life, and also enabling it to establish a complex network of relationships with humans belonging to different groups.

9.6 Early experience and its influence on behaviour

The study of the effects of early experience on the behaviour of the dog has received little attention in recent times. This is unfortunate, because most of the knowledge obtained by Scott and Fuller (1965) represents just one methodological approach to the problem. As they acknowledged, the method of raising large number of animals under controlled conditions resulted in dogs which 'did not develop their maximum capacities' (Scott and Fuller 1965, p. 86), partly because of their restricted experiences. Thus any specific early experience they were given came in addition to living in a relatively impoverished environment, and a broader range of experience could have produced dogs with improved skills.

More recent studies have been based on 'natural' dog populations sharing some or most of their everyday environment with humans, and such dogs may miss out on certain sorts of stimulation, or receive it in excess. This situation offers the chance of looking for correlations between experience and behaviour. In retrospective studies data are collected by questionnaires in order to reconstruct the early rearing environment of the dog and isolate factors that may affect later behaviour. Using this method Serpell and Jagoe (1995) found that many different factors can influence later behaviour. For example, 'dominance-type aggression' reported by the owners was more common in dogs obtained from a pet shop and in dogs that were ill before the age of 14 months. This indicates that restricted social experience during the socialization period can lead to an animal with an overtly agonistic attitude. It is important to remember that such studies are useful in detecting possible risk factors in development and offering hypotheses on early influence, but do not provide causal explanations for behaviour.

More experimental studies aim to find a correspondence between early environment and subsequent performance in certain tests (e.g. Fuchs et al. 2005) or aim at actively influencing the early developmental environment and searching for an effect emerging at some later point in time. Such investigations are of special practical interest because it is assumed that extensive early experience is beneficial for the later capacity of the dog to be successful in training. Pfaffenberg et al. (1976) found that guide dogs for the blind are more likely to pass training if they arrive in their host family shortly after weaning and are not left in kennels for an extended time during the socialization period.

Thus the lack of social experience interferes with later training. Little is known of whether the enrichment of a dog's environment improves training performance or changes its attitude towards its environment. Seksel et al. (1999) varied socialization experience for 6–16-week-old pups by exposing them to different sorts of experience in short sessions. Some dogs were given both handling and early training, other groups received only one of these treatments, and untreated dogs were used as controls. In tests dogs were subjected to different environmental stimulation and training tasks. The lack of major effects of the treatments that was found was explained by the relatively small influence of the socialization experience in comparison to the overall social and environmental stimulation received by the dogs in their home environment.

More specific experience can have an advantageous effect. German shepherd pups that observed their mother searching for and retrieving hidden narcotic sachets when they were 6–12 weeks old responded faster to later training when they became 6 months old (Slabbert and Rasa 1997). This suggests that training for various tasks may benefit from earlier exposure to a skilled conspecific.

9.7 Prediction of behaviour: 'Puppy testing'

Predicting how the behaviour of a puppy will turn out could have a practical application because it could help the breeder to match the puppy to the wishes of prospective owners, and to select puppies for further breeding so that trainers can avoid investing work in 'less talented' individuals. Predicting abilities that are necessary for an animal to become a well-trained working dog could save money and effort (as well as emotion) if it can be ensured that only suitable candidates are enrolled in the training programme. The development of a predictive puppy test became one of the holy grails of dog research, but a review of this literature shows very mixed results. Although there are reports of tests on the basis of which successful prediction of suitability for work or of response to training have been found (e.g. Scott and Bielfelt 1976), negative reports are more frequent. Most problems originate from taking too simplistic a view of development (Box 9.5). When testing for prediction the primary concern is with those aspects of behaviour that are under relatively strong genetic influence and are thus resistant to environmental disturbance. However, some early environmental influence is strong enough to cause long-lasting changes in behaviour which are protected from later modification. In either case we assume that the factors determining the animal's potential happened before the predictive test and no further environmental variation will affect the studied behaviour. Even if this is so, behavioural maturation can interact with the predictability of the test. Although maturation is under strong genetic control, some changes occur 'overnight' while others emerge only gradually. It follows that testing should be done when maturation is near completion, but the timing of this has not been shown for most behaviour patterns, and might be different for different systems.

It is also important to remember that selective breeding affects the structure of development, changing not only the speed of maturation but also the duration of developmental periods, and the sequence of how behaviours emerge. In addition, there is a breed × environment interaction: for

---

Box 9.5 Behavioural development and the problem of puppy tests

Puppy tests are increasingly fashionable because there is a belief that adult behaviour can be predicted on the basis of observing young dogs. Here we present a theoretical framework to illustrate the problems with these tests.

As discussed in the text, perceptual (P) and motor (M) abilities emerge sequentially, and the organism is exposed to various events (E) during development. Any puppy test depends on the ability of the puppy both to perceive certain stimulation, and to show certain patterns of behaviour, neither of which is independent from experience. Puppy tests are usually performed on two or three occasions (dotted box) when a dog is put through a battery of different tests.

In this scheme Test 1 will not measure the effect of E3 at all and the behaviour of the animals will depend on whether their emerging perceptual ability (P2) will precede or be late (P2') with regard to E2. In Test 2 animals developing more rapidly (P3) will have more experience to evaluate E3 than those with a slower rate of development, and it is also not obvious how differential perceptual abilities in Tests 1 and 2

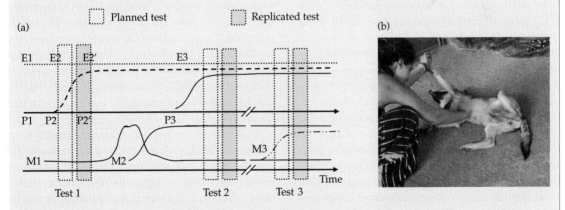

Figure to Box 9.5 (a) A hypothetical schema of behavioural development and the timing of puppy tests. (b) The dominance test is usually part of the puppy test. So far, behaviour during this test has not proved to be predictive. Although the test looks simple (the experimenter puts the dog on its back) there is no generally accepted published version.

continues

Box 9.5 *continued*

affect the relationship between behaviour in these tests. Similar logic could be applied to motor behaviour. Based on this, the development of a useful puppy test might be based on the following considerations. (Note that timings might differ between breeds.)

• *Description of behaviour:* Long-term observational data are needed to describe the development of both perceptual and motor abilities, especially in terms of first emergence, rate of development, and stabilization.
• *Test design:* Depending on the character of interest various behavioural tests, which are supposed to reveal certain abilities, should be tried and also re-tested on the next day.
• *Test battery:* Based on the two previous steps the optimal time for testing should be determined using a combination of tests that are applied throughout the period of development.

example, breeds show differential sensitivity to interaction with humans (Freedman 1958). Such factors could be critical in comparative work on development because using an absolute scale (days after birth) could lead to misleading findings (see Box 9.3).

Another approach to the same problems is to use a battery of tests at a particular age. Although in most cases a series of short tests on a given day seems to be very efficient and time saving, such a practice is contradicted by observations on behavioural development, which suggest that different behavioural systems do not mature in synchrony. Thus testing puppies for sociability, retrieval ability, neophobia, or activity at around 8 weeks of age failed to be predictive as far as suitability for service work was concerned (Wilsson and Sundgren 1998), while various single tests on retrieval (at 8 and 12 weeks) or startle behaviour (at 12 and 16 weeks) provided good predictive value for suitability to become a police dog (Slabbert and Odendaal 1999). The complex relationship between puppy and adult behaviour was also highlighted in studies attempting to predict fearful behaviour in dogs (Goddard and Beilharz 1984). They found that the fear response of dogs changes during development. Before 12 weeks of age dogs reduced their activity in fearful situations, but adult dogs in similar situations became either passive or overtly active. Thus measuring early reactions of fear is not a good predictor for later behaviour. Although such isolated results are not conclusive, they might indicate that the predictive value of a test is likely to be increased if it is done at the 'right' time.

Another important factor is how predictive variable is defined. Many studies rely on a single behavioural variable measured in a test at a particular age, while others obtain 'composite scores'. In the case of the latter researchers combine those variables that are assumed to measure the same trait (which is of course not necessarily the case). For example, in order to predict fearful behaviour, Goddard and Beilharz (1986) defined a 'puppy test index' which consisted of an activity score (at 9 weeks), a fetching score (at 9 weeks), and different scores attributed to fear (reaction to a whistle at 8 weeks, or avoiding objects while walking at 12 weeks). Naturally, from the practical point of view, any kind of measure which proves to have a high predictive value is a valid solution to the problem. But this does not bring us closer to understanding the developmental relationship between the behaviour of the young dog and that of the adult, partly because the predictive value of some of these behavioural variables might apply only to the rearing environment in which they were identified.

As expected, the predictive value of puppy tests increases with age. This was found to be the case in measuring fear in guide dogs for the blind (Goddard and Beilharz 1986) or aggressive behaviour in police dogs (Slabbert and Odendaal 1999). Unfortunately, the prediction often comes too late, when the dog is already participating in the training programme. Late prediction of adult behaviour is also problematic when the aim is to breed for or against some behavioural traits. Despite its being relatively predictive, Goddard and Beilharz (1986) do not recommend selection against fearfulness on the basis of their 'puppy index' because of the uncertainty of the genetic component underlying this trait.

However, even from the practical point of view puppy testing has its advantages, because it exposes the young dog to various physical and social experiences, and the tester can thus gain valuable information on the developmental state of the individual. If the puppy does not perform as expected, corrective measures can be taken to improve its behaviour. Thus regular 'testing' that exposes the puppy to features of its future environment might actually lead to better performance as an adult, despite the tests not being predictive.

9.8 Conclusions for the future

Developmental periods should be viewed as an epigenetic process during which the genetic potential of the organism unfolds in a given environment. Thus a developing dog does not only passively experience environmental stimulation; the organism is built such a way that it actually 'expects' certain kind of stimulation during growth.

Unfortunately, most work on dog development was published more then 30 years ago, and despite considerable research many questions remain unsolved. There is a lack of comparative data both for wolves and dogs, and on different breeds. Comparative studies could widen our understanding how selection and artificial breeding affect the behaviour of the dog by changing the pattern of development.

Little is known about how early stimulation relates to the sensitive periods in development. Such research could reveal species-specific differences as well as the role of different types of stimuli including olfactory, visual, and acoustic influences. It could be important in this regard that the stimuli associated with the development of an affiliative relationship have changed from wolf to dog. Wolves seem to be more dependent on early olfactory stimulation; dogs, by contrast, develop a preference for humans also on the basis of visual cues.

Further reading

Detailed reviews on dog development can be found in Serpell and Jagoe (1995). Lindsay (2001) and Coppinger and Coppinger (2001) provide a range of ideas on the complexity of gene × environment interaction.

CHAPTER 10

Temperament and personality

10.1 Introduction

There is some regularity in individual variation within a species. Although, like any phenotypic trait, individual behaviour is the result of an interaction between genes and environment, some individuals are more similar to each other than to others. This view is probably based on the observation that individuals behave consistently across similar or different situations, and there are limited ways of behaving consistently. For example, upon seeing a novel object crossing its walking path a dog might look at it, follow it, or approach it. If the dog shows a similar behaviour pattern on many occasions towards various novel (or even familiar) objects, we might be led to characterize the dog as being 'curious'. However, what else could the dog have done? There are two more possibilities: the dog could show no interest at all (continue walking along the same path) or it could distance itself from the object by either stopping or changing the direction of its walk. Dogs showing the former pattern would be described as 'not interested'; those showing the latter behaviour would be characterized as 'fearful'. These categorical descriptions are usually regarded as types of behaviour, and the measure(s) (e.g. the distance between dog and object at some point in time after the encounter) that have been used for this categorization are referred to as *traits*. In keeping with this tradition, and to discriminate this sort of behavioural description focusing on the individual from the traditional approach, it seems to be useful to refer to *personality types* and *personality traits*. (Personality traits differ from behavioural traits on the basis that they are usually derived features, that is, they are based on 'weighted' contributions from more than one behavioural trait (see e.g. Jones and

Gosling 2005). Unfortunately, in current research there is great confusion about the term 'personality', partly because ethologists or animal psychologists are afraid to be accused of anthropomorphism. This has led to the blossoming of synonyms including behavioural syndromes, behaviour types, behavioural styles, coping styles, emotional predispositions, and temperament. In dog ethology one encounters mainly two synonyms, *temperament* and *personality*, which, unfortunately, have been used interchangeably by many authors. Recently, many reviews have touched on the topic of personality in dogs, albeit from different perspectives of behavioural genetics (Ruefenacht *et al.* 2002), comparative psychology (e.g. Jones and Gosling 2005) or practical application (Taylor and Mills 2006, Diederich and Giffroy 2006) (Box 10.1).

Here we will take the functional anthropomorphic view and suggest a similarity between the function of personality in humans and animals (Carere and Eens 2005). Thus personality is defined by an array of behavioural traits that are under the influence of selective processes and are the result of some sort of adaptive mechanism. There are perhaps two reasons for preferring the general term 'personality' rather than 'temperament'. First, in the human literature there is a trend to use 'personality' to describe adults, whereas 'temperament' is preferred for developing humans. Second, if fitness consequences are important then it is the more or less stable behavioural pattern of the adult that is of primary importance. Since personality is the product of both genetic and environmental effects, it is expected that it will undergo marked changes from birth to maturation. This process involves complex patterns of genetic activation and environmental stimulation, both of which are expected

Box 10.1 Pavlov and his dogs

Although Pavlov is usually cited as the developer of the laboratory paradigm of associative learning, his contribution to the research on personality was perhaps equally important. He and other workers in his laboratory noted very early on that many dogs showed a specific but consistent behaviour during the training sessions. Importantly, the Pavlovian categorization was not only based on the parameters of the learning process (e.g. number of trials for reaching a criterion, number of trial needed for extinction, etc.); researchers also observed the overall behaviour of the dog before and during the experiment. Dogs were put in three (or four) categories which were assumed to reflect neural properties of the brain ('types of nervous system') (Teplov 1964, Strelau 1997). This categorization, which has some similarities to the Hippocratic–Galenien typology of the four humours, including the problem of objectively assigning a dog to a category, became very popular among dog trainers at that time (and is often referred to today). However, Pavlov's intention was to make this categorization as objective as possible; that is, how dogs reacted to being conditioned in appetitive or aversive situations. In the review of Teplov (1964) the following characteristics were mostly mentioned with regard to these 'types':

- *Weak type (melancholic):* nervous, sensitive (yelp), struggling when restrained, cowardly, inhibited; extreme predominance of inhibitory process
- *Strong–unbalanced type (choleric):* active, lively, prone to being aggressive; moderate predominance of excitatory process
- *Strong–balanced–slow (phlegmatic):* quiet, steady, restrained, moderate predominance of inhibitory process

- *Strong–balanced–mobile (sanguine):* active; reactive to novel stimuli, sleepy in monotonous circumstances; extreme predominance of excitatory process

Most of Pavlov's work received little attention after his time (although Scott and Fuller 1965 mention him in passing), and personality research became dominated by inductive methods (e.g. Cattel *et al.* 1973). In parallel, there has been a long tradition of using the personality (or temperament) of dogs to select them for work (e.g. Humphrey 1934, Pfaffenberg *et al.* 1976, Goddard and Beilharz 1986).

Interestingly, Sheppard and Mills (2002) obtained a two-way categorization of dogs ('negative activation' and 'positive activation') on the basis of questionnaire data that corresponds broadly to the two main types ('weak' and 'strong') in the Pavlovian system.

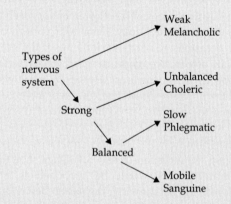

Figure to Box 10.1 Pavlovian typology was developed for dogs first and only later applied to people. However, it is clear that Pavlov also tried to conform to the classic Hippocratic–Galenien human typology. (Redrawn and modified from Strelau 1997).

to decrease as the individual matures. Naturally, this does not mean that personality is resistant to changes after maturation but it is expected that environmental effects in particular have a greater impact before maturity of the individual than after. This also means that the genetic contribution to personality can be seen better at the early stages of development.

Actually, we can turn the distracting synonymous usage of 'personality' and 'temperament' into a useful dichotomy. Temperament could be used for traits that are present at early stages of development (e.g. neonatal, transitional, or social in the case of the dog; see Chapter 9) before maturity is reached. This distinction is also reflected by Goldsmith *et al.* (1987) who distinguish temperament as inherited, early-appearing tendencies that continue throughout life and serve as the foundation for personality. Also note that researchers are inclined to refer to a trait as temperament if it has a more general or broader application (e.g. 'impulsivity', 'boldness', 'activity') and as a personality trait if the character seems to be restricted to a more special context (e.g. 'sociability', 'aggressiveness'). In this respect we might expect that as the animal grows it has to perform in an increasing number of environments, which leads to a more structured pattern of behaviour. For example, the emergence into the social environment is a gradual process which is paralleled by the development of personality traits related to sociality. Many would also argue that temperament is strongly related to the genetic compound of a trait whereas personality is the realized phenotype, the product of a long-lasting gene times environment interaction. Thus we can adopt the definition used by Jones and Gosling (2005): personality represents those characteristics of adult individuals that describe and account for consistent patterns of feeling, thinking and behaving.

10.2 Descriptive approach to personality

A recent review on dog personality (Jones and Gosling 2005) identified various goals of research, such as prediction of behaviour during development, description of behaviour traits to predict behavioural problems or individual suitability for certain training methods, or selection for preferred phenotypes. However, most recent reviews have concluded that many of these aims may be jeopardized by the lack of understanding of (and more attention to) theoretical and methodological problems.

By definition, only individuals can have a personality. In this sense attempts to describe the 'personality profile' of a breed based on expert reports presents an invalid use of the method, even if at the conceptual level some derived traits might correspond (e.g. Draper 1995, Bradshaw and Goodwin 1998). As detailed analysis shows, such breed profiles could be also very different when different populations are compared and also if breeds are

exposed to different selective environments (see below, Svartberg 2005).

10.2.1 'Knowing', observing, or testing

In humans most personality tests consists of a list of questions which usually ask for a judgement of a particular situation. Although such self-reports may seem very subjective, many years of research have established that the responses to these questions do indeed have some or more (statistically significant) relationships with the corresponding behaviour traits of the responder (e.g. Gosling 2001). The practical advantages of this method have led researchers to apply the same questions about dogs. In this case the owner, a familiar person, or an expert is asked to characterize the behaviour of the dog in a series of contexts, without the dog participating in any way in the gathering of the data.

More ethologically oriented methods either involve the observation of the subject in everyday situations or design special behavioural tests in order to reveal special aspects of the behaviour. Observation in natural situations is often very complicated, takes a long time, and is difficult to standardize. Thus researchers prefer to devise *test batteries* in order to describe behavioural traits which could provide the raw material for personality traits. Naturally, to provide a description of the 'whole' personality the test battery should simulate a range of contexts in which different facets of the personality can be revealed. A further aspect of test batteries is that there is a preference for novel (sometimes extreme) stimulation of the dog (e.g. a gunshot) in order to release certain patterns of behaviour. However, these two factors introduce various complications. First, test batteries cannot be extended indefinitely, because dogs cannot be expected to react in the same way over an extended period of time. This limits the number of 'situations' (test units) that can be included in a test battery, which in turn determines the range of behaviours that will be displayed. Even in this case it is impossible to exclude the possibility that the subject's inner state will change during the course of the test battery, which could influence the behaviour. It is thus likely that the test units cannot be regarded as strictly independent measures, especially because this possibility has not been investigated so far. In addition some test units (or aspects of the situation) are repeated within a test battery to provide evidence for internal consistency of behaviour. However, this could be problematic because some carry-over effects of habituation or sensitization can affect the behaviour. For example, in the Dog Mentality Assessment (DMA) test (Svartberg and Forkman 2002) there are two 'play' test units, one in second place of the test battery and one in ninth place. Although play behaviour correlates between the two units, the dogs are subjected to a range of stimulation (metallic noise, 'ghost') before the second play unit. Play behaviour may appear to be resistant to such interventions in a large sample, but in general there could be many hidden factors that affect the behaviour at the second occasion. When testing for aggressive behaviour, Netto and Planta (1997) put a dog through a series of test units lasting for approximately 45 min and included various contexts with the potential to elicit aggressive behaviour on the part of the dog. Although the application of an elaborate testing system was very successful in achieving high criterion validity (dogs with biting history were detected with great success by the test, see below), it also seemed that the test might have sensitized the dogs for this behaviour (dogs got more aggressive towards the end of the test) and the practice of exposing dogs to a stressful situation for such a long time could be also problematic from the welfare point of view.

Although it may be logistically more complex, it is more advantageous that test batteries applied on one occasion test for only a few personality traits, and dogs are then subjected to further testing within a short time, during which changes in personality are not expected.

10.2.2 Describing behaviour: assessment and coding

If the dog does not participate in the testing, the researcher has to rely on the owner's assessment. Such assessment can be based either on human trait rating (e.g. 'energetic', 'anxious'; Gosling and Bonnenburg 1998), human personality inventory (items are 'translated' for being meaningful for dogs; Gosling et al. 2003), or a list of questions drawn from naturalistic life situations (e.g. Serpell and Hsu 2001, Sheppard and Mills 2002).

Questionnaire studies looking for behavioural traits based on owner's reports have provided some evidence that this method can be reliable and valid. Reliability was tested by asking owners to assess their dog and themselves using a human personality inventory, and peers were also requested to assess the focal human and dog (Gosling et al. 2003). The results of this questionnaire study showed that different observers provide similar reports on the same individual (inter-observer reliability), and their judgement is also consistent if they report on a single trait in different contexts (internal consistency). Another questionnaire study (Sheppard and Mills 2002) found that observers are also consistent in their judgements when asked to fill out the same questionnaire after 6 months had elapsed (intra-observer reliability). Although there are some doubts whether owners are unbiased in responding to such questionnaires (e.g. in the case of problem behaviour) (Sheppard and Mills 2002), and each questionnaire to be applied needs to be tested, it seems that reliability can be achieved.

In the case of questionnaires one important way to obtain validity is to look for external criteria that are associated with the trait under investigation. Gosling et al. (2003) asked independent judges to observe owners interacting with the dogs for which the owners had made assessments earlier. There was a considerable agreement between the assessment of the judges and that of the owners. Similarly, Svartberg (2005) reported that there was also an association between owner's assessment in the C-BARQ questionnaire (Serpell and Hsu 2001) and the observed behaviour in the DMA test performed 1–2 years earlier.

The ethological approach to personality prefers the direct measurement of behaviour either in observational situations or in test batteries. This is usually done in two different ways: the observer either rates the behaviour along some scale (similarly to the questionnaire assessment) or there is a detailed ethogram which decomposes the behaviour into elements for which frequency, duration, or latency can be measured (Chapter 2). The behavioural coding is usually assumed to be done by professionals, so good intra-observer and inter-observer reliability is expected. However, in practice (e.g. working field trials, DMA test) several judges observe the behaviour over a long period (in order to obtain large samples) and such reliability is often not achieved or reported (e.g. Svartberg and Forkman 2002, Strandberg *et al.* 2005). This is partly because many of these evaluations take place years after the data were collected; however, there seems to be no reason why in future tests cannot be recorded on video and coded by a small group of trained observers.

In the case of behavioural tests the question of test–retest reliability has long been omitted, although there is now evidence for it, for example for the DMA test for over a range of 2–3 months (Svartberg *et al.* 2005). Importantly, the decrease of aggressiveness between two tests suggest habituation whereas the increase of curiosity and fearlessness scores indicates sensitization to the repeated test situation (see also Ruefenacht *et al.* 2002). The results of the behavioural test may be cross-validated with traits obtained by validated questionnaires on behavioural traits directly or some derived personality traits (see above). A further possibility that is also available for questionnaire-based methods is to look for some other independent measures that predict behavioural differences, such as age or gender.

10.2.3 The construction of personality

Whether assessment or coding of behaviour is used, basically the same statistical methods (factor analysis) are applied to reduce the number of dependent variables and to arrive at a smaller number of derived traits (*factors*) that are independent and can explain the greatest possible amount of variability in the original variables. In practice these factors will be described as personality traits based on the behaviour variables that are associated (*load on*) dominantly with them. In addition, the number of these factors and the relations between them depend crucially on the number of input variables, the nature of the variables, and the correlations between them. Importantly, there is no *a priori* reason to assume that these derived factors make any sense, that is, that they represent a functionally meaningful personality trait.

Test batteries consisting of one or two functionally different situations (e.g. reaction to a threatening stimulus, and a stranger) or one or two functionally similar situations (e.g. play with familiar person or stranger) are likely to reveal only one single factor. This explains the preference for questionnaire studies (and the limits of behaviour testing); hence an adequate set of questions (and a correspondingly large population) could reveal many facets of personality because the analysis of many different variables is likely to show a complex underlying structure of derived variables.

It should be obvious, however, that problems could arise when we confront the personality descriptions derived by these different methods. One source of the difference is that in the case of questionnaire-based methods the evaluation 'happens' in the mind of the observer. Consider the following case: By using a scale with 5 items (scores from 1 = no to 5 = yes) an owner has to indicate whether it is likely that his dog is afraid of vacuum cleaners. They may make an 'intelligent' guess about this trait by combining all the situations in their memory invotving the dog and a vacuum cleaner (and perhaps even other similarly frightening stimuli). This is just a 'mental factor analysis' that is probably not independent of species-specific ('human') subjective elements. In addition, many questions for a trait see a situation from a human perspective. Thus in the present case we may suppose that expected behaviour is 'not being afraid of the vacuum cleaner', which may or may not be true from the dog's perspective. This is in marked contrast to the case when the dog is actually tested with a vacuum cleaner that could be any size or colour, making various noises, etc. and the observer notes 'avoidance behaviour' (on a scale) or latency of approach, looking time, etc. as behaviour recordings. From this it follows that personality traits derived from questionnaires might appear more distinctive and also more familiar to us, partially because the behaviour of the dog was evaluated by a human mind.

In contrast, behaviour-based personality traits could be more difficult to interpret because they cannot simply be projected on to our own personality structure. This seems to be supported by the observation that questionnaire-based personality structures of dogs are more similar to human personality structures (obtained by similar methods) (Gosling *et al.* 2003) than behavioural-based personalities (e.g. Svartberg and Forkman 2002). Naturally, there are some traits that have their equivalents in both types of personality structures, for example 'aggression'. However, it is not easy to find equivalents of the five factors obtained by the DMA test battery (playfulness, curiosity/fearlessness, chase-proneness, sociability, and aggressiveness; Svartberg and Forkman 2002) in the seven personality traits (reactivity, fearfulness, activity, sociability, responsiveness to training, submissiveness, and aggression) suggested in a meta-analysis by Jones and Gosling (2005).

Not forgetting that both methods offer a reliable and valid measure of personality, we might ask how they relate to the underlying biological organization involving genetic components, and neurophysiological control. Again, in the human literature there are examples of both cases; that is, associations with genetic compounds were revealed both in relation to behaviour-derived and questionnaire-derived traits. But in humans this is a within-species situation, and it may not be valid for the dog, or valid only in special circumstances.

10.3 Functional approach to personality

From the perspective of the theory, individuals should behave in optimal way in any given situation, which seems to contradict the idea of personality (Sih *et al.* 2004). Thus, in order to explain the existence of personalities we should be able to show that this form of behavioural organization is adaptive (Dingemanse and Réale 2005), in contrast to a system that shows maximum situation-dependent flexibility. Such questions are rarely asked in the dog personality literature because so far researchers have been not concerned with the question of whether types of personality have differential survival rates. However, this may change as interest grows in understanding the evolutionary transition from wolf to dog.

Sih *et al.* (2004) argued that very variable environments might select for traits that are less flexible, or in other words, more stable over a range of environments, because high flexibility is prone to errors especially if there is little chance to gather adequate information for always behaving optimally. Thus, the 'boldness' trait in a species, which is often associated with exploring novel territories as well as food sources, could be the product of those (broadly similar) selective environments inhabited by individuals of a given population because in this particular case it may pay to be 'always' bolder than to adjust the behaviour on a case-by-case basis to the actual situation. In similar vein, a different environment selects for altered boldness type while in other cases the success of the of personality types is frequency dependent or changes over time (Dingemanse and Réale 2005).

Fox (1972) observed a larger behavioural variability in wolf cubs than in coyotes or foxes. He explained this by assuming that the more complex social system of wolves favours individuals with different behaviour tendencies that fit certain roles in the group. This idea leads to the hypothesis that increasingly complex societies select for more sophisticated personality traits, which determine a finer categorization of personality types. This might explain the superficial observation that the personality trait structure of organisms living in a simpler environment (including social environment) is also simpler.

Many researchers have noted that quite often personality traits extend over different functional units of behaviour. For example, individuals that are bolder in exploring novel environments are often also more aggressive. In this vein Svartberg (2002) argued that shyness–boldness explains a large part of the phenotypic variability that is present in personality traits such as sociability, playfulness, curiosity, and chase-proneness, found in the DMA test. This means that individuals that are more curious (bolder) are also very likely to be more sociable (importantly, this personality trait was measured in the context of reacting to strangers) and playful (with strangers). The simplistic explanation could be that the underlying biological structure overlaps in these traits partly because a limited number of genes affect a large set of phenotypic traits (*pleiotropy*). Thus boldness is determined by common genetic and neurohumoral factors that control the behaviour independently of the particular situation, whether the individual explores an area or a stranger. However, it also seems the case that such correlations between personality traits are not necessarily set. For example, in many species boldness seems to affect aggressive tendencies, but no such relationship was found in the case of the dog. Bolder individuals were not necessarily more aggressive, according to the personality structure described by Svartberg (2002). This means that selection can change the relationship between personality factors in certain environments.

Now we can raise the question of how the process of domestication affected the personality structure of dogs. Importantly, so far no personality model has been put forward for the wolf, and discussion of wolf personalities is confined to single cases or whether assertiveness (the tendency to dominate) is a heritable trait (Packard 2003, but see MacDonald 1987).

One hypothesis would predict that the original wolf and human environment shared many common elements, so selection affected mainly single personality traits by selecting for a different mean value in the population, thus changing the frequency distribution of existing phenotypes. For example, there could have been selection against boldness in dogs because by sharing the anthropogenic environment they had less need to leave the area (the tendency for dispersal to novel areas is often associated with boldness). In addition, certain novel personality types could emerge (i.e. extremely low levels of boldness) that had not been present in the wolf population. This idea is in line with the arguments of Svartberg (2002) and others that the boldness–shyness personality is a trait inherited from the wolf. More interesting consequences can be hypothesized from their other finding that the boldness–shyness trait is independent of the aggressiveness trait in dogs. This suggests that selection for less bold individuals did not necessarily reduce the general level of aggressive behaviour of the population (and vice versa), and more generally selection for aggressive behaviour (in either direction) could be accomplished without affecting the behaviour reflected in the boldness–shyness personality trait. Interestingly, Fox (1972) noted a relationship between aspects of boldness

(prey killing and exploratory behaviour) and dominance in wolf cubs.

Observing young (1–7 months old) wolves MacDonald (1987) noted that fear of objects (the reverse manifestation of boldness) seemed to be independent of their behaviour (attraction) towards humans. This raises the important possibility that selection for a preference towards humans might be (at least partially) independent of being bold or fearful in general (see also Ginsburg and Hiestand 1992). However, it should be noted that in dogs the boldness–shyness personality trait seems to be related to sociability (attraction to strangers, see above), which seems to contradict findings in these wolf cubs.

Thus in both cases mentioned above (boldness × aggression; boldness × sociability) the nature and magnitude of independence remain to be investigated. Nevertheless, one partial answer to these questions can be provided by comparable tests on wolves and dogs with regard to boldness, sociability, or play. In a small population we have found no differences in reaction to novelty in socialized wolves and dogs, but wolves were more aggressive (towards a familiar handler) and less docile (struggling more in the hands of the experimenter) (Győri et al. 2009) (Figure 10.1). Again, this seems to contradict findings in many species like the bighorn sheep in which docile individuals are usually less bold (Dingemanse and Réale 2005).

Recently, Hare and Tomasello (2005) argued that domestication might have affected personality traits, especially those associated with 'fear and aggression'. According to their *emotional reactivity* hypothesis, domestication has affected certain personality traits in a way that has increased the dog's chances of survival in an anthropogenic environment. These ideas are also supported by the selection experiment in foxes (see Chapter 5.6), although there are no data on how this selection affected personality traits in foxes. Although the 'emotional reactivity' hypothesis is a likely candidate to explain behavioural changes from wolf to dog, much work needs to be done in order to achieve greater understanding of the relationship between the different personality traits in both wolves and dogs.

These ideas also gain support from comparison of different breeds in the DMA test battery

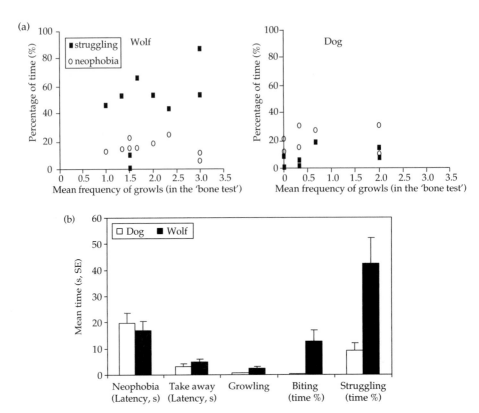

Figure 10.1 (a) In a small sample of wolf cubs (left) and dog puppies (right) there seems to be no clear relationship between a measure of aggressiveness (growling in the bone test) and docility/struggling when restrained (in the hands of the experimenter), and neophobia measured as the latency of approaching a novel object. (b) Wolf cubs struggle more, bite more, growl more than dog pups when interacting with humans. No difference was found in the case of neophobia and the latency to take away a bone. * = significant difference at p < 0.05 (for more detalis see Győri *et al.* 2009).

(Svartberg and Forkman 2002, Svartberg 2006). If traditional breed groups (based on the FCI group-ing) were compared, broad similarity was found (Box 10.2). Most groups showed a similar structure of personality traits, but exceptions occurred (e.g. the sociability and playfulness trait could not be distinguished in the 'retrievers, water dogs, flush-ing dogs' group). A related study did not find dif-ferences in (standardized) values for four personality traits (playfulness, curiosity/fear, socia-bility, aggressiveness) in different groups of dogs (herding dogs, working dogs, gun dogs, and terri-ers). This finding was somewhat surprising because folk knowledge often argues in favour of differ-ences in these traits in these groups of dogs. However, if individual breeds were analysed together by multivariate statistical methods

(cluster analysis) then an interesting four-way grouping resulted, showing a divergent difference in various personality traits. Svartberg (2006) explained these results by providing some evidence that the categorization of breeds in this analysis relates to their present utility and reflects recent selective effects for these new functions. Thus the original (functional) categorization of the breeds refers mainly to morphological similarity but became independent of the underlying behavioural traits because at present many of these breeds fulfil different functions. For example, herding dogs like the Belgian Malinois are now used as police or border guard dogs. Accordingly, Svartberg (2006) argued that dogs (breeds) are under continuous selection by particular human environments (e.g. working dog, herding dog, companion dog)

Box 10.2 A case study for dog personality research: the Dog Mentality Assessment test

Recently, investigations led by Svartberg and others (see text) published a series of studies on dog personality based on the Swedish dog population that was subjected to the Dog Mentality Assessment (DMA) Test from 1997. This data set consists of more than 10 000 dogs belonging to a variety of breeds. Importantly, this test was designed not in order to investigate dog personality but to improve breeding standards in working dogs (Svartberg and Forkman 2002). The utilization of such a large data set has both advantages and disadvantages. The large number of dogs allows detailed statistical analysis of small effects, the use of multivariate methods, quantitative genetic

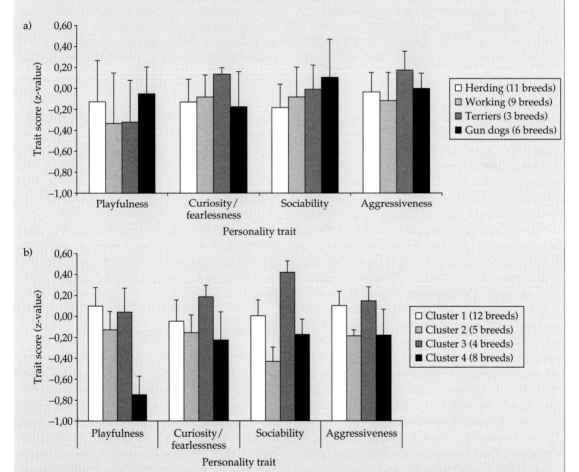

Figure to Box 10.2 (a) The categorization of breeds (herding, working, terriers, gun dogs) does not suggest differences in personality traits (based on 2426 dogs and 29 breeds) (b) A cluster analysis of the same 29 breeds leads to a different grouping (clusters 1–4) which, however, show a more divergent pattern in the personality traits. It is assumed that the breeds in the same cluster had been exposed to the same selective environment which led to similarities in the personality traits. Similar analysis in different countries might have led to other distribution of breeds. (Cluster 1: Australian kelpie, Belgian Tervueren, Rottweiler, Golden retriever, etc.; Cluster 2: Briard, poodle, Belgian Groenendael, etc.; Cluster 3: boxer, Labrador retriever, etc.; Cluster 4: Rough collie, Leonberger, pinscher, English springer spaniel, etc.). Trait scores (mean and standard deviation) normalized for comparative purposes (redrawn from Svartberg 2006).

continues

which can be carried out independently of the morphological traits and historical aspects of the breed. If true, this would also mean that most (if not all) breeds have retained their genetic capacity to fulfil many functions in the human environment, although the effects of such selection may vary. However, it should be noted that the actual pattern (the breeds in the groupings) obtained by Svartberg could be dependent on peculiarities of the Swedish dog population, which for many years was isolated by quarantine laws from most European populations, and/or by the particular attitude of Swedish people in using one or another breed for a given function. In addition, it could be the case that the working breeds are raised in a different environment (which was not controlled for in this arrangement) in comparison to the other breeds. Thus the effect might be not genetic but environmental, and these should be separated experimentally before any final conclusion can be drawn.

If we assume that there is a larger difference between wolf and human environments then selection might have resulted in the emergence of novel personality traits. Similarly, as the chase-proneness trait may be specific for canids in comparison to humans, one might assume the emergence of dog-specific personality traits that could also increase social competence in dogs (see Chapter 8.9). One such candidate trait might be 'playing with humans', which shows no relationship to conspecific play (Svartberg 2005) and might relate to special

aspects of dog–human relationship including a tendency to cooperate (Rooney *et al.* 2001, Naderi *et al.* 2001). Personality tests in dogs do not usually look for cooperativeness (although hunting dogs are tested for such a trait, e.g. Brenoe *et al.* 2002) which may bring in additional traits to the personality structure, as individual dogs vary in this tendency, and some are more independent (e.g. Szetei *et al.* 2003). If such hypotheses were supported then this would provide some argument for the effect of the selective environment on the evolution of personality in general.

10.4 Mechanistic approach

The mechanistic approach to personality is interested in how underlying genetic and physiological processes control or reflect a given personality trait and how the genetic and environmental factors interact during the epigenetic process that determines a certain personality type. Unfortunately, such studies are relatively rare in the case of the dog, especially in concert with investigations on personality traits.

10.4.1 Insights from genetics

The large sample of dogs participating in the DMA test offered the possibility to look for quantitative genetic effects on the personality traits (Saetre *et al.* 2006). Investigating two breeds (German shepherd and Rottweiler) in parallel, the authors found that the pattern of inheritance was very similar in two breeds and it revealed a common underlying genetic factor that was related to the boldness–shyness trait. This trait showed much higher heritability than the individual behavioural traits, which suggest that selection for this trait is possible but also cautions against the practice of selecting on the basis of only a few behavioural traits.

Although quantitative studies are important to verify underlying genetic variance (see Goddard and Beilharz 1984, 1985, Wilsson and Sundgren 1998, 1998, Ruefenacht et al. 2002, van den Berg et al. 2003), and may provide estimates of the number of potential genes, they cannot single out the particular genes that influence personality traits. The wider availability of molecular genetic methods offered the possibility of two different ways to model gene and trait associations. One model assumes that personality traits are under the control of a number of genes (*quantitative trait loci*, QTL) which on their own have a relatively small effect (e.g. Flint et al. 1995). By using polymorphic genetic markers of the nuclear DNA, researchers look for associations between the presence of these markers at a given location on the chromosome and the particular phenotypic trait. So far this method has not been applied to look for association between genetic loci and behavioural traits, but Chase et al. (2002) used this method for analysing the QTL's controlling features of the skeleton. In principle QTL methods could be applied to behaviour; however, these traits may be more variable than skeletal features and more difficult to record, and a large number of genetic markers are needed. Furthermore, significant association between a locus and a trait has to be followed by the search for the gene, which is a very complex task with many pitfalls (e.g. Nadeau and Frankel 2000).

Other genetic models are more hypothesis-driven. Here it is assumed that the phenotypic trait is determined to some extent by genes that have a major effect. The role of such *candidate genes* can be predicted on the basis of neurobehavioural or behaviour genetic studies that show that the modification of certain hormones or transmitter levels (either directly or indirectly) affects personality traits. This kind of analysis assumes that the variability in the phenotypic trait is partially explained by the allele polymorphism in the gene. This means that the manifestation of the trait will depend on the allelic constitution of the individual, because the presence of a certain type of allele predisposes it to a given magnitude of the trait. In recent years this approach has become very popular in humans; for example, certain alleles of the dopamine receptor (DRD4) seem to be associated with novelty seeking or hyperactivity (Castellanos and Tannock 2002). However, this kind of analysis is also not without problems. First, it is not enough to postulate; it has to be shown using independent methods that the present polymorphic alleles indeed cause some measurable biological differences, e.g. change the affinity of the receptor or its distribution in the brain. Second, there is a high chance of getting false positive results. For example, finding that dogs of one breed (characterized by one set of alleles) differ from another breed in a phenotypic trait associated with a candidate gene does not provide evidence for an effect, because other breed-specific background genetic effects can probably account for this. Thus such analysis should be carried out in a single breed in which dogs are derived from a well-described population and are not close relatives (although there are methods that rely on family trees for analysing candidate gene effects) (Box 10.3).

10.4.2 Physiological correlates of personality traits

There has been a continuing interest in investigating the correlation between neurobiological and neuroendocrine variables and personality traits, partly because of a belief that parallels in human and animal models provide support for a homologous origin of these traits. Unfortunately, there has been little systematic research in this area. Most investigations targeted single traits like aggressiveness ('dominance-proneness') or fearfulness ('stress-proneness'). In both cases dogs fit into the broad picture that has been obtained in other mammals including humans, monkeys, and rats.

Dogs that have been characterized as stress-prone by their handlers (Vincent and Mitchell 1996) displayed higher levels of blood pressure and heart

Box 10.3 Human parallels? Candidate gene analysis of the DRD4 receptor in the dog

The gene (*DRD4*) for one type of dopamine receptors expressed at various locations in the brain was one of the first candidate genes for which it was implied that allelic differences are associated with different patterns of behaviour and personality traits in humans. These associations include activity and attention, novelty seeking, and behavioural anomalies such as hyperactivity and attention deficit (Reif and Lesch 2003).

Importantly, a similar polymorphism in the same receptor gene has been revealed in dogs by Japanese researchers who also provided data for the distribution of these alleles among different breeds (Ito *et al.* 2004). The distribution of the most common alleles shows an interesting pattern. Japanese breeds (Akita, Hokkaido, Shiba) are characterized by different set of alleles from European breeds. Most European breeds have the same two versions of the six (or even more) alleles that are known. The one exception is the West Highland terrier, which shares one allele with the

Japanese breeds. A similar pattern is evident for the husky, which is usually regarded as an Asian breed. It also interesting to see that in the case of the German shepherd the samples collected in Japan and Hungary show a similar allele frequency (Table to Box 10.3).

Based on a questionnaire originally designed for measuring the activity and impulsivity traits of human children on the basis of parents' reports, Vas *et al.* (2007) developed a validated method to obtain similar measures for dogs by asking their owners. Using this questionnaire on a population of police dogs (male German shepherds) we found that dogs homozygous for one allele variation of the gene (*DRD4-435*) showed a decreased activity, in contrast to heterozygous dogs or dogs that are homozygous for the other allele (*DRD4-447a*). This finding offers the possibility that, as in humans, this dopamine receptor plays a role in influencing activity levels and/or other aspects of dogs' personality (Héjjas *et al.* 2007).

Table to Box 10.3 The distribution of the DRD4 allele in various dog breeds. All data were collected on individuals living in Japan. Data are from Ito *et al.* (2004), but only a smaller sample of breeds is presented here. In the case of German and Belgian shepherds similar data were collected on a larger sample in Hungary (Héjjas *et al.* 2007#).

Breed	No. of individuals genotyped	Allele types (only the 6 most common)						Total no. of alleles found
		435	447a	447b	486	498	549	
Akita	19	0.00	0.08	0.55	0.13	0.21	0.03	5
Hokkaido	37	0.05	0.01	0.19	0.00	0.45	0.30	5
Shiba	192	0.01	0.10	0.53	0.05	0.05	0.26	7
Beagle	142	0.61	0.35	0.04	0.00	0.00	0.00	5
German shepherd	25	0.58	0.36	0.02	0.00	0.04	0.00	4
German shepherd (#)	294	0.64	0.35	0.00	0.00	0.00	0.00	4
Belgian shepherd (#)	341	0.45	0.55	0.00	0.00	0.00	0.00	3
Golden retriever	174	0.74	0.23	0.03	0.00	0.00	0.00	3
Labrador retriever	134	0.25	0.72	0.03	0.00	0.00	0.00	3
Shetland sheepdog	107	0.16	0.81	0.00	0.00	0.00	0.00	5
Siberian husky	47	0.01	0.33	0.13	0.01	0.51	0.01	6
West Highland white terrier	35	0.00	0.43	0.03	0.00	0.54	0.00	3
Yorkshire terrier	49	0.39	0.49	0.06	0.02	0.01	0.00	6

Box 10.3 *continued*

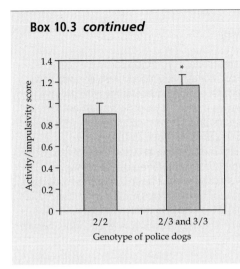

Figure to Box 10.3 Handlers of German shepherd police dogs (N = 72) indicate different levels of activity depending on the allelic constitution of the given individual dog (435: 'short allele'–2; 447; 'long allele'–3). Generally, dogs that are homozygous for the short 435 allele (2) are less active (mean scores obtained on the 'activity impulsivity' trait in the dog ADHD rating scale) (based on data from Hèjjas *et al.* 2007) (* =p<0.05).

rate than dogs that were less sensitive. Stress proneness was described as being unusually fearful and showing difficulties in adapting to novel situations. This is in agreement with findings that stressful stimulation with sudden and novel stimuli increases heart rate in dogs (e.g. Beerda *et al.* 1997). The picture was not so clear when changes in blood cortisol concentration were used as the correlated measure (Box 10.4). Associations with personality traits were not found, and the elevation of cortisol levels seem to be more specific. Various stimuli (noise, electric shock, etc.) did not result in increased cortisol levels (Beerda *et al.* 1997), but in contrast a simulated thunderstorm doubled the cortisol levels (Dreschel *et al.* 2005). Moreover, dogs separated from their conspecific companion and left isolated also reacted by increase in cortisol concentration. Interestingly, humans but not dog companions were effective in reducing these elevated cortisol levels by joining the isolates (Tuber *et al.* 1996, Coppola *et al.* 2006). A similar specific 'calming' (reducing heart rate levels) effect of human presence and petting/grooming was observed in other studies (Kostraczyk and Fonberg 1982, McGreevy *et al.* 2005).

The relationship between physiological correlates, like cortisol, and behaviour is even more complex in the case of aggressiveness. In studies of dominance rank in free-living wolf packs, higher-ranking animals had increased cortisol concentration compared

to lower-ranking companions (Sands and Creel 2004). However, this and other similar observational studies cannot tell what kind of manifestation of the trait helped the individual to reach the top position. It is often assumed that increased aggressiveness is the prerequisite for achieving high rank; however, most observations supporting this idea were done either on young wolves or in wolves living in captivity. Since in nature wolves disperse from their native pack, captive situations might thus be misleading (McLeod *et al.* 1995). Moreover, in such observational cases it is difficult to separate the basal hormone levels, which might reflect the status, from the actual cortisol levels, which can be the result ongoing agonistic interactions.

Recently, we have used a modified version of the 'threatening test' (Vas *et al.* 2005, Figure 8.3) to investigate the relationship between human-directed aggression and cortisol changes in a population of police dogs (male German shepherds) (Horváth *et al.* 2007). Generally the dogs' cortisol levels rose after they were threatened by a human; however, a multivariate analysis revealed that dogs could be categorized as being either bold (showing a tendency to counter-attack), shy (showing a submissive tendency), or ambivalent (displaying passivity and displacement behaviours). These three groups did not differ in pre-test cortisol levels, but the effect of threat in enhancing cortical concentration was largest in the ambivalent group.

Box 10.4 Genetic and physiological aspects of personality

Modern methods make it possible to collect genetic and physiological data in parallel with recording behaviour, in order to understand the neurobiological and neuroendocrine control of personality traits in dogs. Thus the physiological status of the dog can be described by applying non-invasive heart rate measures using portable equipment. Although this measure is sensitive to the bodily movements of the dog, it reliably parallels the dog's reaction to external stimulation (e.g. Beerda *et al.* 1997, Palestrini *et al.* 2005 Maros *et al.* 2007).

So far acute cortisol concentration has usually been determined from blood samples. Saliva cortisol correlates with blood levels (Beerda *et al.* 1997), which offers the possibility of using this non-invasive method by taking a small amount of saliva before and after stimulation (e.g. Dreschel *et al.* 2005, Horváth *et al.* 2007).

A similar method can be used to collect a DNA sample from buccal epithelial cells in the mucous membrane of the dog's mouth by using a cotton swab. These DNA samples offer the possibility of identifying many hundreds of gene polymorphisms (see also Héjjas *et al.* 2007, Overall *et al.* 2006).

(a)  (b)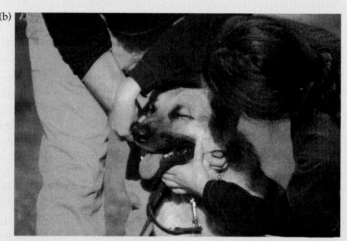

Figure to Box 10. 4 (a) The dog wears a mobile apparatus for measuring heart rate changes in parallel with external stimulation. (b) Collection of saliva or DNA sample takes a few seconds, and although it may feel uncomfortable for the dog, it is not painful.

This suggests that ambivalent dogs, which experienced problems in how to respond to the threatening human, were the most stressed, in contrast to the other dogs which used one or other tactic (attack or withdraw) to resolve the situation.

10.5 Conclusions for the future

Although the investigation of dog personality is one of the oldest subjects in dog behavioural studies, we are only now in the position to build modern personality models of the dog by relying on diverse methodology. Importantly, the unit for obtaining such data should be the individual and

there are some arguments as to why personality trait models should prefer methods that make use of observed behavioural traits recorded under controlled testing conditions.

An important subject could be the comparative and quantitative analysis of dog and wolf personalities, perhaps with an emphasis on early temperamental traits in young animals. It seems that all dog breeds share the same overall personality structure but at the same time they all have the genetic variability to respond to a wide range of anthropogenic environments which might select for certain aspects of a given personality trait without changing the overall structure.

Modern methods offer the possibility of discovering the genetic and neuroendocrine controls of personality traits. In this way the parallels between human and dog personality could be extended because such analysis might reveal not only a behavioural convergence (or homology) but also analogous underlying processes. Such studies could be important for selecting appropriate dogs for various working tasks or identifying potential causes of problem behaviour (Overall 2000).

Further reading

Pavlov's work is important not only for its historic interest but also to judge the advances that have been made. Recent reviews cover this topic from the psychological theory (Jones and Gosling 2005) through methodological issues (Taylor and Mills 2006) to many practicalities (Diederich and Giffroy 2006). A recent comparative review edited by Carere and Eens (2005) provides an ethological perspective.

CHAPTER 11

Afterword: heading towards 21st-century science

11.1 Comparare necesse est!

If the goal of this volume needs to be summarized, it has been to redirect our interest in the comparative behavioural biology of dogs. The foundation of this research was laid down very clearly by Scott and Fuller (1965), but unfortunately these initial efforts have not been continued. It is telling that their book was republished after more than 40 years, and even then there was little experimental research which could have been considered as adding novel knowledge to the topic. However, in recent years important changes have taken place and there is now much hope that the future will bring huge changes in this field.

11.2 Natural model

Especially with regard to population genetic considerations, Scott and Fuller implied that the dog population can be regarded as a model for the human population, but this idea was not reflected either in their experimental design or in their genetic work. A more recent re-invention of this approach comes from behavioural work. First, these researchers not only noted the behavioural parallels between dogs and humans and stressed that dogs and humans share the same environment, but also used these natural populations of dogs for experimental research (e.g. Miklósi et al. 2004, Hare and Tomasello 2005). This also offered the possibility of a direct behavioural comparison between dogs and humans (children), and differed from the method used by Scott and Fuller who based their work on laboratory populations of dogs. Second, behavioural convergence, a similar living environment, and environ-

mental influences make the dog a useful subject to model malformations in human behaviour (Fox 1965, Overall 2000). It is assumed that natural behavioural models have greater construct validity, and provide also a more realistic testing ground than laboratory models which, although well controlled, often fail to be good predictors of what happens in 'real situations'.

11.3 Evolving dogs

Another argument supporting the idea of dogs as a natural model was that researchers have assumed that dogs had to evolve novel aspects of behaviour in order to be successful in the anthropogenic environment. Although this idea was often raised by Scott and Fuller (1965), the experimental programme they carried out did not reflect on aspects of behavioural interaction between humans and dogs, and the comparison of dogs and wolves remained at a very broad and general level.

Today most researchers agree that at the behavioural level not only have convergent processes made dogs fit for the anthropogenic environment, but also that dog and human behaviour actually share some important features. From the evolutionary point of view this means that the changes from an assumed common ancestor (today 'represented' by the extant chimpanzee) towards the Hominidae clade are paralleled by changes that took place during domestication. Importantly, the two processes did not take place on the same time scale. *Homo sapiens* shared its last common ancestor with the chimpanzee approximately 6 million years ago, but the separation of dogs and wolves happened 25 000–50 000 years ago. This means that only the trend of the two processes is comparable, not the result. It is also conceivable that although the selective environment of Hominidae was to some extent different from that of dogs, at the functional level the evolved behavioural pattern has many corresponding elements. In order to emphasize this convergent evolution and make a case for similarities with human evolution, this aspect of dog behaviour could be called the *convergent dog behaviour complex* (Topál *et al.* 2009) (Figure 11.1). This line of research suggests that an understanding of the nature of dog–wolf differences should shed some light on how early hominid species must have differed from the last common ancestor (Hare and Tomasello 2005).

One interesting difference between dog and human evolution is that the human common ancestor lived at a relatively well-determined location somewhere in East Africa, whereas the ancestor of the dog was a very successful species that ranged over half the world. With regard to the wolf, also taking into account the evolutionary history of Canidae, we know that it proved to be a very adaptable species that was just about to become the top predator in the northern hemisphere when humans migrated into this area. Thus we could hypothesize that the wolf was a very potent species for being domesticated

because it had the genetic potential to invade novel niches. However, the overall behavioural similarity of other *Canis* species does not exclude the possibility that they could also have been domesticated if the historical situation had been different.

The achievements of modern genetics make a comparative study of the *Canis* species possible, not only on the basis of phylogenetic relationships but also at the level of functional differences and similarities of the genes. Similarly, as has now been done for the human and chimpanzee genomes, the comparative genomics of wolf, dog, coyote, jackal, etc., could actually point to critical differences.

Under the effect of human influence, dogs evolved an even greater diversity than had been present in wolves. The variety of human environments allowed the survival of genotypes which would have no chance in the 'wilderness'. Modern dogs (breeds) also mirror nicely the mosaic evolution of various phenotypic traits which probably contributed to the success of the Canidae over many millions of years. Scott and Fuller (1965) also noted that the wolf phenotype seems to be present in discrete pieces in the various breeds. This process did not, however, lead to the evolution of a 'super wolf'; on the contrary, these dogs could never rival wolves in nature, apart from locations lacking the ancestor (e.g. Australia).

With regard to the human environment, comparative experiments have also shown that dogs have a greater behavioural (phenotypic) plasticity, which seems to parallel the human case. This means that the dog shows a more variable pattern of behaviour over a range of environments than the wolf. This sort of phenotypic plasticity at the population level, which is reflected in the possible ranges of the phenotype (*reaction norm*), should not be confused with the capacity of a genotype to interact with the environment and produce different phenotypes. Dogs as a species could be said to be more plastic than wolves when put in the same anthropogenic environment. It would be interesting to understand how selection led to such increased phenotypic plasticity, as well as understanding the underlying genetic basis of this feature. Any information on these topics could also be revealing for human evolution.

Although recent genetic studies on wolves and dogs noted that present genetic variability is greater

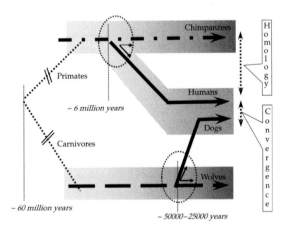

Figure 11.1 A schematic evolutionary relationship between the Hominidae and Canidae. With respect to many social behavioural skills there is a convergent relationship between dogs and humans that live in the same anthropogenic niche. There seems to be a correspondence between behavioural changes that occurred during domestication of dogs and at the split of the linkages leading to extant chimpanzee/human clades.

than expected, this could soon change through the extermination of wolves and the closed breeding practices of dogs. Neither small wolf populations nor the 400 or more dog breeds can retain the genetic variability that was present in the species. Today dogs are subject to a dangerous 'game' which involves irresponsible playing with one tiny aspect of their phenotype: the form. This leads to two important problems. Breeders are encouraged to inbreed in order to fulfil the requirements which lead to genetically homozygous populations, and the absence of selection for behaviour leads to the disappearance of breed-specific traits. Thus this trend brings nothing good for dogs in terms of their evolution because genotypes are being lost and genetic variability is decreasing. McGreevy and Nicholas (1999) made many recommendations that were aimed not only at improving dog welfare but also maintaining genetic variability. For example, they advocated that breeds should not be considered as 'closed' populations, and dogs from other breeds should be crossed in. The example of the Pharaoh dog shows that any form or type of dog can be constructed by using a good 'mixture' from the available breeds. It may be surprising, but one could make a 'Labrador retriever lookalike' in a few generations without using any of the hunting breeds. The recent fashion for cross-breeds (e.g. the labradoodle) does not solve the problem if these dogs are not subsequently allowed to breed. Naturally, mongrel dogs can provide some hope in maintaining genetic diversity, but we know very little about them at the population level. Studies comparing genetic variability in breeds with different populations of feralized dogs are needed. At one extreme one might also consider the 'domestication' of wolves originating from different populations, which was done to some extent in the case of the foxes. Such an experiment would take many years, but apart from providing valuable scientific insights it could also contribute to the increased genetic variability of dogs.

11.4 Behavioural modelling

Using arguments based on functional anthropomorphism, dogs show a high level of social competence in their interactions with humans. This is undoubtedly the result of a selective process, because such competence is not shown by wolves raised in intensive social contact with humans. At the mechanistic level we need to find out more about the changes in genetic control which led to the emergence of the convergent dog behaviour complex. Non-exclusive behaviour models have argued for three kinds of changes, with the common insight that the dog has obtained a more open behavioural system which is in many ways more reactive to the environmental stimulations than the wolf. First, Frank (1980) hypothesized that dogs possess a merged information-processing system which is less constrained in learning, and is more open to making associations between stimuli and behaviour. Second, selection could have affected emotional aspects of 'aggression' and 'fear' (Hare and Tomasello 2005); and third, there are also arguments in favour of a direct selection of certain behavioural traits (e.g. looking into the face of humans, Miklósi *et al.* 2003).

In order to make progress towards a refined model of dog behaviour, the systematic collection of more data is inevitable. However, this is often compromised by the poor design of comparative experiments. The unified look of dogs belonging to the same breed often deceives researchers (and lay people) into believing that this resemblance also reflects genetic and environmental similarity. However, there is no reason to think that two members of the same breed behave more similarly to each other in all respects just because they share some aspects of form. Many workers would deny such thinking, but actually this error is committed when one tests for breed differences (in the sense of genetic difference) without accounting for the possible environmental (and other genetic but not breed-specific) differences which could potentially account for the observed effect. Because in the case of dogs (just as in the case of humans) we are working with natural populations and not laboratory-raised animals, the right sampling and control is very important. In the end such questions can only be solved by large-scale comparative studies.

In the case of sociocognitive behaviours it could be useful to adopt a distributed approach (Johnson 2001) which emphasizes the study of simple behavioural actions that occur among interacting individuals. The distributed approach stresses that the emerging social cognitive structures are the outcomes of the interaction between the individual and its social environment. At the moment we know very little about the interactions between humans and dogs or within the dog community in natural situations. The process of *ontogenetic ritualization* could be very revealing, and studies on the diversity of play signals in dog–human play can be regarded as a step in this direction (e.g. Mitchell and Thompson 1991, Rooney *et al.* 2001). Similarly, tests aimed at measuring performance should be complemented by reports on other aspects of the behaviour during the test. For example, when testing dogs in the two-way choice test using the pointing gesture, it is advisable to give an overall behavioural description and not only report the performance.

The distributed approach makes it obligatory to provide a developmental perspective of the dog–human interaction. We know very little about the effect of humans on the cognitive development of dogs. This is most obvious in the case of dog learning. Despite a huge literature on the practical application of dog training methods (e.g. Lindsay 2001) this field has received very little attention from researchers. Many see dog training as a simple mechanistic method that makes the animal do things in the presence of certain signals. Interestingly, parents (and nurses or teachers) avoid referring to 'training' when they teach infants or young children. The reason is that they also feel that such 'training' should take place in an interactive situation and should not be based on mechanistic application of rules of conditioning ('You get a chocolate if you do your homework!'). In the case of both children and dogs such training should preferably be part of daily interaction and should be based on the rich social tool set that is available both for the human and the dog. Similarly, as in the case of children, any success will depend also on the interacting partners, including the genetic constitution of the dog and the human's social skills. At present we know very little about how various training methods affect the cognitive abilities of dogs.

There should be also some advance in the study of developmental mechanisms in dogs. Although any such programme depends crucially on the cooperation of dog breeders, eventually such information could also help in determining the most advantageous developmental environments for dogs. The study of perinatal learning of olfactory cues (Wells and Hepper 2006) represents such an approach, but similar research involving a range of early stimulation could also reveal the role of early experience on the behavioural (and cognitive) skills of the adult. In parallel, early developmental differences would help to give another perspective on dog–wolf differences. It is likely that a difference in the sensitivity to olfactory cues, especially species-specific cues, is one important starting point for the developmental divergence of the two species in the human environment.

Novel genetic tools open up a new horizon for behaviour genetic studies in dogs. Such investigations could be very informative because the phenotypic variability in natural dog populations should be comparable to that found in humans. The solution is not to put the dog in a laboratory setting but to develop methods which are sensitive enough to detect gene–behaviour associations under these natural conditions. Similar research with regard to personality traits is also required.

11.5 Ethical implications and researchers' mission

Researchers, including us, have found that it is not very helpful to talk about 'experimenting with dogs' because for most dog owners the word 'experiment' has a negative connotation. What it normally means is that a researcher makes an observation under controlled conditions, but most people think (not without reason) that animal experiments bring mostly pain and suffering to the subjects and often end with their death. Fortunately this is not the case in the experiments we propose. In most cases the experiments are designed in such a way that they are part of the daily life of the dog and very often they provide the dog (and the owner) with an enjoyable new experience.

Thus one can see a close parallel between research on dogs and human infants. To put it simply, everything that is ethically acceptable in the case of an infant could also be considered as acceptable in the case of dogs, but nothing more. Relying on this simple guideline, dog researchers could strengthen a positive side of science which does not aim to gather novel information at the cost of destroying lives unnecessarily. Instead of using methods that are or can be harmful to our subjects, dog researchers should promote the application of alternative, non-invasive methods. Although at present such methods have a limited capacity to solve complex methodological problems, dog research could be one driving force for the invention of such techniques.

11.6 Dog genome and bioinformatics

In genetics there is a long tradition of sharing and publishing information. With the characterization of genes in canine disorders and the publication of the dog genome, this revolution in information exchange has also reached dog researchers. Various databases enable interested researchers or experts to search for such information.

Similar databases might be useful for collecting phenotypic information. For example, there are calls to make available the digitized versions of skeletons which have been described in published studies. Since dog research is in its infancy this could be the right time to move in this direction. In the case of behavioural phenotypes a short video clip could be regarded as the unit. Behavioural coding software allows for the notation of video images on a frame-by-frame basis, so that a viewer can see in parallel the description of the behaviour. To start with there would have to be discussion on how behaviour should be described, but in the end researchers would arrive at a system that could be used by everyone who contributes to such a phenotypic database.

On a different note, a closer relationship with robotic technology would also be fruitful. AIBO, the robotic dog, is perhaps an oversimplified recreation of a dog (Kubinyi *et al.* 2004), but through the construction and testing of such robots important insights can be gained about the organization of behaviour. Also, such robots could be also useful in teaching people how to interact with living beings.

11.7 'Paws in hands'

At the moment it seems that the future of dogs and humans is tightly coupled. Although there are no data, it is likely that the human population boom has also been paralleled by a huge increase in the dog population. A very crude estimate would put the size of the dog population somewhere between 0.5 and 1 billion individuals. A large part of the human economy, including veterinary medicine and the production of dog food, is devoted to the support of dogs. Sharing the environment not only means social contact but also that both species are exposed to the same negative effects, like air pollution. Thus it is not so surprising that humans and dogs share many diseases, including not only cancer and various inherited eye diseases but also certain forms of psychiatric conditions. Ageing affects both the human and the dog population.

Similarly, recent changes in the human living, which include lessening of social contacts and leading a very individualistic lifestyle, affect not only human relationships but also our relationship with dogs. Despite arguments that animals should be allowed to live a full life in 'nature', many dogs are prevented from living a natural life in human communities because they spend most of their time alone or at the end of a leash. In families where adults have little time to provide a socially rich environment for their children, dogs will also lack such experience.

In this sense the job of dog ethologists is the same as that of teachers and child psychologists: using all means available to teach humans in modern societies to keep up family life, which has always been essential for providing the appropriate social environment for both our children and our best friends.

References

Abler, W. (1997). Gene, language, number: The particulate principle in nature. *Evolutionary Theory*, 11, 237–248.

Adams, G.J. and Johnson, K.G. (1995). Guard dogs: sleep, work and the behavioural responses to people and other stimuli. *Applied Animal Behaviour Science*, 46, 103–115.

Adler, L.L. and Adler, H.E. (1977). Ontogeny of observational learning in the dog (*Canis familiaris*). *Developmental Psychobiology*, 10, 267–271.

Agnetta, B., Hare, B., and Tomasello, M. (2000). Cues to food locations that domestic dogs (*Canis familiaris*) of different ages do and do not use. *Animal Cognition*, 3, 107–112.

Ainsworth, M.D.S. (1969). Object relations, dependency and attachment: a theoretical review of the infant-mother relationship. *Child Development*, 40, 969–1025.

Albert, A. and Bulcroft, K. (1987). Pets and urban life. *Anthrozoös*, 1, 9–23.

Albert, A. and Bulcroft, K. (1988). Pets, families, and the life course. *Journal of Marriage and the Family*, 50, 543–552.

Alberch, P., Gould, S.J. Oster, G.F., and Wake, D.B. (1979). Size and shape in ontogeny and phyilogeny. *Paleobiology*, 5, 296–317.

Allen, K.M., Blascovich, J., Tomaka, J., and Kelsey, R.M. (1991). Presence of human friends and pet dogs as moderators of automatic responses to stress in women. *Journal of Personality and Social Psychology*, 61, 582–589.

Anderson, J.R., Sallaberry, P., and Barbier, H. (1995). Use of experimenter-given cues during object choice tasks by capuchin monkeys. *Animal Behaviour*, 49, 201–208.

Arant, B.S. and Gooch, W.M. (1982). Developmental changes in the mongrel canine brain during postnatal life. *Human Development*, 7, 179–194

Arkow, P.S. and Dow, S. (1984). Ties that do not bind: a study of human-animal bonds that fail. In: Anderson, R.K., Hart, B.L., and Hart, L.A., eds. *The pet connection: Its influence on our health and quality of life*, pp. 348–354. University of Minneapolis Press, Minneapolis.

Baerends, G.P. (1976). The functional organization of behaviour. *Animal Behaviour*, 24, 726–738.

Baker, P.J., Robertson, C.P.J., Funk, S.M., and Harris, S. (1998). Potential fitness benefits of group living in the red fox, *Vulpes vulpes*. *Animal Behaviour*, 56, 1411–1424.

Baldwin, D.A. and Baird, J.A. (2001). Discerning intentions in dynamic human action. *Trends in Cognitive Sciences*, 5, 171–178.

Balserio, S.C. and Correira, H.R. (2006). Is olfactory detection of human cancer by dogs based on major histocompatibility complex dependent odour components? – A possible cure and a precocious diagnosis of cancer. *Medical Hypotheses*, 66, 270–272.

Bánhegyi, P. (2005). *Rank-order dependent social learning in detour and manipulation tasks in dogs*. (In Hungarian). Dissertation for the fulfilment of MSc in Zoology, Szent István University, Budapest.

Banks, M.R. and Banks, W.A. (2005). The effects of group and individual animal assisted therapy on loneliness in residents of long-term care facilities. *Anthrozoös*, 18, 358–378.

Baranyiova, E., Holub, A., Tyrlik, M., Janackova, B., and Ernstova, M. (2005). The influence of urbanization on the behaviour of dogs in the Czech Republic. *Acta Veterinaria Brno*, 74, 401–409.

Bardeleben, C., Moore, R.L., and Wayne, R.K. (2005). Isolation and molecular evolution of the selenocysteine tRNA (Cf TRSP) and RNase PRNA (Cf RPPH1) genes in the dog family Canidae. *Molecular Biology and Evolution*, 22, 347–359.

Barrett, L. and Henzi, P. (2005). The social nature of primate cognition. *Proceedings of the Royal Society of London Series B*, 272, 1865–1875.

Bartosiewicz, L. (1994). Late Neolithic dog exploitation: chronology and function. *Acta Archeologica Academiae Scientiarum Hungaricae*, 46, 59–71.

Bartosiewicz, L. (2000). Metric variability in Roman period dogs in Pannonia provinvia and the Barbaricum, Hungary. In: Crockford, S.J. ed. *Dogs through time: an archaeological perspective*, pp. 181–192. Archeopress, London.

Bateson, P. (1981) Control of sensitivity to the environment during development. In: Immelmann, K., Barlow, G.W., Petrinovich, L., and Main, M., eds. *Behavioral development: The Bielefeld Interdisciplinary Project,* pp 432–453. Cambridge University Press, Cambridge.

Bateson, P.J.B. and Horn, G. (1994). Imprinting and recognition memory: a neural net model. *Animal Behaviour,* 48, 695–715.

Beaver, B.V. (2001). A community approach to dog bite prevention. *Journal of the American Veterinary Medical Association,* 218, 1732–1749.

Beck, A.M. (1973). *The ecology of stray dogs.* York Press, Baltimore, MD.

Becker, F.R., King, J.E., and Markee, J.E. (1962). Olfactory studies on olfactory discrimination in dogs: II. Discriminatory behaviour in a free environment. *Journal of Comparative Physiological Psychology,* 5, 773–780.

Beerda, B., Schilder, M.B.H., Van Hooff, J.A., and De Vries, H.W. (1997). Manifestation of chronic and acute stress in dogs. *Applied Animal Behaviour Science,* 52, 307–319.

Bekoff, M. (1974). Social play and soliciting by infant canids. *American Zoologist,* 14, 323–340.

Bekoff, M. (1977). Social communication in Canids: evidence for the evolution of stereotyped mammalian display. *Science,* 197, 1097–1099.

Bekoff, M. (1995a). Play signals as punctuation: the structure of social play in canids. *Behaviour,* 132, 419–429.

Bekoff, M. (1995b). Cognitive ethology and the explanation of nonhuman animal behavior. In: Roitblat, H.L. and Meyer, J.A., eds. *Comparative approaches to cognitive science,* pp. 119–150. MIT Press, Cambridge, MA.

Bekoff, M. (1996). Cognitive ethology, vigilance, information gathering, and representation: who might know what and why? *Behavioural Processes,* 35, 225–237.

Bekoff, M. (2000). Naturalizing the bonds between people and dogs. *Anthrozoös,* 13, 11–12.

Bekoff, M. (2001). Observations of scent-marking and discriminating self from other by a domestic dog (*Canis familiaris*): tales of displaced yellow snow. *Behavioural Processes,* 55, 75–79.

Bekoff, M. and Allen, C. (1998). Intentional communication and social play: how and why animals negotiate and agree to play. In: Bekoff, M., Byers, J.A., eds. *Animal play: evolutionary, comparative, and ecological perspectives,* pp. 97–114. Cambridge University Press, Cambridge.

Bekoff, M. and Byers, J.A. (1981). A critical reanalysis of the ontogeny of mammalian social and locomotor play. An ethological hornet's nest. In: Immelman, K., Barlow, G.W. Petrinovich, L., and Main, M., eds. *Behavioural development, The Bielefeld Interdisciplinary Project,* pp. 296–337. Cambridge University Press, New York.

Bekoff, M. and Jamieson, D. (1991). Reflective ethology, applied philosophy, and the moral status of animals. *Perspectives in Ethology,* 9, 1–47.

Bekoff, M. and Meaney, C.A. (1997). Interactions among dogs, people, and the environment in Boulder, Colorado: a case study. *Anthrozoös,* 10, 23–29.

Belyaev, D.K. (1979). Destabilizing selection as a factor in domestication. *Journal of Heredity,* 70, 301–308.

Belyaev, D.K. Plyusnina, I.Z., and Trut, L.N. (1985). Domestication in the silver fox (*Vulpes fulvus*): changes in physiological boundaries of the sensitive period of primary socialization. *Applied Animal Behaviour Science,* 13, 359–370.

Benecke, N. (1992). Archäolzoologische Studien zur Entwicklung der Haustierhaltung. Akademie Verlag, Berlin.

Bering, J.M. (2004). A critical review of the 'enculturation hypothesis': the effects of human rearing on great ape social cognition. *Animal Cognition,* 7, 201–213.

Beritashvili, J.S. (1965). *Neural mechanisms of higher vertebrate behaviour.* J. &A. Churchill, London.

Bernstein, P.L., Friedmann, E., and Malaspina, A. (2000). Animal-assisted therapy enhances resident social interaction and initiation in long-term care facilities. *Anthrozoös,* 13, 213–224.

Biben, M. (1982). Object play and social treatment of prey in bush dogs and crab-eating foxes. *Behaviour,* 79, 201–211.

Bitterman, M. E. 1965. Phyletic differences in learning. *American Psychology,* 20, 396–410.

Björnerfeldt, S., Webster, M., and Vila, C. (2006). Relaxation of selective constraint on dog mitochondrial DNA following domestication. *Genome Research,* 16, 990–993.

Bleicher, N. (1963). Physical and behavioural analysis of dog vocalisations. *American Journal of Veterinary Research,* 24, 415–427.

Bloom, P. (2004). Can a dog learn a word? *Science,* 304, 1605–1606.

Blough, D.S. and Blough, P.M. (1977). Animal psychophysics. In: Honig, W.K. and Staddon, J.E.R., eds. *Handbook of operant behavior.* Prentice-Hall, Englewood Cliffs, NJ.

Blumberg, M.S. and Wasserman, E.A. (1995). Animal mind and the argument from design. *American Psychologist,* 50, 133–144.

Boehm, T. and Zufall, F. (2006). MHC peptides and the sensory evaluation of genotype. *Trends in Neurosciences,* 29, 100–107.

Boitani, L. (1982). Wolf management in intensively used areas of Italy. In: Harrington, F.H. and Paquet, P.C., eds. *Wolves of the world: perspectives of behavior, ecology and conservation,* pp. 158–172. Noyes Publications, Park Ridge, NJ.

Boitani, L. (1983). Wolf and dog competition in Italy. *Acta Zoologica Fennica,* 174, 259–264.

Boitani, L. (2003). Wolf conservation and recovery. In: Mech, D. and Boitani, L., eds. *Wolves: Behaviour, ecology and conservation*, pp. 317–340. University of Chicago Press, Chicago.

Boitani, L. and Ciucci, P. (1995). Compative social ecology of feral dogs and wolves. *Ethology, Ecology and Evolution*, 7, 49–72.

Boitani, L., Franscisci, P., Ciucci, P., and Andreoli, G. (1995). Population biology and ecology of feral dogs in central Italy. In: Serpell, J., ed. *The domestic dog: its evolution, behaviour and interactions with people*, pp. 217–244. Cambridge University Press, Cambridge.

Bökönyi, S. (1974). The dog. In: Bökönyi, S., ed. *History of domesticated mammals in central and eastern Europe*, pp. 313–333, Akadémiai Kiadó, Budapest.

Bolhuis, J.J. (1991). Mechanisms of avian imprinting. A review. *Biological Review*, 66, 303–345.

Bolk, L. (1926). *Das Problem der Menschenwerdung*. Gustav Fischer, Jena.

Bolwig, N. (1962). Facial expression in primates with remarks on a parallel development in certain carnivores. *Behaviour*, 22, 167–191.

Bonas, S., McNicholas, J., and Collis, G.M. (2000). Pets in the network of family relationships: an empirical study. In: Podberscek, A.L., Paul, E.S., and Serpell, J.A., eds. *Companion animals & us: exploring the relationships between people and pets*, pp. 209–236. Cambridge University Press, Cambridge.

Bowlby, J. (1972). *Attachment*. Penguin, London.

Boyd, C.M. Fotheringham, B., Litchfield, C., McBryde, I., Metzer, J.C. Scanlon, P. *et al.* (2004). Fear of dogs in a community sample. *Anthrozoös*, 17, 146–166.

Bradshaw, J.W.S. and Goodwin, D. (1998). Determination of behavioural traits of pure-bred dogs using factor analysis and cluster analysis: a comparison of studies in the USA and UK. *Research in Veterinary Science*, 66, 73–76.

Bradshaw, J.W.S. and Lea, A.M. (1993). Dyadic interactions between domestic dogs during exercise. *Anthrozoös*, 5, 234–253.

Bradshaw, J.W.S. and Nott, H.M.R. (1995). Social communication and behaviour of companion dogs. In: Serpell, J. ed. *The domestic dog*, pp. 116–130, Cambridge University Press, Cambridge.

Bräuer J., Call, J., and Tomasello, M. (2004). Visual perspective taking in dogs (*Canis familiaris*) in the presence of barriers. *Applied Animal Behaviour Science*, 88, 299–317.

Brenoe, Ú.T., Larsgard, A.G., Johannessen, K.R., and Uldal, S.H. (2002). Estimates of genetic parameters for hunting performance in three breeds of gun hunting dogs in Norway. *Applied Animal Behavioral Science*, 77, 209–215.

Brewer, D., Clark, T., and Phillips, A. (2001). *Dogs in antiquity: Anubis to Cerberus The origin of the domestic dog*. Aris & Phillips, Oxford.

Brisben Jr., I.L., Austad, S.N. (1991). Testing the individual odour theory of canine olfaction. *Animal Behaviour*, 42, 63–69.

Brodgen, W.J. (1942). Imitation and social facilitation in the social conditioning of forelimb flexion in dogs. *American Journal of Psychology*, 55, 72–83.

Bryant, B.K. (1990). The richness of the child-pet relationship: a consideration of both benefits and costs of pets to children. *Anthrozoös*, 3, 253–261.

Burghardt, G.M. (1985). Animal awareness. Current perceptions and historical perspective. *American Psychologist*, 40, 905–919.

Burghardt, G.M. and Gittleman, J.L. (1990). Comparative behaviour and phylogenetic analyses: new wine, old bottles. In: Bekoff, M. and Jamieson, D., eds. *Interpretation and explanation in the study of animal behaviour*, pp. 192–225. Westview Press, Boulder, CO.

Butler, J.R.A., du Toit, J.T., and Bingham, J. (2004). Free-ranging domestic dogs (*Canis familiaris*) as predators and prey in rural Zimbabwe: threats of competition and disease to large wild carnivores. *Biological Conservation*, 115, 369–378.

Buytendijk, F.J.J. (1935). *The mind of a dog*, translated by Lillian. A. Clare. Allen & Unwin, London.

Buytendijk, F.J.J. and Fischel, W. (1936). Über die reaktionen des Hundes auf menschliche Wörter. *Archives de Physiologie*, 19, 1–19.

Byrne, R.W. (1995). *The thinking ape. The evolution of intelligence*. Oxford University Press, Oxford.

Cain, A.O. (1985). Pet as family members. In: Sussman, A., ed. *Pets and the family*, pp. 5–10. Haworth Press, New York.

Cairns, R.B. and Werboff, J. (1967). Behavior development in the dog: an interspecific analysis. *Science*, 158, 1070–1072.

Call, J. (2001). Chimpanzee social cognition. *Trends in Cognitive Sciences*, 5, 388–393.

Call, J. (2004). Inferences about the location of food in the great apes (*Pan paniscus, Pan troglodytes, Gorilla gorilla*, and *Pongo pygmeus*). *Journal of Comparative Psychology*, 118, 232–241.

Call, J., Bräuer, J., Kaminski, J., and Tomasello, M. (2003). Domestic dogs (*Canis familiaris*) are sensitive to attentional state of humans. *Journal of Comparative Psychology*, 117, 257–263.

Cameron-Beaumont, C., Lowe, S.E., and Bradshaw, J.W.S. (2002). Evidence suggesting preadaptation to domestication throughout the small Felidae. *Biological Journal of the Linnean Society*, 75, 361–366.

Candland, D. K. (1993). *Feral children and clever animals*. Oxford University Press, New York.

Carbone, C., Mace, G.M., Roberts, S.C., and Macdonald, D.W. (1999). Energetic constraints on the diet of terrestrial carnivores. *Nature*, 402, 286–287.

Carere, C. and Eens, M. (2005). Unravelling animal personalities: how and why individual consistently differ. *Behaviour,* 1149–1287.

Caro, T.M. and Bateson, P. (1986). Organisation and ontogeny of alternative tactics. *Animal Behaviour,* 34, 1483–1499.

Casinos, A., Bou, J., Castiella, M.J., and Viladiu, C. (1986). On the allometry of long bones in dogs *(Canis familiaris). Journal of Morphology,* 190, 73–79.

Castellanos, F.X. and Tannock, R. (2002). Neuroscience of attention-deficit/hyperactivity disorder: the search for endophenotypes. *Nature Reviews. Neuroscience,* 3, 617–628.

Cattel, R.B., Bolz, C.R., and Korth, B. (1973). Behavioral types in purebred dogs objectively determined by taxonome. *Behavior Genetics,* 3, 205–216.

Cavalli-Sforza, L.L. and Feldman, M.W. (2003). The application of molecular genetic approaches to the study of human evolution. *Nature Genetics,* 33, Suppl., 266–275.

Cenami Spada, E. (1996). Amorphism, mechanomorphism, and anthropomorphism. In: Mitchell, R.W. Thompson, N.S., and Miles, H.L., eds. *Anthromorphism, anecdotes and animals,* pp. 254–276. State University of New York Press, New York.

Chaix, L. (2000). A preboreal dog from the northern Alps (Savoie, France). In: Crockford, S.J. ed. *Dogs through time: an archaeological perspective,* pp. 49–59. British Archaeological Reports International Series 889, Oxford.

Chalmers, N.R. (1987). Developmental pathways in behaviour. *Animal Behaviour,* 35, 659–674.

Chapuis, N. and Varlet, C. (1987). Short cuts by dogs in natural surroundings. *Quarterly Journal of Experimental Psychology,* 39B, 49–64.

Chapuis, N., Thinus-Blanc, C., and Poucet, B. (1983). Dissociation of mechanisms involved in dogs' oriented displacements. *Quarterly Journal of Experimental Psychology,* 35B, 213–219.

Chase, K., Carrier, D.R., Adler, F.R. *et al.* (2002). Genetic basis for systems of skeletal quantitative traits: principal component analysis of the canid skeleton. *Proceedings of the National Academy of Science of the USA,* 99, 9930–9935.

Cheney, D. and Seyfarth, R. (1990). *How monkeys see the world: inside the mind of another species.* University of Chicago Press, Chicago.

Chisholm, K., Carter, M., Ames, E., and Morison, S. (1995). Attachment security and indiscriminately friendly behavior in children adopted from Romanian orphanages. *Development and Psychopathology,* 7, 283–294

Ciucci, P., Boitani, L., Lucchini, V., and Randi, E. (2003). Dew claw on wolves as an indication of hybridization with dogs. *Canadian Journal of Zoology,* 81, 2077–2081.

Clark, A.B. and Ehlinger, T.J. (1987). Pattern and adaptation in individual differences. In: Bateson, P.P.G. and Klopfer, P.H.P., eds. *Perspectives in Ethology* vol. 7, pp.1–47. Plenum Press, New York.

Clark, G. (1997) Osteology of the Kuri Maori: the prehistoric dog of New Zealand. *Journal of Archaeological Science,* 24, 113–126.

Clark, M.M. and Galef, B.G. (1980). Effects of rearing environment on adrenal weights, sexual development and behaviour in Gerbils: An examination of Richter's domestication hypothesis. *Journal of Comparative and Physiological Psychology,* 94, 857–863.

Clark, T. (2001). The dogs of the ancient Near East. In: Brewer, D., Clark, T., and Phillips, A., eds. *Dogs in antiquity,* pp. 49–80. Aris & Phillips, Oxford.

Clutton-Brock, J. (1984). Dog. In: Mason, I.L. ed. *Evolution of domesticated animals,* pp. 198–210. Longman, London.

Clutton-Brock, J. (1995). Origin of the dog: domestication and early history. In: Serpell, J., ed. *The domestic dog: its evolution, behaviour and interactions with people,* pp.7–20. Cambridge University Press, Cambridge.

Clutton-Brock, J. and Hammond, N. (1994). Hot dogs: comestible canids in preclassic Maya culture at Cuello, Belize. *Journal of Archaeological Science,* 21, 819–826.

Clutton-Brock, J. and Noe-Nygaard, N. (1990). New osteological and C-isotope evidence on Mesolithic dogs: companions to hunters and fishers at Star Carr, Seamer Carr and Kongemose. *Journal of Archaeological Science,* 17, 643–53.

Coren, J.A. and Fox, M.W. (1976). Vocalization in wild canids and possible effects of domestication. *Behavioral Processes,* 1, 77–92.

Coren, S. (1994). The intelligence of dogs. Canine consciousness and capabilities. Free Press, New York.

Coren, S. (2005). How dogs think. Understanding the canine mind. Pocket Books, London.

Coile, D.C. Pollitz, C.H., and Smith, J.C. (1989). Behavioral determination of critical flicker fusion in dogs. *Physiology and Behavior,* 45, 1087–1092.

Collier, S. (2006). Breed-specific legislation and the pit bull terrier: are the laws justified? *Journal of Veterinary Behaviour Clinical Applications and Research,* 1, 17–22.

Collis, G.M. (1995). Health benefits of pet ownership: attachment vs. psychological support. In: *Animals, health and quality of life,* p. 7. VIIth International Conference on Human Animal Interactions.

Colombo, M. and D'Amato, M.R. (1986). A comparison of visual and auditory short-term memory in monkeys *(Cebus apella). Quarterly Journal of Experimental Psychology,* 38, 425–448.

Compton, J.M. and Scott, J.P. (1971). Allomimetic behaviour system: Distress vocalization and social facilitation of feeding in Telomian dogs. *Journal of Psychology,* 78, 165–179.

Cooper, J.J., Ashton, C., Bishop, S., West, R., Mills, D.S., and Young, R.J. (2003). Clever hounds: social cognition in the domestic dog (*Canis familiaris*). *Applied Animal Behaviour Science*, 81, 229–244.

Coppinger, R.P. and Coppinger, L. (2001). *Dogs: a new understanding of canine origin, behavior and evolution*. University of Chicago Press, Chicago.

Coppinger, R. and Schneider, R. (1995). Evolution of working dogs. In: Serpell, J., ed. *The domestic dog*, pp. 22–47. Cambridge University Press, Cambridge.

Coppinger, R.P. and Smith, K.C. (1990). A model for understanding the evolution of mammalian behaviour. In: Genoways, H.H., ed. *Current mammalogy*, pp. 335–374. Plenum Press, New York.

Coppinger, R., Glendinning, J., Torop, E., Matthay, C., Sutherland, M., and Smith, C. (1987). Degree of behavioral neoteny differentiates canid polymorphs. *Ethology*, 75, 89–108.

Coppola, C.L., Grandin, T., and Enns, R.M. (2006). Human interaction and cortisol: can human contact reduce stress for shelter dogs? *Physiology and Behavior*, 87, 537–541.

Corbett, L.K. (1988). Social dynamics of a captive dingo pack: population regulation by dominant female infanticide. *Ethology*, 78, 177–198.

Corbett, L.K. (1995). *The dingo in Australia and Asia*. Comstock/Cornell University Press, Ithaca, NY.

Corbett, L.K. and Newsome, A. (1975). Dingo society and its maintenance: a preliminary analysis: In: Fox, W.M., ed. *The wild canids: Their systematics, behavioural ecology and evolution*, pp 369–379. Van Nostrand Reinhold, New York.

Covert, A.M., Whiren, A.P., Keith, J., and Nelson, C. (1985). Pets, early adolescents, and families. In: Sussman, M., ed. *Pets and the family*, pp. 95–107. Haworth Press, Binghampton, NY.

Cox, R.P. (1993). The human/animal bond as a correlate of family functioning. *Clinical Nursing Research*, 2, 224–231.

Cracknell N.R., Mills D.S., Kaulfuss P. (2007). Can stimulus enhancement explain the apparent success of the model-rival training technique in dogs? Landsberg., G., Mattiello S., Mills D. (eds) *Proceedings of the 6th International Veterinary Behaviour Meeting and European College of Veterinary Behavioural Medicine- Companion Animals and European Society of Veterinary Clinical Ethology*. pp. 108–109. Fondazionne Iniziative Zooprofilattiche e Zootechniche, Brescia.

Crawford, E.K., Worsham, N.L., and Swinehart, E.R. (2006). Benefits derived from companion animals, and the use of the term 'attachment'. *Anthrozoos*, 19, 98–112.

Crockford, S.J. (ed.) (2000). *Dogs through time: an archaeological perspective*. British Archaeological Reports International Series 889, Oxford.

Crockford, S.J. (2006). *Rhythms of life*. Trafford Publishing, Victoria, Canada.

Csányi, V. (1988). Contribution of the genetical and neural memory to animal intelligence. In: Jerison, H. and Jerison, I., eds. *Intelligence and evolutionary biology*, pp. 299–318. Springer-Verlag, Berlin.

Csányi, V. (1989). *Evolutionary systems and society: a general theory*. Duke University Press, Durham, NC.

Csányi, V. (1993). How genetics and learning make a fish an individual: a case study on the paradise fish. In: Bateson, P.P.G, Klopfer, P.H. and Thompson, N.S., ed. *Perspectives in Ethology, Behaviour and Evolution*, pp. 1–52. Plenum Press, New York.

Csányi, V. (2000). The 'human behaviour complex' and the compulsion of communication: key factors in human evolution. *Semiotica*, 128, 45–60.

Csányi, V. (2005). *If dogs could talk*. North Point Press, New York.

Custance, D.M. Whiten, A., and Bard, K.A. (1995). Can young chimpanzee (*Pan troglodytes*) imitate arbitrary actions? Hayes & Hayes (1952) revisited. *Behaviour*, 132, 837–857.

Cutt, H., Giles-Cortia, B., Knuimana, M., and Burkeb, V. (2007). Dog ownership, health and physical activity: A critical review of the literature. *Health and Place*, 13, 261–272.

Dalziel, D.J., Uthman, B.M., Mcgorray, S.P., and Reep, R.L. (2003). Seizure-alert dogs: a review and preliminary study. *Seizure*, 12, 115–120.

Daniels, T.J. and Bekoff, M. (1989). Feralization: the making of wild domestic animals. *Behavioural Processes*, 19, 79–94.

Darimont, C.T., Reimchen, T.E., and Paquet, P.C. (2003). Foraging behaviour by gray wolves on salmon streams in coastal British Columbia. *Canadian Journal of Zoology*, 81, 349–353.

Darwin, C. (1872). The expressions of the emotions in man and animals. John Murray. London.

Davis, S.J.M,. and Valla, F.R. (1978). Evidence for domestication of the dog 12,000 years ago in the Natufian of Israel. *Nature*, 276, 608–610.

Dawkins, R. (1986). *The blind watchmaker*. Longman, London.

Dayan, T. and Galili, E. (2000). A preliminary look at some new domesticated dogs from submerged Neolithic sites off the Carmel coast. In: Crockford, S.J., ed. *Dogs through time: an archaeological perspective*, pp. 29–33. British Archaeological Reports International Series 889, Oxford.

De Palma, C., Viggiano, E., Barillari, E. *et al.* (2005). Evaluating the temperament in shelter dogs. *Behaviour*, 142, 1307–1328.

de Waal, F.B.M. (1989). *Peacemaking among primates*. Harvard University Press, Cambridge.

de Waal, F.B.M. (1991). Complementary methods and convergent evidence in the study of primate social cognition. *Behaviour*, 118, 297–320.

de Waal, F.B.M. (1996). *Good natured. The origins of right and wrong in humans and other animals*. Harvard University Press, Cambridge, MA.

Derix, R., Van Hoof, J., DeVries, H., and Wensing, J. (1993). Male and female mating competition in wolves: female suppression vs. male intervention. *Behaviour*, 127, 141–174.

Devenport, L.D. and Devenport, J.A. (1990). The laboratory animal dilemma: A solution in the backyards. *Psychological Science*, 1, 215–216.

Diederich, C. and Giffroy, J.M. (2006). Behavioural testing in dogs: A review of methodology in search for standardisation. *Applied Animal Behaviour Science*, 97, 51–72.

Dingemanse, N.J. and Réale, D. (2005). Natural selection and animal personality. *Behaviour*, 142, 1159–1185.

Doogan, S. and Thomas, G.V. (1992). Origins of fear of dogs in adults and children: the role of conditioning processes and prior familiarity with dogs. *Behaviour Research and Therapy*, 30, 387–394.

Doré, Y.F. and Goulet, S. (1998). The comparative analysis of object knowledge. In: Langer, J. and Killen, M., eds. *Piaget, evolution and development*, pp. 55–72. NJ. Lawrence Erlbaum Associates, Mahwah, NJ.

Doty, R.L. and Dunbar, I.F. (1974). Attraction of beagles to conspecific urine, vaginal and anal sac secretion odours. *Physiology and Behavior*, 12, 825–833.

Draper, T.W. (1995). Canine analogs of human personality factors. *Journal of General Psychology*, 122, 241–252.

Dreschel, N.A., Douglas, A., and Granger, D.A. (2005). Physiological and behavioral reactivity to stress in thunderstorm-phobic dogs and their caregivers. *Applied Animal Behaviour Science*, 95, 153–168.

Dumas, C. and Paré, D.D. (2006). Strategy planning in dogs (*Canis familiaris*) in a progressive elimination task. *Behavioural Processes*, 73, 22–28.

Dunbar, I.F. (1977). Olfactory preferences in dogs: the response of male and female beagles to conspecific odors. *Behavioral Biology*, 20, 471–481.

Dunbar, I.F., Buehler, M., and Beach, F.A. (1980). Developmental and activational effects of sex hormones on the attractiveness of dogs' urine. *Physiology and Behavior*, 24, 201–204.

Dyer, F.C. (1998). Spatial cognition: lesson from central-place foraging insects. In: Balda, R.P., Pepperberg, I.M., and Kamil, A.C., eds. *Animal cognition in nature*, pp. 119–155. Academic Press, San Diego, CA.

Eckstein, G. (1949). Concerning a dog's word comprehension. *Science*, 13, 109.

Edenburg, N., Hart, H., and Bouw, J. (1994). Motives for acquiring companion animals. *Journal of Economic Psychology*, 15, 191–206.

Edinger, L. (1915). Zur Methodik in der Tierpsychologie. *Zeitschrift für Physiologie*, 70. 101–124.

Edney, A. (1993). Dogs and human epilepsy. *The Veterinary Record*. 132, 337–8

Egenvall, A., Hedhammar, A., Bonnett, B.N., and Olson, P. (1999). Survey of the Swedish dog population. Age, sex, breed, location and enrolment in animal insurance. *Acta Veterinaria Scandinavica*, 40, 231–240.

Egenvall, A., Bonnett, B.N., Olsson, P., and Hedhammar, A. (2000). Gender, age, breed and distribution of morbidity and mortality in insured dogs in Sweden during 1995 and 1996. *Veterinary Record*, 29, 519–525.

Elliot, O. and Scott, J.P. (1961). The development of emotional distress reactions to separation, in pups. *Journal of Genetic Psychology*, 99, 3–22.

Emery, N.J. and Clayton, N.S. (2004). The mentality of crows: convergent evolution of intelligence in corvids and apes. *Science*, 306, 1903–1907.

Erdőhegyi, Á., Topál, J., Virányi, Zs., and Miklósi, Á. (2007). Dog-logic: inferential reasoning in a two-way choice task and its restricted use. *Animal Behaviour*, 74, 725–737.

Fabrigoule, C. (1987). Study of cognitive processes used by dogs in spatial tasks. In: Ellen, P. and Thinus-Blanc, C., eds. *Cognitive processes and spatial orientation in animal and man*, pp. 114–123. Aix-en-Provence, France.

Feddersen-Petersen, D. (1991) The ontogeny of social play and agonistic behaviour in selected canid species. *Bonner zoologische Beiträge*, 42, 97–114.

Feddersen-Petersen, D. (2000). Vocalisation of European wolves (*Canis lupus*) and various dog breeds (*Canis l. familiaris*). *Archives für Tierzucht, Dummerstorf*, 43, 387–397.

Feddersen-Petersen, D. (2001a). Zur Biologie der Aggression des Hundes. *Deutsche Tierärztliche Wochenschrift*, 108, 94–101.

Feddersen-Petersen, D. (2001b). *Hunde und ihre Menschen*. Kosmos Verlag, Stuttgart.

Feddersen-Petersen, D. (2004). *Hundepsychologie. Sozialverhalten und Wesen. Emotionen und Individualität*. Kosmos Verlag, Stuttgart.

Fentress, J.C. (1967). Observations on the behavioral development of a hand-reared male timber wolf. *American Zoologist*, 7, 339–351.

Fentress, J. (1976). Dynamic boundaries of patterned behaviour: Interaction and self-organisation. In: Bateson, P.P.G. and Hinde, R.A., eds. *Growing points in ethology,* pp. 135–169. Cambridge University Press, Cambridge.

Fentress, J. (1993). The covalent animal. In: Davis, H. and Balfour, D., eds. *The inevitable bond,* pp. 44–72. Cambridge University Press, Cambridge.

Fentress, J.C. and Gadbois, S. (2001). The development of action sequences. In: Blass, E.M., ed. *Handbooks of behavioral neurobiology,* Volume 13: *Developmental psychobiology, developmental neurobiology and behavioral ecology: mechanisms and early principles.* Kluwer Academic Publishers, New York.

Fentress, J.C. and Ryon, J. (1982). A long-term study of distributed pup feeding in captive wolves. In: Harrington, F.H. and Paquet, P.C., eds. *Wolves of the world: perspectives of behavior, ecology and conservation,* pp. 238–261. Noyes Publications, Park Ridge, NJ.

Fentress, J.C., Ryon, J., McLeod, P.J. and Havkin, G.Z. (1987). A multidimensional approach to agonistic behavior in wolves. In: Frank, H., ed. *Man and wolf: advances, issues, and problems in captive wolf research,* pp. 253–275. Junk Publishers, Dordrecht.

Finlayson, C. (2005). Biogeography and evolution of the genus *homo. Trends in Ecology and Evolution,* 20, 457–463.

Fischel, W. (1933). Das Verhalten von Hunden bei doppelter Handlungsmöglichkeit. *Zeitschrift für Physiologie,* 19, 170–182.

Fischel, W. (1941). Tierpsychologie und Hundeforschung. *Zeitschrift für Hundeforschung,* 17, 1–71.

Fiset, S., Gagnon, S., and Beaulieu, C. (2000). Spatial encoding of hidden objects in dogs (*Canis familiaris*). *Journal of Comparative Psychology,* 114, 315–324.

Fiset, S., Beaulieu, C., and Landry, F. (2003). Duration of dog's (*Canis familiaris*) working memory in search for disappearing objects. *Animal Cognition,* 6, 1–10.

Fiset, S., Landry, F., and Ouellette, M. (2006). Egocentric search for disappearing objects in domestic dogs: evidence for a geometric hypothesis of direction. *Animal Cognition,* 9, 1–12.

Fisher, J.A. (1990). The myth of anthropomorphism. In: Bekoff, M. and Jamieson, D., ed. *Interpretation and explanation in the study of animal behaviour,* pp. 96–116. Westview Press, Boulder, CO.

Fisher, P.M. (1983). On pigs and dogs: Pets as produce in three societies. In Katcher, A.H. and Beck, A.M., eds., *New perspectives on our lives with companion animals,* pp. 132–137. University of Pennsylvania Press, Philadelphia.

Fitch, W.M. (2000). Homology. A personal view on some of the problems. *Trends in Genetics,* 16, 227–231.

Flint, J., Corley, R., DeFries, J.C., Fulker, D.W., Gray, J.A., and Miller, S. (1995). A simple genetic basis for a complex psychological trait in laboratory mice. *Science,* 269, 1432–1435.

Fondon, J.W. and Garner, H.R. (2004). Molecular origins of rapid and continuous morphological evolution. *Proceedings of the National Academy of Sciences of the USA,* 28, 18058–18063.

Fox, M.W. (1965). *Canine behavior.* C.C. Thomas, Springfield, IL.

Fox, M.W. (1970). A comparative study of the development of facial expressions in canids; wolf, coyote and foxes. *Behaviour,* 36, 49–73.

Fox, M.W. (1971). Behavioral effects of rearing dogs with cats during the 'critical period of socialization'. *Behaviour,* 35, 273–280.

Fox, M.W. (1972). Socio-ecological implications of individual differences in wolf litters: A developmental perspective. *Behaviour,* 41, 298–313.

Fox, M.W. (1975). *The wild canids: Their systematics, behavioural ecology and evolution.* Van Nostrand Reinhold, New York.

Fox, M.W. (1978). *The dog: its domestication and behavior.* Garland STPM Press, New York.

Fox, M.W. (1990). Sympathy, empathy, and understanding animal feelingsmdashand feelings for animals In: Bekoff, M. and Jamieson, D., eds. *Interpretation and explanation in the study of animal behaviour,* pp. 420–434. Westview Press, Boulder, CO.

Frank, H. (1980). Evolution of canine information processing under conditions of natural and artificial selection. *Zeitschrift für Tierpsychologie,* 59, 389–399.

Frank, H. and Frank, M.G. (1982). On the effects of domestication on canine social development and behaviour. *Applied Animal Ethology,* 8, 507–525.

Frank, M.G. and Frank, H. (1988). Food reinforcement versus social reinforcement in timber wolf pups. *Bulletin of the Psychonomic Society,* 26, 467–468.

Frederickson, E. (1952). Perceptual homeostasis and distress vocalization in the puppy. *Journal of Personality,* 20, 472–477.

Freedman, D.G. (1958). Constitutional and environmental interactions in rearing of four breeds of dogs. *Science,* 127, 585–586.

Freedman, D.G., King, J.A., and Elliot, O. (1961). Critical period in the social development of dogs. *Science,* 133, 1016–1017.

Friedmann, E. (1995). The role of pets in enhancing human well-being: physiological effects. In: Robinson, I., ed. *The Waltham book of human–animal interaction: benefits and responsibilities of pet ownership,* pp. 33–53. Pergamon, London.

Friedmann, E., Katcher, A.H., Thomas, S.A., and Lynch, J.J. (1980). Animal companions and one-year survival of patients after discharge from a coronary care unit. *Public Health Reports*, 95, 307–312.

Fuchs, T., Gaillard, C., Gebhardt-Henrich, S., Ruefenacht, S., and Steiger, A. (2005). External factors and reproducibility of the behaviour test in German shepherd dogs in Switzerland. *Applied Animal Behaviour Science*, 94, 287–301.

Fukuzawa, M., Mills, D.S., and Cooper, J.J. (2005). More than just a word: non-semantic command variables affect obedience in the domestic dog (*Canis familiaris*). *Applied Animal Behaviour Science*, 91, 129–141.

Fuller, T.K. Mech, L.D., and Cochrane, J.F. (2003). Wolf population dynamics. In: Mech, D. and Boitani, L., eds. *Wolves: behaviour, ecology and conservation*, pp. 161–191. University of Chicago Press, Chicago.

Furman, W. and Burhmester, D. (1985). Children's perceptions of the personal relationships in their social networks. *Developmental Psychology*, 21, 116–1024.

Furton, K.G. and Myers, L.J. (2001). The scientific foundation and efficacy of the use of canines and chemical detectors for explosives. *Talanta*, 43, 487–500.

Gácsi, M. (2003). The ethological study of attachment behaviour of dogs toward their owner. (In Hungarian). PhD dissertation. Eötvös Lóránd University, Budapest.

Gácsi, M., Topál, J., Miklósi, Á., Dóka, A. and Csányi, V. (2001). Attachment behaviour of adult dogs (*Canis familiaris*) living at rescue centres: forming new bonds. *Journal of Comparative Psychology*, 115, 423–431.

Gácsi, M., Miklósi, Á., Varga, O., Topál, J., and Csányi, V. (2004). Are readers of our face readers of our minds? Dogs (*Canis familiaris*) show situation-dependent recognition of human's attention. *Animal Cognition*, 7, 144–153.

Gácsi, M., Győri, B., Miklósi, Á., *et al.* (2005). Species-specific differences and similarities in the behavior of hand raised dog and wolf pups in social situations with humans. *Developmental Psychobiology*, 47, 111–122.

Gácsi, M., Kara, E., Belényi, B., Topál, J., and Miklósi, Á. (2008). Critical considerations on the comprehension of human pointing in dogs: Developmental and methodological concerns *Animal Cognition* (in press).

Gácsi, M., Topál, J., and Miklósi, A. (2009). Effects of selection for cooperation and attention? New perspectives on evaluating dogs' performance in human pointing tests (submitted).

Gadbois, S. (2002). The socioendocrinolgy of aggression-mediated stress in timber wolves (*Canis lupus*). PhD dissertation, Dalhousie University, Halifax, NS.

Gagnon, S. and Doré, F.Y. (1992). Search behavior in various breeds of adult dogs (*Canis familiaris*): object permanence and olfactory cues. *Journal of Comparative Psychology*, 106, 58–68.

Gagnon, S. and Doré, F.Y. (1993). Search behavior of dogs (*Canis familiaris*) in invisible displacement problems. *Animal Learning and Behavior*, 21, 246–254.

Gagnon, S. and Doré, F.Y. (1994). Cross-sectional study of object permanence in domestic pups (*Canis familiaris*). *Journal of Comparative Psychology*, 108, 220–232.

Galik, A. (2000). Dog remains from the late Hallstatt period of the chimney cave Durezza, near Villach (Carinthia, Austria). In: Crockford, S.J., ed. *Dogs through time: an archaeological perspective*, pp. 129–137. British Archaeological Reports International Series 889, Oxford

Gallistel, C.R. (1990). *The organization of learning*. MIT Press, Cambridge, MA.

Garcia, J. and Koelling, R.A. (1966). Relation of cue to consequence in avoidance learning. *Psychonomic Science*, 5, 123–124.

Gazit, I. and Terkel, J. (2003). Domination of olfaction over vision in explosives detection by dogs. *Applied Animal Behaviour Science*, 82, 65–73.

Gazit, I., Goldblatt, A., and Terkel, J. (2005). The role of context specificity in learning: The effects of training context on explosives detection in dogs. *Animal Cognition*, 8, 143–150.

Gergely, G. and Csibra, G. (2006). Sylvia's recipe: The role of imitation and pedagogy in the transmission of human culture. In Enfield, N.J. and Levinson, S.C., eds., *Roots of human sociality: culture, cognition, and human interaction*, pp. 229–255. Berg Publishers, Oxford.

Gese, E.M. and Mech, L.D. (1991). The dispersal of wolves (*Canis lupus*) in northeastern Minnesota, 1969–1989. *Canadian Journal of Zoology*, 69, 2946–2955.

Ginsberg, J.R. and Macdonald, D.W. (1990). *Foxes, wolves, jackals and dogs. An action plan for the conservation of canids*. IUCN World Conservation Union, Gland.

Ginsburg, B.E. (1975). Non-verbal communication: The effect of affect on individual and group behaviour. In: Pliner, P., Krames, L. and Alloway, T., eds. *Non-verbal communication of aggression*, pp. 161–173. Plenum Press, New York.

Ginsburg, B.E. (1987). The wolf pack as a socio-genetic unit. In: Frank, H., ed. *Man and wolf: Advances, issues and problems in captive wolf research*, pp. 401–413. Dr W. Junk Publishers, Dordrecht.

Ginsburg, B.E. and Hiestand, L. (1992). Humanity's 'best friend': the origins of our inevitable bond with dogs. In: Davis, H. and Balfour, D., eds. *The inevitable bond*, pp. 93–108. Cambridge University Press, Cambridge.

Gittleman, J.L. (1986). Carnivore life history patterns: allometric, phylogenic, and ecological associations. *American Naturalist*, 127, 744–771.

Goddard, M.E. and Beilharz, R.G. (1984). A factor analysis of fearfulness in potential guide dogs. *Applied Animal Behaviour Science*, 12, 253–265.

Goddard, M.E. and Beilharz, R.G. (1985). Individual variation in agonistic behaviour in dogs. *Animal Behaviour*, 33, 1338–1342.

Goddard, M.E. and Beilharz, R.G. (1986). Early prediction of adult in potential guide dogs. *Applied Animal Behaviour Science*, 15, 247–260.

Goldsmith, H., Buss, A., Plomin, R. *et al.* (1987). Roundtable: what is temperament? Four approaches. *Child Development*, 58, 505–529.

Gomez, J.C. (1996) Nonhuman primate theories of (nonhuman primate) minds: some issues concerning the origins of mindreading. In: Carruthers, P. and Smith, P.K., eds. *Theories of theories of mind*, pp. 330–343. Cambridge University Press, Cambridge.

Gomez, J.C. (2004). *Apes, monkeys, children and the growth of the mind*. Harvard University Press, Cambridge, MA.

Gomez, J.C. (2005). Species comparative studies and cognitive development. *Trends in Cognitive Sciences, 9*, 118–125.

Goodwin, M., Gooding, K.M. and Regnier, F (1979). Sex pheromone in the dog. *Science*, 203, 559–61.

Goodwin, D., Bradshaw, J.W.S., and Wickens, S.M. (1997). Paedomorphosis affects visual signals of domestic dogs. *Animal Behaviour*, 53, 297–304.

Goring-Morris, A.N. and Belfer-Cohen A. (1998). The articulation of cultural processes and Late Quaternary environmental changes in CisJordan. *Paleorient*, 23, 71–93.

Gosling, S.D. (2001). From mice to men: what can we learn about personality from animal research? *Psychological Bulletin, 127*, 45–86.

Gosling, S.D. and Bonnenburg, A.V. (1998). An integrative approach to personality research in anthrozoology: ratings of six species of pets and their owners. *Anthrozoös*, 11, 148–156.

Gosling, S.D., Kwan, V.S.Y., and John, O.P. (2003). A dog's got personality: A cross-species comparative approach to evaluating personality judgments. *Journal of Personality and Social Psychology*, 85, 1161–1169.

Gould, S.J. and Lewontin, R.C. (1979). The sprandels of San Marco and the Panglossian paradigm: a critique of the adaptationist programme. *Proceedings of the Royal Society of London B*, 205, 581–598.

Gould, S.J. and Vbra, E.S. (1982). Exaptation—a missing term in the science of form. *Paleobiology*, 8, 4–15.

Graham, L., Wells, D.L., and Hepper, P.G. (2005). The influence of olfactory stimulation on the behaviour of dogs housed in a rescue shelter. *Applied Animal Behaviour Science*, 91, 143–153.

Grayson, D.K. (1988). *Danger Cave, Last Supper Cave and Hanging Rock Shelter: the faunas.* Anthropological papers of the American Museum of Natural History 66.

Griffin, D. (1976). *The question of animal awareness.* Rockefeller University Press, New York.

Griffin, D.R. (1984). *Animal minds.* University of Chicago Press. Chicago.

Gromko, M.H., Briot, A., Jensen, S.C and Fukui, H.H. (1991). Selection on copulation duration in *Drosophila melanogaster*: Predictability of direct response versus unpredictability of correlated response. *Evolution*, 45, 69–81.

Grzimek, B. (1941). Über einen zahlenverbellenden Artistenhund. *Zeitschrift für Tierpsychologie*, 4, 306–310.

Grzimek, B. (1942). Weitere Vergleichsversuche mit Wolf und Hund. *Zeitschrift für vergleichende Physiologie*, 5, 59–73.

Gubernick, D.J. (1981). Parent and infant attachment in mammals. In: Gubernick, D.J. and Klopfer, P.H., eds. *Parental care in mammals*, pp. 243–300. Plenum Press, London.

Guy, N.C., Luescher, U.A., Dohoo, S.E. *et al.* (2001a). A case series of biting dogs: characteristics of the dogs, their behaviour, and their victims. *Applied Animal Behaviour Science*, 74, 43–57.

Guy, N.C., Luescher, U.A., Dohoo, S.E. *et al.* (2001b). Risk factor for dog bites to owner in a general veterinary caseload. *Applied Animal Behaviour Science*, 74, 29–42.

Guy, N.C., Luescher, U.A., Dohoo, S.E. *et al.* (2001c). Demographic and aggressive characteristics of dogs in a general veterinary caseload. *Applied Animal Behaviour Science*, 74, 15–28.

Győri, B., Gácsi, M., Kubinyi, E., Virányi., Zs, Topál, J., and Miklósi, Á. (2009). Comparative investigation of early behavioural traits in hand-reared wolves and differently socialized dogs. submitted.

Hall, E.R. and Kelson, K.R. (1959). *The mammals of North America*, Vol II. Ronald Press, New York.

Handley, B.M. (2000). Preliminary results in determining dog types from prehistoric sites in the northeastern United States. In: Crockford, S.J., ed. *Dogs through time: an archaeological perspective*, pp. 205–217. Archeopress, London.

Harcourt, R.A. (1974). The dog in prehistoric and early historic Britain. *Journal of Archaeological Science*, 1, 151–175.

Hare, B. and Tomasello, M. (1999). Domestic dogs (*Canis familiaris*) use human and conspecific social cues to locate hidden food. *Journal of Comparative Psychology*, 113, 1-5.

Hare, B. and Tomasello, M. (2005). Human-like social skills in dogs? *Trends in Cognitive Sciences*, 9, 405–454.

Hare, B. and Tomasello, M. (2006). Behaviour genetics of dog cognition: Human-like social skills in dogs are heritable and derived. In: Ostrander, E.A., Giger, U., Lindbladh, K. (eds), *The Dog and its Genome*, pp. 497–515. Cold Spring Harbor Press: Woodbury, New York.

Hare, B., Call, J and Tomasello, M. (1998). Communication of food location between human and dog (*Canis familiaris*). *Evolution of Communication*, 2, 137-159.

Hare, B., Brown, M., Williamson, C., and Tomasello, M. (2002). The domestication of social cognition in dogs. *Science*, 298, 1634–1636.

Hare, B., Plyusnina, I., Ignacio, N. *et al.* (2005). Social cognitive evolution in captive foxes is a correlated by-product of experimental domestication. *Current Biology*, 15, 226–230.

Harrington, F.H. and Asa, C.S. (2003). Wolf communication. In: Mech, D. and Boitani, L., eds. *Wolves: behaviour, ecology and conservation*, pp. 66–103. University of Chicago Press, Chicago.

Harrington, F.H. and Mech, L.D. (1978). Wolf vocalisation. In: Hall, R.L. and Sharp, H.S., eds. *Wolf and man. Evolution in parallel*, pp. 109–133. Academic Press, New York.

Harrington, F.H. and Paquet, P.C. (1982). *Wolves of the world*. Noyes Publications, Park Ridge, NJ.

Hart, B.L. and Miller, M.F. (1985). Behavioral breed profiles: A quantitative approach. *Journal of the American Veterinary Medical Association*, 168, 11175–1180.

Hart, L.A. (1995). Dogs as human companions: a review of the relationship. In: Serpell, J., ed. *The domestic dog*, pp. 161–178. Cambridge University Press, Cambridge.

Hauser, M.D. (1996) *The evolution of communication*. MIT Press, Cambridge, MA.

Hauser, M.D. (2000). A primate dictionary? Decoding the function and meaning of another species vocalizations. *Cognitive Science*, 24, 445–475.

Hayes, R.D., Bear, A.M. and Larsen, D.G. (1991). Population dynamics and prey relationships of an exploited and recovering wolf population in the southern Yukon. Yukon Fish and Wildlife Branch Final Report. TR-91–1.

Healy, S. (1998). *Spatial representation in animals*. Oxford University Press. New York.

Heffner, H.E. (1983). Hearing in large and small dogs: absolute thresholds and size of the tympanic membrane. *Behavioral Neuroscience*, 97, 310–318.

Heffner, H.E. (1998). Auditory awareness. *Applied Animal Behaviour Science*, 57, 259–268.

Heffner, H.E. and Heffner, R.S. (2003). Audition. In: Davis, S., ed. *Handbook of research methods in experimental psychology*, pp. 413–440. Blackwell, Oxford.

Heffner, R.S., Koay, G., and Heffner, H.E. (2001). Sound localization in a new-world frugivorous bat, *Artibeus jamaicensis*: acuity, use of binaural cues, and relationship to vision. *Journal of the Acoustical Society of America*, 109, 412–421.

Heimburger, N. (1962). Beobachtungen an handaufgezogenen Wildcaniden (Wölfin und Schakalin) und Versuche über ihre Gedächtnisleistungen. *Zeitschrift für Tierpsychologie*, 18, 265–284.

Héjjas, K., Vas, J., Topál, J. *et al.* (2007). Association of the dopamine D4 receptor gene polymorphism and the 'activity-impulsivity' endophenotype in dogs. *Animal Genetics*. (in press).

Helton, W.S. (2005). Animal expertise, conscious or not. *Animal Cognition*, 8, 67–74.

Hemmer, H. (1990). *Domestication: The decline of environmental appreciation*. Cambridge University Press, Cambridge.

Hennessy, W.M., Davis, H.M., Williams, M.T., *et al* (1997). Plasma cortisol levels of dogs at a county animal shelter. *Physiology and Behavior*, 62, 485–490.

Hennessy, M.B., Williams, M.T., Miller, D.D., *et al* (1998). Influence of male and female petters on plasma cortisol and behaviour: can human interaction reduce the stress of dogs in a public animal shelter? *Applied Animal Behaviour Science*, 61, 63–77.

Henshaw, R.E. (1982). Can the wolf be returned to New York? In: Harrington, F.H. and Paquet, P.C., eds.. *Wolves of the world: perspectives of behavior, ecology and conservation*, pp. 395–422. Noyes Publications, Park Ridge, NJ.

Hepper, P.G. (1988). The discrimination of human odour by the dog. *Perception*, 17, 549–554.

Hepper, P.G. (1994). Long-term retention of kinship recognition established during infancy in the domestic dog. *Behavioral Processes*, 33, 3–14.

Hepper, P.G. and Wells, D.L. (2005). How many footsteps do dogs need to determine the direction of an odour trail? *Chemical Senses*, 30, 291–298.

Herman, L.M. (2002). Vocal, social and self-imitation by bottlenosed dolphins. In: Dautenhahn, K. and Nehaniv, C.L., eds. *Imitation in animals and artifacts*, pp. 63–108. MIT Press, Cambridge, MA.

Herre, W. and Röhrs, M. (1990). *Haustiere—zoologisch gesehen*. Gustav Fischer, Stuttgart.

Heyes, C.M. (1993). Anecdotes, training, trapping and triangulating. *Animal Behaviour*, 46, 177–188.

Heyes, C. (2000). Evolutionary psychology in the round. In: Heyes, C. and Ludwig, H., eds. *The evolution of cognition*, pp. 3–22. MIT Press, Cambridge, MA.

Hirsch-Pasek, K. and Treiman, R. (1981). Doggerel: motherese in a new context. *Journal of Child Language*, 9, 229–237.

Ho, S.Y.W. and Larson, G. (2005). Molecular clocks: when times are a-changin'. *Trends in Genetics*, 22, 79–83.

Holland, P.C. (1990). Forms of memory in pavlovian conditioning. In: McGaugh, J.L., Weinberger, N.M., and Lynch, G., eds. *Brain organization and memory: cells, systems, and circuits*, pp. 78–105. Oxford Science Publications, Oxford.

Hood, B. (1995). Gravity rules for 2–4 olds? *Cognitive Development*, 10, 577–598.

Hood, B.M., Hauser, M.D., Anderson, L. and Santos, L.R. (1999). Gravity biases in a non-human primate? *Developmental Sciences*, 2, 35–41.

Horisberger, U., Stark, K.D.C., Rüfenacht, J.C., Pillonel, C. and Steiger, A. (2005). The epidemiology of dog bite injuries in Switzerland-characteristics of victims, biting dogs and circumstances. *Anthrozoös*, 17, 320–339.

Horváth, Zs., Igyártó, B.Z., Magyar, A. and Miklósi, Á (2007). Three different coping styles in police dogs exposed to a short-term challenge. *Hormones and Behavior* 2, 621–630.

Houpt, K.A. (2006) Terminology think tank: Terminology of aggressive behavior. *Journal of Veterinary Behaviour: Clinical Applications and Research*, 1, 39–41.

Hsu, Y. and Serpell, J.A. (2003). Development and validation of a questionnaire for measuring behavior and temperament trait in pet dogs. *Journal of the American Veterinary Medical Association*, 223, 1293–1300.

Hubel, D.H. and Wiesel, T.N. (1998). Early exploration of the visual cortex. *Neuron*, 20, 401–412.

Hulse, S.H. Flower, H., and Honig, W.K. (1978). *Cognitive processes in animal behavior*. Lawrence Erlbaum Associates, Hillsdale, NJ.

Humphrey, E.S. (1934). 'Mental tests' for shepherd dogs. *Journal of Heredity*, 25, 129–135.

Irion, D.N., Schaffer, A.L., Famula, T.R., Eggleston, M.L., Hughes, S.S., and Pedersen, N.C. (2003). Analysis of genetic variation in 28 dog breed populations with 100 microsatellite markers. *Journal of Heredity*, 94, 81–87.

Ishiguro, N., Okumura, N., Matsui, A., and Shigehara, N. (2000). Molecular genetic analysis of ancient Japanese dogs. In: Crockford, S.J., ed. *Dogs through time: an archaeological perspective*, pp. 287–292. British Archaeological Reports International Series 889, Oxford.

Itakura S., Agnetta B., Hare B., and Tomasello M. (1999). Chimpanzee use of human and conspecific social cues to locate hidden food. Developmental Sciences 2, 448–456.

Ito, H., Nara, H., Inouye-Mrayama, M. *et al.* (2004). Allele frequency distribution of the canine dopamine receptor D4 gene exon III and I in 23 breeds. *Journal of Veterinary Medical Science*, 66, 815–820.

Jacobs, G.H., Deegan, J.F., Crognale, M.A., and Fenwick, J.A. (1993). Photopigments of dogs and foxes and their implications for canid vision. *Visual Neuroscience*, 10, 173–180.

Jagoe, A. and Serpell, J. (1996). Owner characteristics and interactions and the prevalence of canine behaviour problems. *Applied Animal Behaviour Science*, 47, 31–42.

Jedrzejewski, W., Jedrzejewska, B., Okarma, H., Schmidt, K., Zub, K., and Musiani, M. (2000). Prey selection and predarion by wolves in Białowieża primeval forest, Poland. *Journal of Mammalogy*, 81, 197–212.

Jedrzejewski, W., Schmidt, H., Theuerkauf, J. *et al.* (2002). Kill rates and predation by wolves on ungulate populations in Białowieża primeval forest (Poland). *Ecology*, 83, 1341–1356.

Jenkins, H.M., Barrera, F.J., Ireland, C., and Woodside, B. (1978). Signal centred action patterns of dogs in appetitive classical conditioning. *Learning and Motivation*, 9, 272–296.

Jhala, Y.V. and Giles, R.H. (1991). The status and conservation of the wolf in Gujarat and Rajasthan, India. *Conservation Biology*, 5, 476–483.

Johnson, C.M. (2001). Distributed primate cognition: a review. *Animal Cognition*, 4,167–183.

Johnson, H.M. (1912). The talking dog. *Science*, 35, 749–751.

Johnston, B. (1997). Harnessing thought. Guide dog—a thinking animal with a skilful mind. Queen Anne Press, London.

Jolicoeur, P. (1959). Multivariate geographic variation in the wolf, *Canis lupus* L. *Evolution*, 13, 283–299.

Jones, A.C. and Gosling, S.D. (2005). Temperament and personality in dogs (*Canis familiaris*): a review and evaluation of past research. *Applied Animal Behaviour Science*, 95, 1–53.

Jordan, P.A. Shelton, P.C., and Allen, D.L. (1967). Numbers, turnover, and social structure of the Isle Royale wolf population. *American Zoologist*, 7, 233–252.

Kamil, A.C. (1988). A synthetic approach to the study of animal intelligence. In: Leger, D.W., ed. *Comparative study in modern psychology*, pp. 230–257. Nebraska Symposium On Motivation, vol. 35, University of Nebraska Press, Lincoln, Ne.

Kamil, A.C. (1998). On the proper definition of cognitive ethology. In: Balda, R.P. Pepperberg, I.M., and Kamil, A.C., eds. *Animal cognition in nature*, pp. 1–29. Academic Press, San Diego, CA.

Kaminski, J., Call, J., and Fischer, J. 2004. Word learning in a domestic dog: evidence for 'fast mapping'. *Science*, 304, 1682–1683.

Kaminski, J., Call, J., and Tomasello, M. (2004). Body orientation and face orientation: two factors controlling apes' begging behaviour from humans. *Animal Cognition*, 7, 216–224.

Katcher, A.H. and Beck, A.M. (1983). *New perspective on our lives with companion animals.* University of Pennsylvania Press, Philadelphia, PA.

Kazdin, A.E. (1982). *Single-case research designs.* Oxford University Press, Oxford.

Kemencei, Z. (2007). *The development of comprehension of human gestural signals in cats.* (In Hungarian). Dissertation for the fulfilment of MSc in Zoology. Szent István University, Budapest.

Kerepesi, A., Jonsson, G.K., Miklósi, Á., Topál, J., Csányi, V., and Magnusson, M.S. (2005). Detection of temporal patterns in dog-human interaction. *Behavioural Processes,* 70, 69–79.

Kim, K.S., Tanabe, Y., Park, C., and Kha, J.H. (2001). Genetic variability in east Asian dogs using microsatellite loci analysis. *Journal of Heredity,* 92, 398–403.

Kleiber, M (1961). *The fire of life.* John Wiley and Sons, New York.

Kleiman, D.G. and Eisenberg, J.F. (1973). Comparisons of canid and felid social systems from an evolutionary perspecive. *Animal Behaviour,* 21, 637–659.

Klingenberg, C.P. (1998). Heterochrony and allometry: the analysis of evolutionary change in ontogeny. *Biological Reviews,* 73, 79–123.

Klinghammer, E. and Goodman, P.A. (1987). Socialization and management of wolves in captivity. In: Frank, H., ed. *Man and wolf,* pp. 31–61. Junk Publishers, Dordrecht.

Koch, S.A. and Rubin, L.F. (1972). Distribution of cones in retina of the normal dog. *American Journal of Veterinary Research,* 33, 361–363.

Koda, N. (2001). Development of play behaviour between potential guide dogs for the blind and human raisers. *Behavior* 53, 41–46.

Köhler, O. (1917). *Intelligenzprüfungen an Menschenaffen.* Springer, Berlin (translated as: *The mentality of apes,* Routledge and Kegan Paul, London, 1925).

Koler-Matznick, J. (2002). The origin of the dog revisited. *Anthrozoös,* 15, 98–118.

Koler-Matznick, J., Brisbin, I.L., and McIntyre, J.K. (2000). The New Guinea singing dog: a living primitive dog. In: Crockford, S.J., ed. *Dogs through time: an archaeological perspective,* pp. 239–247. Archeopress, London.

Koler-Matznick, J., Brisbin, I.L. Feinstein, M., and Bulmer, S. (2003). An updated description of the New Guinea singing dog (*Canis hallstromi,* Troughton 1957). *Journal of Zoology,* 261, 109–118.

Koop B.F., Burbidge, M., Byun, A., Rink, U., and Crockford, S.J. (2000). Ancient DNA evidence of a separate origin for North American indigenous dogs. In: S. J. Crockford (ed). *Dogs Through Time: An Archaeological Perspective,* pp. 271–286. British Archaeological Series 889 Oxford.

Koskinen, M.T. and Bredbacka, P. (2000). Assessment of the population structure of five Finnish dog breeds with microsatellites. *Animal Genetics,* 31, 310–317.

Kostraczyk, E. and Fonberg, E. (1982). Heart-rate mechansims in instrumental conditioning reinforced by petting dogs. *Physiology and Behavior,* 28, 27–30.

Kotrschal, K., Bromundt, V., and Föger, B. (2004). *Faktor Hund.* Czernin Verlag, Wien.

Kowalska, D.M., Kusmierek, P., Kosmal, A., and Mishkin, M. (2001). Neither perirhinal/entorhical nor hippocampal lesions impair short-term auditory recognition memory in dogs. *Neuroscience,* 104, 965–978.

Kreeger, T.J. (2003). The internal wolf: physiology, pathology, and pharmacology. In: Mech, D. and Boitani, L., eds. *Wolves: behaviour. ecology and conservation,* pp. 317–340. University of Chicago Press, Chicago.

Krestel, D., Passe, D., Smith, J.C., and Jonsson, L. (1984). Behavioral determination of olfactory thresholds to amylacetate in dogs. *Neuroscience and Biobehavioral Reviews,* 8, 169–174.

Kruska, D. (1988). Mammalian domestication and its effect on brain structure and behaviour. *NATO ASI Series on Intelligence and Evolutionary Biology,* G17, 211–249.

Kruska, D.C.T. (2005). On the evolutionary significance of encephalization in some eutherian mammals: effects of adaptive radiation, domestication and feralization. *Brain, Behaviour and Evolution,* 65, 73–108.

Kubinyi, E., Virányi, Zs, and Miklósi, Á. (2007). Comparative social cognition: From wolf and dog to humans. *Comparative Cognition and Behavior Reviews,* 2, 26–46. Retrieved from http.psych.queensu.ca/ccbr/index.html

Kubinyi, E., Miklósi, Á., Topál, J., and Csányi, V. (2003a). Social anticipation in dogs: a new form of social influence. *Animal Cognition,* 6, 57–64.

Kubinyi, E., Topál, J., Miklósi, Á., and Csányi, V. (2003b). The effect of human demonstrator on the acquisition of a manipulative task. *Journal of Comparative Psychology,* 117, 156–165.

Kubinyi, E., Miklósi, Á., Kaplan, F., Gácsi, M., Topál, J., and Csányi, V. (2004). Can a dog tell the difference? Dogs encounter AIBO, an animal-like robot in a neutral and in a feeding situation. *Behavioural Processes,* 65, 231–239.

Kukekova, A.V., Acland, G.M., Oskina, I.N. *et al.* (2005). The genetics of domesticated behaviour in canids: What can dogs and silver foxes tell us about each other? In: Ostrander, E.A., Giger, U., and Lindbladh, K., eds. *The dog and its genome,* pp. 515–538. Cold Spring Harbour Press, Woodbury, NY.

Kurtén, B. (1968). *Pleistocene mammals of Europe.* Aldine Press, Chicago.

Kurtén, B. and Anderson, E. (1980). *Pleistocene mammals of North America.* Columbia University Press, New York.

Lakatos, G., Soproni, K., Dóka, A., and Miklósi, Á. (2009). A comparative approach to dogs' (*Canis familiaris*) and human infants' understanding of various forms of pointing gestures. *Animal Cognition* (in press).

Laland, K.N. (2004). Social learning strategies. *Learning and Behaviour,* 32, 4–14.

Laska, M., Wieser, A., Bautista, R.M.R., Teresa, L., and Salazar, H. (2004). Olfactory sensitivity for carboxylic acids in spider monkeys and pigtail macaques. *Chemical Senses,* 29, 101–109.

Lawicka, W. (1969). Differing effectiveness of auditory quality and location cues in two forms of differentiation learning. *Acta Biologica,* 29, 83–92.

Lawrence, B. and Reed, C.A. (1983). The dogs of Jarmo. In: Braidwood, L.S. *et al.,* eds. *Prehistoric archaeology along the Zargos flanks,* pp. 485–494. Oriental Insitute of the University of Chicago, Chicago.

Le Boeuf, B.J. (1967). Interindividual association in dogs. *Behaviour,* 29, 268–295.

Lehman, N.E., Eisenhawer, E.A., Hansen, K. *et al.* (1991). Introgression of coyote mitochondrial DNA into sympatric North American gray wolf populations. *Evolution,* 45, 104–119.

Lehman, N.E., Clarkson, E.P., Mech, L.D., Meier, T.J., and Wayne, R.W. (1992). A study of the genetic relationship within and among wolf packs using DNA fingerprinting and mitochondrial DNA. *Behavioral Ecology and Sociobiology,* 30, 83–94.

Lehner, P.N. (1996). *Handbook of ethological methods.* Cambridge University Press, Cambridge.

Leonard, J.A., Wayne, R.K. Wheeler, J., Valadez, R., Guillén, S., and Vilá, C. (2002). Ancient DNA evidence for Old World origin of New World Dogs. *Science,* 298, 1613–1616.

Leonard, J.A., Vilá, C., and Wayne, R.K. (2005). Legacy lost: genetic variability and population size of extirpated US grey wolves (*Canis lupus*). *Molecular Ecology,* 14, 9–17.

Levinson, B.M. (1969). *Pet-oriented child psychotherapy.* C.C. Thomas, Springfield, IL.

Lichtenstein, P.E. (1950). Studies of anxiety: the production of a feeding inhibition in dogs. *Journal of Comparative and Physiological Psychology,* 43, 16–29.

Lim, K., Fisher, M., and Burns-Cox, C.J. (1992). Type 1 diabetics and their pets. *Diabetic Medicine,* 9, S3–S4.

Lindberg, J., Björnerfeldt, S., Saetre, P. *et al.* (2005). Supplemental data: selection for tameness has changed brain gene expression in silver foxes. *Current Biology,* 15, 915–916.

Lindblad-Toh, K., Wade, C.M., Mikkelsen, T.S. *et al.* (2005). Genom sequence, comparative analysis and haplotype structure of the domestic dog. *Nature,* 438, 803–819.

Lindsay S. (2001). *Handbook of applied dog behavior and training,* Volume 1: *Adaptation and learning.* Iowa University Press, Ames, IA.

Lindsay S. (2005). *Handbook of Applied Dog Behavior and Training,* Volume 3: *Procedures and Protocols.* Blackwell Publishing.

Line, S. and Voith, V. (1986). Dominance aggression of dogs towards people: behavior profile and response to treatment. *Applied Animal Behaviour Science,* 16, 77–83.

Lockwood, R. (1979). Dominance in wolves: useful construct or bad habit. In: Klinghammer, E., ed. *The behavior and ecology of wolves,* pp. 225–243. Garland STPM Press, New York.

Lorenz, K. (1950). The comparative method in studying innate behaviour patterns. *Symposia of the Society for Experimental Biology,* 4, 221–268.

Lorenz, K. (1954). *Man meets dog.* Houghton Mifflin, Boston.

Lorenz, K. (1969). The innate basis of learning. In: Pribram, K., ed. *On the biology of learning,* pp. 13–93. Harcourt, Brace and World, New York.

Lorenz, K. (1974). Analogy as a source of knowledge. *Science,* 185, 229–234.

Lorenz, K. (1981). *The foundations of ethology.* Springer-Verlag, Wien.

Lubbock, H. (1888). *The senses, instincts and intelligence of animals.* Kegan Paul & Co., London.

Macdonald, D.W. (1983). The ecology of carnivore social behaviour. *Nature,* 301, 379–384.

Macdonald, D.W. and Carr, G.M. (1995). Variation in dog society: between resource dispersion and social influx. In: Serpell, J., ed. *The domestic dog,* pp. 199–216. Cambridge University Press,Cambridge.

Macdonald, D.W. and Sillero-Zubiri, C. (2003). *The biology and conservation of wild canids.* Oxford University Press, Oxford.

MacDonald, K. (1987). Development and stability of personality characteristics in pre-pubertal wolves: implications for pack organisation and behaviour. In: Frank, H., ed. *Man and wolf: advances, issues, and problems in captive wolf research,* pp. 293–312. Junk Publishers, Dordrecht.

MacDonald, K.B. and Ginsburg, B.E. (1981). Induction of normal prepubertal behaviour in wolves with restricted rearing. *Behavioural and Neural Biology,* 33, 133–162.

Mader, B., Hart, L.A., and Bergin, B. (1989). Social acknowledgements for children with disabilities: effects of service dogs. *Child Development,* 60, 1529–1534.

Magnusson, M.S. (2000). Discovering hidden time patterns in behaviour: T-patterns and their detection. *Behavior Research Methods, Instruments and Computers,* 32, 93–110.

Malm, K. and Jensen, P. (1997). Weaning and parent-offspring conflict in the domestic dog. *Ethology*, 103, 653–664.

Manaserian, N.H. and Antonian, L. (2000). Dogs of Armenia. In: Crockford, S.J., ed. *Dogs through time: an archaeological perspective*, pp.181–192. British Archaeological Reports International Series 889, Oxford.

Mandairon, N., Stack, C., Kiselycznyk, C., and Linster, C. (2006). Broad activation of the olfactory bulb produces long-lasting changes in odor perception. *Proceedings of the National Academy of Sciences of the USA*, 103, 13543–13548.

Maragliano, L., Ciccone, G., Fantini, C., Petrangeli, C., Saporito, G., Di Traglia, M., and Natoli, E (2007). Biting dogs in Rome (Italy). International Journal of Pest Management, 53, 329–334.

Maros, K., Dóka, A., and Miklósi, Á. (2008). Behavioural correlation of heart rate changes in family dogs. *Applied Animal Behaviour Science*, 109, 329–341.

Maros, K., Pongrácz, P., Bárdos, Gy., Molnár Cs., Faragó, T., and Miklósi, Á. (2008). Dogs can discriminate barks from different situations. *Applied Animal Behavior Science* (in press).

Marston, L.C. and Bennett, P.C. (2003). Reforging the bond—towards successful canine adoption. *Applied Animal Behaviour Science*, 83, 227–245.

Marston, L.C., Bennett, P.C. and Coleman, G.J. (2004). What happens to shelter dogs? An analysis of data for 1 year from three Australian shelters. *Journal of Applied Animal Behaviour Welfare Science*, 7, 27–47.

Marston, L.C., Bennett, P.C., and Coleman, G.J. (2005a). Factors affecting the formation of a canine-human bond. *IWDBA Conference Proceedings*, 132–138.

Marston, L.C, Bennett, P.C. and Coleman, G.J. (2005b). What happens to shelter dogs? Part 2. Comparing three Melbourne welfare shelters for nonhuman animals. *Journal of Applied Animal Behaviour Welfare Science*, 8, 25–45.

Martin, P. and Bateson, P. (1986). *Measuring behaviour*. Cambridge University Press, Cambridge.

Matas, L., Arend, R.A., and Sroufe, L.A. (1978). Continuity of adaptation in second year: The relationship between quality of attachment and later competence. *Child Development*, 49, 547–556.

Mayr, E. (1963). *Animal species and evolution*. Harvard University Press, Cambridge, MA.

Mayr, E. (1974). Behaviour programs and evolutionary strategies. *American Science*, 62, 650–659.

Mazzorin, J. and Tagliacozzo, A. (2000). Morphological and osteological changes in the dog from the Neolithic to the roman period in Italy. In: Crockford, S.J.,

ed. *Dogs through time: an archaeological perspective*, pp.141–161. British Archaeological Reports International Series 889, Oxford.

McBride, A. (1995). The human-dog relationship. In: Robinson, I., ed. *The Waltham book of human–animal interactions*, pp. 99–112. Pergamon, Oxford.

McConnell, P.B. (1990). Acoustic structure and receiver response in domestic dogs, *Canis familiaris*. *Animal Behaviour*, 39, 897–904.

McConnell, P.B. and Baylis, J.R. (1985). Interspecific communication in cooperative herding: Acoustic and visual signals from shepherds and herding dogs. *Zeitschrift für Tierpsychologie*, 67, 302–328.

McGreevy, P.D. and Nicholas, F.W. (1999). Some practical solutions to welfare problems in dog breeding. *Animal Welfare*, 8, 329–341.

McGreevy, P.D. Grassi, T.D., and Harman, A.M. (2004). A strong correlation exists between the distribution of retinal ganglion cells and nose length in the dog. *Brain, Behavior and Evolution*, 63, 13–22.

McGreevy, P.D., Righetti, J., and Thomson, C. (2005). The reinforcing value of physical contact and the effect of grooming in different anatomic areas. *Anthrozoös*, 18, 236–244.

McKinley, J. and Sambrook, T.D. (2000). Use of human-given cues by domestic dogs (*Canis familiaris*) and horses (*Equus caballus*). *Animal Cognition*, 3, 13-22.

McKinley, S. and Young, R.J. (2003). The efficacy of the model-rival method when compared with operant conditioning for training domestic dog to perform a retrievel task. *Applied Animal Behaviour Science*, 81, 357–365.

McLeod, P.J. (1996). Developmental changes in associations among timber wolf (*Canis lupus*) postures. *Behavioural Processes*, 38, 105–118.

McLeod, P.J. and Fentress, J.C. (1997). Developmental changes in the sequential behavior of interacting timber wolf pups. *Behavioural Processes*, 39, 117–136.

McLeod, P.J., Moger, W.H., Ryon, J., Gadbois,S., and Fentress, J.C. (1995). The relation between urinary cortisol levels and social behaviour in captive timber wolves. *Canadian Journal of Zoology*, 74, 209–216.

McPhail, E.M. and Bolhuis, J.J. (2001). The evolution of intelligence: adaptive specialisation versus general process. *Biological Reviews*, 76, 341–364.

Mech, L.D. (1966). Hunting behaviour of timber wolves in Minnesota. *Journal of Mammalogy*, 47, 347–348.

Mech, L.D. (1970). *The wolf: The ecology and behaviour of an endangered species*. Natural History Press, New York.

Mech, L.D. (1995). A ten year history of the demography and productivity of an arctic wolf pack. *Arctic*, 48, 329–332.

Mech, L.D. (1999) Alpha status, dominance, and division of labor in wolf packs. *Canadian Journal of Zoology, 77,* 1196–1203.

Mech, L.D. and Boitani, L. (2003). Wolf social ecology. In: Mech, D. and Boitani, L., eds. *Wolves: behavior, ecology and conservation,* pp. 1–34. University of Chicago Press, Chicago.

Mech, L.D. and Peterson, R.O. (2003). The wolf as a carnivore. In: Mech, D. and Boitani, L., eds. *Wolves: behavior, ecology and conservation,* pp.104–130. University of Chicago Press, Chicago.

Mech, L.D., Adams, L.G., Burch, J.W., and Dale, B.W. (1998). *The wolves of Denali.* University of Minnesota Press, Minneapolis, MN.

Mech, L.D., Wolf, P.C., and Packard, J.M. (1999). Regurgitative food transfer among wild wolves. *Canadian Journal of Zoology, 77,* 1–4.

Medjo, D.C. and Mech, L.D. (1976). Reproductive activity in nine- and ten-month-old wolves. *Journal of Mammalogy, 57,* 406–408.

Megitt, M.J. (1965). The association between Australian Aborigines and Dingoes. In: Leeds, A. and Vayda, A.P., eds. *Man, culture and animals,* pp. 7–26. AAAS Publications, Washington, DC.

Meisterfeld, C.W. and Pecci, E. (2000). *Dog and human behavior: amazing parallels, similarities.* M R K Publishing, Petaluma, CA.

Mekosh-Rosenbaum, V., Carr, W.J., Goodwin, J.L., Thomas, P.L., D'Ver, A., and Wysocki, C.J. (1994). Age-dependent responses to chemosensory cues mediating kin recognition in dogs (*Canis familiaris*). *Physiology & Behavior, 55,* 495–499.

Menault, E. (1869). *The intelligence of animals.* Cassel, Petter & Galpin, London.

Mendelsohn, H. (1982). Wolves of Israel. In: Harrington, F.H. and Paquet, P.C., eds. *Wolves of the world: perspectives of behavior, ecology and conservation,* pp 173–195. Noyes Publications, Park Ridge, NJ.

Menzel, R. (1936). Welpe und Umwelt. *Zeitschrift für Hundeforschung, 3,* 1–72.

Mertens, P.A. and Unshelm, J. (1996). Effects of group and individual housing on the behaviour of kennelled dogs in animal shelters. *Anthrozoös, 9,* 40–51.

Miklósi, Á. and Soproni, K. (2006). A comparative analysis of the animals' understanding of the human pointing gesture. *Animal Cognition, 9,* 81–94.

Miklósi, Á., Polgárdi, R., Topál, J., and Csányi, V. (1998). Use of experimenter-given cues in dogs. *Animal Cognition, 1,* 113–121.

Miklósi, Á., Polgárdi, R., Topál, J., and Csányi, V. (2000). Intentional behaviour in dog-human communication: An experimental analysis of 'showing' behaviour in the dog. *Animal Cognition, 3,* 159–166.

Miklósi, Á., Kubinyi, E., Topál, J., Gácsi, M., Virányi, Zs., and Csányi, V. (2003). A simple reason for a big difference: wolves do not look back at humans but dogs do. *Current Biology, 13,* 763–766.

Miklósi, Á., Topál, J., and Csányi, V. (2004). Comparative social cognition: What can dogs teach us? *Animal Behaviour, 67,* 995–1004.

Miklósi, Á., Topál, J., and Csányi, V. (2007). Big thoughts in small brains? Dogs as model for understanding human social cognition. *NeuroReport, 18,* 467–471.

Milgram, N.W., Adams, B., Callahan, H. *et al.* (1999). Landmark discrimination learning in the dog. *Learning and Memory, 6,* 54–61.

Milgram, N.W., Head, E., Muggenburg, B. *et al.* (2002). Landmark discrimination learning in the dog: effects of age, an antioxidant fortified food, and cognitive strategy. *Neuroscience and Behavioral Reviews, 26,* 679–695.

Miller, M. and Lago, D. (1990). Observed pet-owner in-home interactions: species differences and association with the pet relationship scale. *Anthrozoös, 4,* 49–54.

Miller, P.E. and Murphy, C.J. (1995). Vision in dogs. *Journal of the American Veterinary Medical Association, 207,* 1623–1634.

Millot, J.L., Filiatre, J.C., Eckerlin, A., Gagnon, A.C., and Montagner, H. (1987). Olfactory cues in the relation between children and their pets. *Applied Animal Behaviour Science, 17,* 189–195.

Mills, D.S. (2005). What's in a word? A review of the attributes of a command affecting the performance of pet dogs. *Anthrozoös, 18,* 208–221.

Mills, D.S., Ramos, D., Estelles, M.G., and Hargrave, C. (2006). A triple blind placebo-controlled investigation into the assessment of the effect of dog appeasing pheromone (DAP) on anxiety related behaviour of problem dogs in the veterinary clinic. *Applied Animal Behaviour Science, 98,* 114–126.

Mitchell, R.W. (2001). Americans' talk to dogs: similarities and differences with talk to infants. *Research on Language and Social Interactions, 34,* 183–210.

Mitchell, R.W. and Hamm, M. (1997). The interpretation of animal psychology: anthropomorphism or behavior reading? *Behaviour, 134,* 173–204.

Mitchell, R.W. and Thompson, N.S. (1991). Projects, routines and enticements in dog-human play. In: Bateson, P.P.G. and Klopfer, R.H., eds. *Perspectives in ethology,* Vol. 9, pp. 189–216. Plenum Press, New York.

Molnár, Cs. (2008). (In Hungarian). PhD Dissertation. Eötvös University, Budapest.

Molnár, Cs., Pongrácz, P., Dóka, A., and Miklósi, Á. (2006). Can humans discriminate dogs individually by acoustic parameters of barks? *Behavioral Processes, 73,* 76–83.

Molnár, Cs., Kaplan, F., Roy, P., Pachet, F., Pongrácz, P., and Miklósi, Á. (2008). A machine learning approach to the classification of dog (*Canis familiaris*) barks. *Animal Cognition*, 11, 389–400.

Morey, D.F. (1992). Size, shape and development in the evolution of the domestic dog. *Journal of Archaeological Science*, 19, 181–204.

Morey, D.F. (2006). Burying key evidence: the social bond between dogs and people. *Journal of Archaeological Science*, 33, 158–175.

Morey, D.F. and Aaris-Sorensen, K. (2002). Paleoeskimo dogs of the eastern Arctic. *Arctic*, 55, 44–56.

Morgan, C.L. (1903). *An introduction to comparative psychology*. Walter Scott, London.

Morton, E. (1977). On the occurrence and significance of motivation-structural rules in some bird and mammal sounds. *American Naturalist*, 111, 855–869.

Murie, A. (1944). *The wolves of Mount McKinley*. US National Park Service Fauna Series No. 5.

Musil, R. (2000). Evidence for the domestication of wolves in central European Magdalenian sites. In: Crockford, S.J., ed. *Dogs through time: an archaeological perspective*, pp. 21–28. British Archaeological Reports International Series 889, Oxford.

Nadeau, J.H. and Frankel, W.N. (2000). The roads from phenotypic variation to gene discovery: mutagenesis versus QTLs. *Nature Genetics*, 25, 381–384.

Naderi, Sz., Csányi, V., Dóka, A., and Miklósi, Á. (2001). Cooperative interactions between blind persons and their dog. *Applied Animal Behaviour Science*, 74, 59–80.

Nagel, T. (1974). What is it like to be a bat? *Philosophical Review*, 4, 435–450.

Neff, M.W., Robertson, K.R., Wong, A.K. *et al.* (2004). Breed distribution and history of canine mdr1–1delta, a pharmacogenetic mutation that marks the emergence of breeds from the collie lineage. *Proceedings of the National Academy of Sciences of the USA*, 101, 11725–11730.

Nelson, G.S. (1990). Human behaviour and the epidemiology of helminth infections. In: Barnard, C.J. and Behnke, J.M., eds. *Parasitism and host behaviour*, pp. 234–263. Taylor & Francis, London.

Netto, W.J. and Planta, D.J.U. (1997). Behavioural testing for aggression in the domestic dog. *Applied Animal Behaviour Science*, 52, 243–263.

Neuhaus, W. and Regenfuss, E. (1967). Über die Sehschärfe des Haushundes bei verschiedenen Helligkeiten. *Zeitschrift für Vergleichende Physiologie*, 57, 137–146.

Ney, J.A.J. (1999). Social learning in canids: an ecological perspective. In: Box, H.O. and Gibson, K.R., eds. *Mammalian social learning*, pp. 259–277. Cambridge University Press, Cambridge.

Nobis, G. (1979). Der älteste Haushund lebte vor 14.000 Jahren. *Umschau*, 19, 610.

Normando, S., Stefanini, C., Meers, L., Adamelli, S., Coultis, D., and Bono, G. (2006). Some factors influencing adoption of sheltered dogs. *Anthrozoös*, 19, 211–225.

Notari, L. and Goodwin, D. (2006). A survey of behavioural characteristics of pure-bred dogs in Italy. *Applied Animal Behaviour Science*, 30, 1–13.

Nowak, R.M. (2003). Wolf evolution and taxonomy. In: Mech, D. and Boitani, L., eds. *Wolves: behavior, ecology and conservation*, pp. 239–258. University of Chicago Press, Chicago.

Nunez, E.A., Becker, D.V., Furth, E.D., Belshaw, B.E., and Scott, J.P. (1970). Breed differences and similarities in thyroid function in purebred dogs. *American Journal of Physiology*, 218, 1337–1341.

Odendaal, J.S.J. (1996). An ethological approach to the problem of dogs digging holes. *Applied Animal Behaviour Science*, 52, 299–305.

Okarma, H. (1995). The tropic ecology of wolves and their predatory role in ungulate communities of forest ecosystems in Europe. *Acta Theriologica*, 40, 335–386.

Okarma, H. and Buchalczyk, T. (1993). Craniometrical characteristics of wolves *Canis lupus* in Poland. *Acta Theriologica*, 38, 253–262.

Okarma, H., Jędrzejewski, W., Schmidt, K., Śnieżko, S., Bunevich, A.N. and Jędrzejewska, B. (1998). Home ranges of wolves in Białowieża Primeval Forest, Poland, compared with other Eurasian populations. *Journal of Mammalogy*, 79, 842–85

Olsen, S.J. (1985). The fossil ancestry of *Canis*. In: Olsen, S.J., ed. *Origins of the domestic dog: the fossil record*, pp. 1–29. University of Arizona Press, Tucson, AZ.

Olsen, S.J. and Olsen, J.W. (1977). The Chinese wolf, ancestor of New World dogs. *Science*, 3, 533–535.

Olsen, S.L. (2001). The secular and sacred roles of dogs at Botai, North Kazakhstan. In: Crockford, S.J., ed. *Dogs through time: an archaeological perspective*, pp.71–92. British Archaeology Reports International Series 889, Oxford.

Orihel, J.S., Ledger, R.A., and Fraser, D. (2005). A survey and management of inter-dog aggression. *Anthrozoös*, 18, 273–287.

Osadchuk, L.V. (1992a). Endocrine gonadal function in silver fox under domestication. *Scientifur*, 16, 116–121.

Osadchuk, L.V. (1992b). Some peculiarities in reproduction in silver fox males under domestication. *Scientifur*, 16, 285–288.

Osadchuk, L.V. (1999). Testosterone, estradiol and cortisol responses to sexual stimulation wit reference to mating activity in domesticated silver fox males. *Scientifur*, 23, 215–220.

Osthaus, B., Slater, A.M., and Lea, S.E.G. (2003). Can dogs defy gravity? A comparison with the human infant and a non-human primate. *Developmental Science*, 6, 489–497.

Osthaus, B., Lea, S.E.G., and Slater, A.M. (2005). Dogs (*Canis lupus familiaris*) fail to show understanding of means end connections in a string pulling task. *Animal Cognition*, 8, 37–47.

Ostrander, E.A. and Wayne, R.K. (2005). The canine genome. *Genome Research*, 15, 1706–1716.

Ostrander, E.A., Giger, U., and Lindblad-Toh, K. (eds.) (2006). *The dog and its genome*. Cold Spring Harbor Monograph Series 44. Cold Spring Harbor Laboratory Press.

Overall, K. (2000). Natural animal models of human psychiatric conditions: Assessment of mechanism and validity. *Progress in Neuropsychopharmacology Biology Psychiatry*, 24, 727–776.

Overall, K.L. and Love, M. (2001). Dog bites to humans demography, epidemiology, injury, and risk. *Journal of the American Veterinary Medical Association*, 218, 1923–1934.

Overall, K.L., Hamilton, S.P., and Chang, M.L. (2006). Understanding the genetic basis of canine anxiety: phenotyping dogs for behavioural, neurochemical and genetic assessment. *Journal of Veterinary Behavior: Clinical Applications and Research*, 1, 124–141.

Packard, J.M. (2003). Wolf behaviour: reproductive, social and intelligent. In: Mech, D. and Boitani, L., eds. *Wolves: behavior, ecology and conservation*, pp. 35–65. University of Chicago Press, Chicago.

Packard, J.M. and Mech, L.D. (1980). Population regulation in wolves. In: Cohen, M.N., Malpass, R.S., and Klein, H.G., eds. *Biosocial mechanisms of population regulation*, pp 135–150. Yale University Press, New Haven, CT.

Packard, J.M. Seal, U.S., Mech, L.D., and Plotka, E.D. (1985). Causes of reproductive failure in two family groups of wolves (*Canis lupus*). *Zeitschrift für Tierpsychologie*, 68, 24–40.

Packard, J.M., Mech, L.D., and Ream, R.R. (1992). Weaning in an arctic wolf pack: behavioral mechanisms. *Canadian Journal of Zoology*, 70, 1269–1275.

Pageat, P. and Gaultier, E. (2003). Current research in canine and feline pheromones. *Veterinary Clinic of North America (Small Animal Practice)*, 33, 187–211.

Pal, S.K. (2003). Reproductive behaviour of free-ranging rural dogs (*Canis familiaris*) in relation to mating strategy, season and litter production. *Acta Theriologica*, 48, 271–281.

Pal, S.K. (2004). Parental care in free-ranging dogs, *Canis familiaris*. *Applied Animal Behaviour Science*, 90, 31–47.

Pal, S.K., Ghosh, B., and Roy, S. (1998). Agonistic behaviour of free-ranging dogs (*Canis familiaris*) in relation to season, sex, and age. *Applied Animal Behaviour Science*, 59, 331–348.

Palestrini, C., Prato-Provide, E., Spiezio, C., and Verga, M. (2005). Heart rate and behavioural responses of dogs in the Ainsworth's strange situation: a pilot study. *Applied Animal Behaviour Science*, 94, 75–88.

Paquet, P.C. and Harrington, F.H. (1982). *Wolves of the world: perspectives of behavior, ecology and conservation*. Noyes Publications, Park Ridge, NJ.

Parker, G.A. (1974). Assessment strategy and the evolution of animal conflicts. *Journal of Theoretical Biology*, 47, 223–243

Parker, H.G. and Ostrander, E.A. (2005). Canine genomics and genetics: running with the pack. PLoS *Genetics*, 1, 507–513.

Parker, H.G., Kim, L.V., Sutter, N.B. *et al.* (2004). Genetic structure of the purebred domestic dog. *Science*, 304, 1160–1164.

Parthasarathy, V. and Crowell-Davis, S.L. (2006). Relationship between attachment to owners and separation anxiety in pet dogs (*Canis lupus familiaris*) *Journal of Veterinary Behavior: Clinical Applications and Research*, 1, 109–120.

Patronek, G.J. and Glickman, L.T. (1994). Development of a model for estimating the size and dynamics of the pet dog population. *Anthrozoös*, 7, 25–42.

Patronek, G.J. and Rowan, A.N. (1995). Determining dog and cat number and population dynamics. *Anthrozoös*, 8, 199–205.

Pavlov, I.P. (1927). *Lectures on conditioned reflexes*. Oxford University Press, Oxford.

Pavlov, I.P. (1934). An attempt at physiological interpretation of obsessional neurosis and paranoia. *Journal of Mental Science*, 80, 187–197.

Paxton, D.W. (2000). A case for a naturalistic perspective. *Anthrozoös*, 13, 5–8.

Pederson, S. (1982). Geographic variation in Alaskan wolves in Carbyn, L.N., ed. *Wolves in Alaska and Canada: their status, biology and management.* pp. 345–361. Canadian Wildlife Service Report Series Number 45, Canadian Wildlife Service, Ottawa.

Peichl, L. (1992). Morphological types of ganglion cells in the dog and wolf retina. *Journal of Comparative Neurology*, 324, 590–602.

Pepperberg, I.M. (1991). Learning to communicate: the effects of social interaction. In: Bateson, P.J.B. and Klopfer, P.H., eds. *Perspectives in ethology*, pp. 119–164 Plenum Press, New York.

Pepperberg, I.M. (1992). Social interaction as a condition for learning in avian species: a synthesis of the disciplines of ethology and psychology. In: Davis, H. and Balfour, D., eds. *The inevitable bond*, pp. 178–205. Cambridge University Press, Cambridge.

Peters, R. (1978). Communication, cognitive mapping and strategy in wolves and hominids. In: Hall, L. and Sharp, H.S., eds. *Wolf and man: evolution in parallel*, pp. 95–107. Academic Press, New York.

Peterson, R.O. and Ciucci, P. (2003). The wolf as a carnivore. In: Mech, D. and Boitani, L., eds. *Wolves: behavior, ecology and conservation*, pp. 104–130. University of Chicago Press, Chicago.

Peterson, R.O., Woolington, J.D., and Bailey, T.N. (1984). Wolves of the Kenai Peninsula, Alaska. *Wildlife Monographs*, 88, 1–52.

Peterson, R.O., Jacobs, A.K., Drummer, T.D., Mech, D.L., and Smith, D.W. (2002). Leadership behaviour in relation to dominance and reproductive status in grey wolves. *Canadian Journal of Zoology*, 80, 1405–1412.

Pettijohn, T.F., Wont, T.W., Ebert, P.D., and Scott, J.P. (1977). Alleviation of separation distress in 3 breeds of young dogs. *Developmental Psychobiology*, 10, 373–381.

Pfaffenberg, C.J., Scott, J.P., Fuller, J.L., Binsburg, B.E., and Bilfelt, S.W. (1976). *Guide dogs for the blind: their selection, development and training*. Elsevier, Amsterdam.

Pfungst, O. (1907). Das Pferd des Herr von Osten (der Kluge Hans), eine Beitrag zur experimentellen Tier- und Menschpsychologie. Barth, Leipzig.

Pfungst, O. (1912). Über 'sprechende' Hunde. In: Schumann, ed. *Bericht über den V. Kongress für experimentelle Psychologie*, pp. 241–245. Barth, Leipzig.

Pickel, D., Manucy, G.P., Walker, D.B., Hall, S.B., and Walker, J.C. (2004). Evidence for canine olfactory detection of melanoma. *Applied Animal Behavior Science*, 89, 107–114.

Pigliucci, M. (2005). Evolution of phenotypic plasticity: Where are we going now? *Trends in Ecology and Ecolution*, 20, 481–486.

Podberscek, A.L. (2006). Positive and negative aspects of our relationship with companion animals. *Veterinary Research Communications*, 30, 21–27.

Podberscek, A.L. (2007). Dogs and cats as food in Asia. In: Bekoff, M., ed. *The encyclopedia of human-animal interactions*. Greenwood Press, Westport, CT.

Podberscek, A.L. and Blackshaw, J.K. (1993). A survey of dog bites in Brisbane. *Australia. Australian Veterinary Practitioner*, 23, 178–183.

Podberscek, A.L. and Serpell, J.A. (1996). The English cocker spaniel: preliminary findings on aggressive behaviour. *Applied Animal Behaviour Science*, 47, 75–89.

Podberscek, A.L. and Serpell, J.A. (1997). Environmental influences on the expression of aggressive behaviour in English cocker spaniels. *Applied Animal Behaviour Science*, 52, 215–227.

Podberscek, A., Paul, E., and Serpell, J. (eds.) (2000). *Companion animals and us*. Cambridge. Cambridge University Press.

Pongrácz, P., Miklósi, Á., and Csányi, V. (2001). Owners' beliefs on the ability of their pet dogs to understand human verbal communication. A case of social understanding. *Current Cognitive Psychlogy*, 20, 87–107.

Pongrácz, P., Miklósi, Á., Kubinyi, E., Gurobi, K., Topál, J., and Csányi, V. (2001). Social learning in dogs: The effect of a human demonstrator on the performance of dogs (*Canis familiaris*) in a detour task. *Animal Behaviour*, 62, 1109–1117.

Pongrácz, P., Miklósi, Á., Dóka, A., and Csányi, V. (2003). Successful application of video-projected human images for signaling to dogs. *Ethology*, 109, 809–821.

Pongrácz, P., Miklósi, Á.,Timár-Geng, K., and Csányi, V. (2003). Preference for copying unambiguous demonstrations in dogs. *Journal of Comparative Psychology*. 117, 337–343.

Pongrácz, P., Miklósi, Á., Timár-Geng, K., and Csányi, V. (2004). Verbal attention getting as a key factors in social learning between dog (*Canis familiaris*) and human. *Journal of Comparative Psychology*, 118, 375–383.

Pongrácz, P., Miklósi, Á., Molnár, Cs., and Csányi, V. (2005). Human listeners are able to classify dog barks recorded in different situations. *Journal of Comparative Psychology*, 119, 136–144.

Poresky, R. H., Hendrix, C., Mosier, J. E., and Samuelson, M. L. (1987). The companion animal bonding scale: Internal reliability and construct validity. *Psychological Reports*, 60, 743–746.

Povinelli, D.J. (2000). *Folk physics for apes*. Oxford University Press, Oxford.

Povinelli, D.J., Bierschwale, D.T., and Cech, C.G. (1999). Comprehension of seeing as a referential act in young children, but not juvenile chimpanzees. *British Journal of Developmental Psychology*, 17, 37–60.

Povinelli, D.J. and Vonk, J. (2003). Chimpanzee minds: Suspiciously human? *Trends in Cognitive Science*, 7, 157–160.

Povinelli, D.J., Nelson, K.E., and Boysen, S.T. (1990). Inferences about guessing and knowing by chimpanzees (*Pan troglodytes*). *Journal of Comparative Psychology*, 104, 203–210.

Prato-Previde, E., Custance, D.M. Spiezio, C., and Sabatini, F. (2003). Is the dog-human relationship an attachment bond? An observational study using Ainsworth's strange situation. *Behaviour*, 140, 225–254.

Pretterer, G., Bubna-Littitz, H., Windischbauer, G., Gabler, C., and Griebel, U. (2004). Brightness discrimination in the dog. *Journal of Vision*, 4, 241–249.

Price, E.O. (1984). Behavioral aspects of animal domestication. *Quarterly Review of Biology*, 59, 2–32.

Prothmann, A., Bienert, M., and Ettrich, C. (2006). Dogs in child psychotherapy: effects on state of mind. *Anthrozoös*, 19, 265–277.

Pullianen, E. (1965). Studies of wolf (*Canis lupus*) in Finland. *Annales Zoologica Fennica*, 2, 215–259.

Quignon, P., Kirkness, E., Cadieu, E. *et al.* (2003). Comparison of the canine and human olfactory receptor gene repertoires. *Genome Biology,* 4, 80–88.

Rabb, G.B., Woolpy, J.H., and Ginsburg, B.E. (1967). Social relationships in a group of captive wolves. *American Zoology,* 7, 305–312.

Radinsky, L.B. (1973). Evolution of the canid brain. *Brain, Behaviour, Evolution,* 7, 169–202.

Radovanovic, I. (1999). 'Neither person nor beast': dogs in the burial practice of the Iron Gates Mesolithic. *Documenta Praehistorica,* 26, 71–87.

Rajecki, D.W., Lamb, M.E., and Obmascher, P. (1978). Toward a general theory of infantile attachment: a comparative review of aspects of the social bond. *Behavioral and Brain Sciences,* 3, 417–464.

Randi, E. and Lucchini, V. (2002). Detecting rare introgression of domestic dog genes into wild wolf (*Canis lupus*) populations by Bayesian admixture analysis of microsatelitte variation. *Conservation Genetics,* 3, 31–45.

Randi, E., Lucchini, V., and Fransisci, F. (1993). Allozyme variability in the Italian wolf (*Canis lupus*) population. *Heredity,* 71, 516–522.

Randi, E., Lucchini, V., Christensen, M.F. *et al.* (2000). Mitochondrial DNA variability in Italian and east European wolves: detecting the consequences of small population size and hybridization. *Conservation Biology,* 14, 464–473.

Range, F., Aust, U., Steurer, M., *et al.* (2008). Visual categorization of natural stimuli by domestic dogs. *Animal Cognition.* (in press).

Rasmussen, J.L. and Rajecki, D.W. (1995). Differences and similarities in humans' perceptions of the thinking and feeling of a dog and a boy. *Society and Animals,* 3, 117–137.

Reif, A. and Lesch, K.-P. (2003). Toward a molecular architecture of personality. *Behavioural Brain Research,* 139, 1–20

Reynolds, P.C. (1993). The complementation theory of language and tool use. In: Gibson, K.R. and Ingold, T., eds. *Tool Use, language and cognition in human evolution.* Cambridge University Press, Cambridge.

Reznick, D.N. and Ghalambor, C.K. (2001). The population ecology of contemporary adaptations: what empirical studies reveal about the conditions that promote adaptive evolution. *Genetica,* 112, 183–198.

Richter, C. (1959). Rats, man and the welfare state. *American Psychologist,* 14, 18–28.

Riedel, J., Buttelmann, D., Call, J., and Tomasello, M. (2006). Domestic dogs (*Canis familiaris*) use a physical marker to locate hidden food. *Animal Cognition,* 9, 27–35.

Riegger, M.H. and Guntzelman, J. (1990). Prevention and amelioration of stress and consequences of interaction between children and dogs. *Journal of the American Veterinary Medical Association,* 196, 1781–1785.

Ristau, C.A. (1991). Aspects of the cognitive ethology of an injury-feigning bird, the piping plover. In Ristau, C.A., ed. *Cognitive ethology,* pp. 91–126. Lawrence Erlbaum Associates, Hillsdale, NJ.

Robinson, I. (1995). *The Waltham book of human–animal interaction: benefits and responsibilities of pet ownership,* Pergamon, Oxford.

Rockman, M.V., Hahn, M.W., Soranzo, N., Zimprich, F., Goldstein, D.B., and Wray, G.A. (2005). Ancient and recent positive selection transformed opioid *cis*-regulation in humans. *PLoS Biology,* 3, 2208–2219.

Roitblat, H.L. Bever, T.G., and Terrace, H.S. (1984). *Animal cognition.* Lawrence Erlbaum Associates, Hillsdale, NJ.

Romanes, G.J. (1882a). Foxes, wolves, jackals, etc. In: Romanes, G.J., ed. *Animal intelligence,* pp. 426–436. Trench & Co., London.

Romanes, G.J. (1882b). Monkeys, apes, and baboons. In: Romanes, G.J., ed. *Animal intelligence,* pp. 471–498. Trench & Co., London.

Rooney, N.J. and Bradshaw, J.W.S. (2003). Links between play and dominance and attachement dimensions of dog-human relationships. *Journal of Applied Animal Welfare Science,* 6, 67–94.

Rooney, N.J., Bradshaw, J.W.S., and Robinson, I.H. (2000). A comparison of dog-dog and dog-human play behaviour. *Applied Animal Behaviour Science,* 235–248.

Rooney, N.J., Bradshaw, J.W.S., and Robinson, I.H. (2001). Do dogs respond to play signals given by humans? *Animal Behaviour,* 61, 715–722.

Ross, S and Ross, J.G. (1949). Social facilitation of feeding behaviour in dogs: I. Group and solitary feeding. *Journal of Genetic Psychology,* 74, 97–108.

Ross, S., Scott, J.P., Cherner, M., and Denenberg, V.H. (1960). Effects of restraint and isolation on yelping in pups. *Animal Behaviour,* 8, 1–5.

Roy, M.S., Geffen, E., Smith, D., Ostrander, E.A., and Wayne, R.K. (1994). Patterns of discrimination and hybridisation in North American wolflike Canids, revealed by analysis of microsatellite loci. *Molecular Biology and Evolution,* 11, 533–570.

Ruefenacht, S., Gebhardt-Henrich, S., Miyake, T., and Gaillard, C. (2002). A behavior test on German shepherd dogs: heritability of seven different traits. *Applied Animal Behavior Science,* 79, 113–132.

Sablin, M.V. and Khlopachev, A.A. (2002). The earliest Ice Age dogs: evidence from Eliseevichi. *Current Anthropology,* 43, 795–799.

Saetre, P., Lindberg, J., Leonard, J.A. *et al.* (2004). From wild wolf to domestic dog: gene expression changes in the brain. *Molecular Brain Research,* 126, 198–206

Saetre, P., Strandberg, E., Sundgren, P., Pettersson, E.U., Jazin, E., and Bergström, T.F. (2006). The genetic contribution to canine personality. *Genes, Brain, Behaviour,* 5, 240–248.

Salmon, P.W. and Salmon, I.M. (1983). Who owns who? Psychological research into the human-pet bond in Australia. In: Kachter, A.H. and Beck, A.M., eds. *New perspective on our lives with companion animals,* pp. 244–265. University of Pennsylvania Press, Philadelphia, PA.

Salzinger, K. and Waller, M.B. (1962). The operant control of vocalization in the dog. *Journal of the Experimental Analysis of Behavior,* 5, 383–389.

Sands, J. and Creel, S. (2004). Social dominance, aggression and faecal glucocorticoid levels in a wold population of wolves, *Canis lupus. Animal Behaviour,* 67, 387–396.

Sarris, E.G. (1937). Die individuellen Unterschiede bei Hunden. *Zeitschrift für angewandte Psychologie und Charakterkunde,* 52, 257–309.

Savage-Rumbaugh, E.S. and Lewin, R. (1994). *Kanzi, the ape at the brink of the human mind.* John Wiley & Sons, New York.

Savage-Rumbaugh, E., Murphy, J., Sevcik, R.A., Brakke, K.E., Williams, S.L., and Rumbaugh, D.M. (1993). Language comprehension in ape and child. *Monographs of the Society for Research in Child Development,* 58, 1–221.

Savishinsky, J.S. (1983). Pet ideas: The domestication of animals, human behavior and human emotions. In Katcher, A.H. and Beck, A.M., eds. *New perspectives on our lives with companion animals,* pp 112–131. University of Pennsylvania Press, Philadelphia, PA.

Savolainen, P. (2006). mtDNA studies of the origin of dogs. In: Ostrander, E.A. Giger, U., and Lindbladh, K., eds. *The dog and its genome,* pp. 119–140. Cold Spring Harbor Laboratory Press, New York.

Savolainen, P., Zhang, Y., Luo, J., Lundeberg, J., and Leitner, T. (2002). Genetic evidence for an east Asian origin of domestic dogs. *Science,* 298, 1610–1613.

Savolainen, P., Leitner, T., Wilton, A., Matisoo-Smith, E., and Lundeberg, J. (2004). A detailed picture of the origin of the Australian dingo, obtained from the study of mitochondrial DNA. *Proceedings of the National Academy of Sciences of the USA,* 17, 12387–12390.

Schaller, G.B. and Lowther, G.R. (1969). The relevance of carnovore behaviour to the study of early hominids. *Southwestern Journal of Anthropology,* 25, 307–341.

Schassburger, R.M. (1993). *Vocal communication in the timber wolf,* Canis lupus, Linnaeus. Advances in Ethology, No. 30. Paul Parey, Berlin.

Schenkel, R. (1947). Ausdrucks-Studien an Wölfen. *Behaviour,* 81–129.

Schenkel, R. (1967). Submission: Its features and function in the wolf and dog. *American Zoologist,* 7, 319–329.

Schleidt, W.M. (1973). Tonic communication: continual effects of discrete signs in animal communication systems. *Journal of Theoretical Biology,* 42, 359–386.

Schleidt, W.M. and Shalter, M.D. (2003). Co-evolution of humans and canids. *Evolution and Cognition,* 9, 57–72.

Schmidt, P.A. and Mech, L.D. (1997). Wolf pack size and food acquisition. *American Naturalist,* 150, 513–517.

Schneirla, T. (1959). An evolutionary and developmental theory of biphasic processes underlying approach and withdrawal. In: Jones, M., ed. *Nebraska Symposium on Motivation,* University of Nebraska Press, Lincoln, NE.

Schoon, G.A.A. (1996). Scent identification lineups by dogs (*Canis familiaris*): experimental design and forensic application. *Applied Animal Behaviour,* 49, 257–267.

Schoon, G.A.A. (1997). The performance of dogs in identifying humans by scent. PhD disseration, Rijksuniveristeit Leiden.

Schoon, G.A.A. (2004). The effect of the ageing of crime scene objects on the results of scent identification lineups using trained dogs. *Forensic Science International,* 147, 43–47.

Schotté, C.S. and Ginsburg, B.E. (1987). The wolf pack as a socio-genetic unit. In: Frank, H., ed. *Man and wolf,* pp. 401–413. Junk Publishers, Dordrecht.

Schwab, C., and Huber, L. (2006). Obey or not obey? Dogs (*Canis familiaris*) behave differently in response to attentional states of their owners. *Journal of Comparative Psychology,* 120, 169–175.

Schwarz, M. (2000). The form and meaning of Maya and Mississippian dog representations. In: Crockford, S.J., ed. *Dogs through time: an archaeological perspective,* pp. 271–285. British Archaeological Reports International Studies 889, Oxford.

Scott, J.P. (1945). Social behaviour, organisation and leadership in a small flock of domestic sheep. *Comparative Psychology Monograph,* 18, 1–29.

Scott, J.P. (1962). Critical periods in behaviour development. *Science,* 138, 949–958.

Scott, J.P. (1986). Critical periods in organisational processes. In: Falkner, F. and Tanner, J.M., eds. *Human growth,* pp. 181–196. Plenum Press, New York.

Scott, J.P. (1992). The phenomenon of attachment in human-nonhuman relationships. In: Davis, H. and Balfour, D., eds. *The inevitable bond,* pp. 72–92. Cambridge University Press, Cambridge.

Scott, J.P. and Bielfelt, S.W. (1976). Analysis of the puppy testing program. In: Pfaffenberger, C.J. *et al.,* eds. *Guide dogs for the blind: their selection, development and training,* pp. 39–75. Elsevier, Amsterdam.

Scott, J.P. and Fuller, J.L. (1965). *Genetics and the social behaviour of the dog.* University of Chicago Press, Chicago.

Seddon, J.M. and Ellegren, H. (2002). MHC Class II genes in European wolves: A comparison with dogs. *Immunogenetics,* 54, 490–500.

Séguinot, V., Cattet, J., and Benhamou, S. (1998). Path integration in dogs. *Animal Behaviour,* 55, 787–797.

Seksel, K., Mazurski, E.J., and Taylor, A. (1999). Puppy socialisation programs: short and long term behavioural effects. *Applied Animal Behaviour Science,* 62, 335–349.

Seligman, M.E.P., Maier, S.F., and Geer, J.H. (1965). Alleviation of learned helplessness in the dog. *Journal of Abnormal Psychology,* 73, 256–262.

Senay, E.C. (1966). Toward an animal model of depression: a study of separation behaviour in dogs. *Journal of Psychiatric Research,* 47, 65–71.

Serpell, J. (1996). Evidence for association between pet behaviour and owner attachment levels. *Applied Animal Behaviour Science,* 47, 49–60.

Serpell, J. (ed.) (1995). *The domestic dog. Its evolution: behaviour, and interactions with people.* Cambridge University Press, Cambridge.

Serpell, J. and Jagoe, J.A. (1995). Early experience and the development of behavior. In: Serpell, J. ed. *The domestic dog, its evolution, behavior, and interactions with people,* pp. 79–175. Cambridge University Press, Cambridge.

Serpell, J. and Hsu, Y. (2001). Development and validation of a novel method for evaluating behaviour and temperament in guidedogs. *Applied Animal Behaviour Science,* 72, 347–364.

Serpell, J.A. and Hsu, Y. (2005). Effects of breed, sex and neuter status on trainability in dogs. *Anthrozoös,* 18, 196–207.

Sharma, D.K., Maldonado, J.E., Jhala, Y.V., and Fleischer, R.C. (2003). Ancient wolf lineages in India. *Proceeding of the Royal Society, Biology Letters,* 271, S1–S4.

Sharp, H.S. (1978). Comparative ethnology of the wolf and chipewyan. In: Hall, R.L. and Sharp, H.S., eds. *Wolf and man: evolution in parallel,* pp. 55–79. Academic Press, New York.

Sheldon, J.W. (1988). Wild dogs: the natural history of the nondomestic Canidae. Academic Press, San Diego, CA.

Sheppard, G. and Mills, D.S. (2002). The development of a psychometric scale for the evaluation of the emotional predispositions of pet dogs. *Journal of Comparative Psychology,* 15, 201–222.

Sheppard, G. and Mills, D.S. (2003). Evaluation of dog appeasing pheromone as a potential treatment for dogs fearful of fireworks. *Veterinary Record,* 152, 432–436.

Sherman, C.K., Reisner, I.R., Taliaferro, L.A., and Houpt, K.A. (1996). Characteristics, treatment and outcome of 99 cases of aggression between dogs. *Applied Animal Behavior Science,* 47, 91–108.

Shettleworth, S.J. (1972). Constraints on learning. *Advances in the Study of Behaviour,* 4, 1–68.

Shettleworth, S.J. (1998). *Cognition, evolution and behaviour.* Oxford University Press, Oxford.

Shigehara, N. and Hongo, H. (2000). Ancient remains of Jomon dogs from Neolithic sites in Japan. In: Crockford, S.J., ed. *Dogs through time: an archaeological perspective,* pp. 61–67. British Archaeological Reports International Studies 889, Oxford.

Sih, A., Bell, A.M., and Johnson, J.C. (2004). Behavioral syndromes: an ecological and evolutionary overview. *Trends in Ecology and Evolution,* 19, 372–378

Silk, J.B. (2002). Using the 'F'-word in primatology. *Behaviour,* 139, 421–446.

Slabbert, J.M. and Odendaal, J.S.J. (1999). Early prediction of adult police dog efficiency—longitudinal study. *Applied Animal Behaviour Science,* 64, 269–288.

Slabbert, J.M. and Rasa, O.A.E. (1997). Observational learning of an acquired maternal behaviour pattern by working dog pups: an alternative training method? *Applied Animal Behaviour,* 53, 309–316.

Slater, P.J.B. (1978). Data collection. In: Colgan, P.W., ed. *Quantitative ethology,* pp. 7–15. John Wiley & Sons, New York.

Smith, B.D. (1998). *The emergence of agriculture.* Scientific American Library, New York.

Sober, E.R. and Wilson, D.S. (1998). *Unto others: the evolution and psychology of unselfish behavior.* Harvard University Press, Cambridge, MA.

Solomon, R.L. and Wynne, L.C. (1953). Traumatic avoidance learning: acquisition in normal dogs. *Psychological Monographs: General and Applied,* 67, 1–19.

Solomon, R.L., Turner, L.H., and Lessac, M.S. (1968). Some effects of delay of punishment on resistance to temptation in dogs. In: Walters, R.H., Cheyne, J.A., and Banks, R.K., eds. *Punishment,* pp. 124-135. Penguin, London.

Soproni, K., Miklósi, Á., Topál, J., and Csányi, V. (2001). Comprehension of human communicative signs in pet dogs. *Journal of Comparative Psychology,* 115, 122–126.

Soproni, K., Miklósi, Á., Topál, J., and Csányi, V. (2002). Dogs' (*Canis familiaris*) responsiveness to human pointing gestures. *Journal of Comparative Psychology,* 116, 27–34.

Stahler, D.R. Smith, D.W., and Landis, R. (2002). The acceptance of a new breeding male into a wild wolf pack. *Canadian Journal of Zoology,* 80, 360–365.

Stanley, W.C. and Elliot, O. (1962). Differential human handling as reinforcing events and as treatments influencing later social behaviour in basenji puppies. *Psychological Reports,* 10, 775–788.

Steen, J.B. and Wilsson, E. (1990). How do dogs determine the direction of tracks? *Acta Physiologica Scandinavica*, 139, 531–534.

Steward, M. (1983). Loss of a pet—loss of a person: a comparative study of bereavement. In Katcher, A.H. and Beck, A.M., eds. *New perspectives on our lives with companion animals*, pp. 390–406. University of Pennsylvania Press. Philadelphia, PA.

Strandberg, E., Jacobsson, J., and Seatre, P. (2005). Direct genetic, maternal and litter effects on behaviour in German shepherd dogs in Sweden. *Livestock Production Science*, 93, 33–42.

Street, M. (1989). Jager und Schamen: Bedburg-Königshoven ein Wohnplatz am Niederrhein vor 10.000 Jahren. Römisch-Germanischen Zentralmuseums, Main.

Strelau, J. (1997). The contribution of Pavlov's typology of CNS properties to personality research. *European Psychologist*, 2, 125–138.

Studdert-Kennedy, M. (1998). The particulate origins of language generativity: From syllable to gesture. In Hurford, J., Studdert-Kennedy, M., and Knight, C., eds. *Approaches to the evolution of language: social and cognitive bases*, pp. 202–221. Cambridge University Press, Cambridge.

Sundqvist, A.K., Björnerfeldt, S., Leonard, J.A. *et al.* (2006). Unequal contribution of sexes in the origin of dog breeds. *Genetics*, 172, 1121–1128.

Svartberg, K. (2002). Shyness–boldness predicts performance in working dogs. *Applied Animal Behaviour Science*, 79, 157–174.

Svartberg, K. (2005). A comparison of behaviour in test and in everyday life: evidence of three consistent boldness-related personality traits in dogs. *Applied Animal Behaviour Science*, 91, 103–128.

Svartberg, K. (2006). Breed-typical behaviour in dogs—historical remnants or recent constructs? *Applied Animal Behaviour Science*, 96, 293–313.

Svartberg, K. and Forkman, B. (2002). Personality traits in the domestic dog (*Canis familaris*). *Applied Animal Behavior Science*, 79, 133–155.

Svartberg, K., Tapper, I., Temrin, H., Radesater, T., and Thorman, S. (2005). Consistency of personality traits in dogs. *Animal Behaviour*, 69, 283–291.

Szetei, V., Miklósi, Á., Topál, J., and Csányi, V. (2003). When dogs seem to loose their nose: an investigation on the use of visual and olfactory cues in communicative context between dog and owner. *Applied Animal Behavior Science*, 83, 141–152.

Tapp, P.D. Siwak, C.T. Estrada, J., Holowachuk, D., and Milgram, N.W. (2003). Effects of age on measures of complex working memory span in the beagle dog (*Canis familaris*) using two versions of a spatial list learning paradigm. *Learning & Memory*, 10, 148–160.

Taylor, H., Williams, P., and Gray, D. (2004). Homelessness and dog ownership: an investigation into animal empathy, attachment, crime, drug use, health and public opinion. *Anthrozoös*, 17, 353–368.

Taylor, K.D. and Mills, D.S. (2006). The development and assessment of temperament tests for adult companion dogs. *Journal of Veterinary Behavior: Clinical Applications and Research*, 1, 94–108.

Tchernov, E. and Horwitz, L.K. (1991). Body size diminution under domestication: unconscious selection in primeval domesticates. *Journal of Anthropological Archaeology*, 10, 54–75.

Tchernov, E. and Valla, F.F. (1997). Two new dogs, and other Natufian dogs, from the southern Levant. *Journal of Archaeological Science*, 24,65–95.

Tembrock, G. (1976). Canid vocalisation. *Behavior Processes*, 1, 57–75.

Templer, D.I., Salter, C.A., Dickery, S., Baldwin,R., and Veleber, D.M. (1981). The construction of a pet attitude scale. *Psychological Record*, 31, 343–348.

Teplov, B.M. (1964). Problems in the study of general types of higher nervous activity in man and animal. In: Gray, J.A., ed. *Pavlov's typology*, pp. 3–141. Pergamon Press, London.

Theberge, J.B. and Falls, J.B. (1967). Howling as a means of communication in timber wolves. *American Zoologist*, 7, 331–338.

Thesen, A., Steen, J.B. and Doving, K.B. (1993). Behaviour of dogs during olfactory tracking. *Journal of Experimental Biology*, 180, 247–251.

Thompson, P.C., Rose, K., and Kok, N.E. (1992). The behavioural ecology of dingoes in North-western Australia: V. Population dynamics and variation in the social system. *Wildlife Research*, 19, 565–584.

Thorndike, E.L. (1911). *Animal intelligence*. Macmillan, New York.

Timberlake, W. (1994). Behavior systems, associationism, and Pavlovian conditioning. *Psychonomic Bulletin and Review*, 1, 405–420.

Tinbergen, N. (1963). On aims and methods of ethology. *Zeitschrift für Tierpsychologie*, 20, 410–433.

Toates, F. (1998). The interaction of cognitive and stimulus-response processes in the control of behavior. *Neuroscience and Biobehavioral Reviews*, 22, 59–83.

Tomasello, T. and Call, J. (1997). *Primate cognition*. Oxford University Press, New York.

Topál, J., Miklósi, Á., and Csányi, V. (1997). Dog-human relationship affects problem solving behavior in the dog. *Anthrozoös*, 10, 214–224.

Topál, J., Miklósi, Á., Csányi, V., and Dóka, A. (1998). Attachment behavior in dogs (*Canis familaris*): a new application of Ainsworth's (1969) strange situation test. *Journal of Comparative Psychology*, 112, 219–229.

Topál, J., Gácsi, M., Miklósi, Á., Virányi, Z., Kubinyi, E., and Csányi, V. (2005a). The effect of domestication and socialization on attachment to human: a comparative study on hand reared wolves and differently socialized dog pups. *Animal Behaviour*, 70, 1367–1375.

Topál, J., Kubinyi, E., Gácsi, M., *et al.* (2005b). Obeying Social Rules: A Comparative Study on Dogs and Humans. *Journal of Cultural and Evolutionary Psychology*, 3, 213–238.

Topal, J., Erdőhegyi, Á., Mányik, R., *et al.* (2006a). Mindreading in a dog: an adaptation of a primate 'mental attribution' study. *International Journal of Psychology and Psychological Therapy*, 6, 365–379.

Topal, J., Byrne, R.W., Miklósi, Á., *et al.* (2006b). Reproducing human actions and action sequences: "Do as I do!" in a dog. *Animal Cognition*, 9, 355–367

Topal, J., Miklósi, Á., Gácsi, M., *et al.* (2008). Dog as a complementary model for understanding human social behaviour. Submitted to *Advances in the Study of Behaviour.*

Triana, E. and Pasnak, R. (1981). Object permanence in cats and dogs. *Animal Learning and Behaviour*, 9, 135–139.

Tripp, A.C. and Walker, J. (2003). The great chemical residue detection debate: dog versus machine. In: Harmon, R.S. Holloway, J.H. and Broach, J.T., eds. *Detection and remediation technologies for mines and minelike targets VIII*, pp. 983–990. Proceedings of SPIE, Orlando, FL.

Trut, L.N. (1980). The genetics and phenogenetics of domestic behaviour. In: Trut, L.N., ed. *Problems in general genetics*, pp. 123–137. MIR, Moscow.

Trut, L.N. (1999). Early canid domestication: the farm-fox experiment. *American Scientist*, 87, 160–168.

Trut, L.N. (2001). Experimental studies in early canid domestication. In: Ruvinsky, A. and Sampson, J., eds. *The genetics of the dog*, pp. 15–41. CABI Publishing, Wallingford.

Trut, L.N. Naumenko, E.V., and Belyaev, D.K. (1972). Change in the pituary-adrenal function of silver foxes during selection according to behaviour. *Genetika*, 8, 35–40.

Tuber, D.S., Henessy, M.B., Sanders, S., and Miller, J.A. (1996). Behavioral and glucocorticoid responses of adult domestic dogs (*Canis familiaris*) to companionship and social separation. *Journal of Comparative Psychology*, 110, 103–108.

Turnbull, P.F. and Reed, C.A. (1974). The fauna from the terminal Pleistocene of Palegawra Cave, a Zarzian occupation site in northeastern Iraq. *Fieldiana Anthropology*, 63, 81–146.

Uexküll, J. (1909). *Umwelt und Innerleben der Tiere*. Berlin.

Ujfalussy, D., Kulcsár, Zs., and Miklósi, Á. (2007). Numerical competence in dogs and wolves. Unpublished.

Valadez, R. (2000). Prehispanic dog types in middle America. In: Crockford, S.J., ed. *Dogs through time: an archaeological perspective*, pp. 210–222. British Archaeological Reports International Series 889, Oxford.

van den Berg, L., Schilder, M.B.H., and Knol, B.W. (2003). Behavior genetics of canine aggression: behavioural phenotyping of golden retrievers by means of an aggression test. *Behavior Genetics*, 33, 469–483.

Van Hooff, J.A.R.A.M. Wensing, J. (1987). Dominance and its behavioral measures in a captive wolf pack. In: H. Frank (ed.), *Man and Wolf: Advances, Issues and Problems in Captive Wolf Research*, pp. 219–252. Junk, Dordrecht.

Van Valkenburgh, B., Sacco, T., and Wang, X. (2003). Pack hunting in Miocene borophagine dogs: evidence from craniodental morphology and body size. In: Flynn, L., ed. *Vertebrate fossils and their context: contributions in honor of Richard H. Tedford*, pp. 147–162. Bulletin of the American Museum of Natural History, Tedford.

Vas, J., Topál, J., Gácsi, M., Miklósi, Á., and Csányi, V. (2005). A friend or an enemy? Dogs' reaction to an unfamiliar person showing behavioural cues of threat and friendliness at different times. *Applied Animal Behaviour Science*, 94, 99–115.

Vas, J., Topál, J., Pech, É., and Miklósi, Á. (2007). Measuring attention deficit and activity in dogs: a new application and validation of a human ADHD questionnaire. *Applied Animal Behaviour Science*, 103, 105–117.

Verginelli, F., Capelli, C., Coia, V. *et al.* (2005). Mitochondrial DNA from prehistoric canids highlights relationships between dogs and south-east European wolves. *Molecular Biology and Evolution*, 22, 2541–2551.

Vila, C., Savolainen, P., Maldonado, J.E. *et al.* (1997). Multiple and ancient origins of the domestic dog. *Science*, 276, 1687–1689.

Vila, C., Amorim, I.R. Leonard, J.A. *et al.* (1999). Mitochondrial DNA phylogeography and population history of the grey wolf *Canis lupus*. *Molecular Ecology*, 8, 2089–2103.

Vila, C., Walker, C., Sundqvist, A.K. *et al.* (2003). Combined use of maternal, paternal and bi-parental genetic markers for the identification of wolf–dog hybrids. *Heredity*, 90, 17–24.

Vila, C., Seddon, J.M., and Ellegren, H. (2005). Genes of domestic mammals augmented by backcrossing with wild ancestors. *Trends in Genetics*, 21, 214–218.

Vincent, I.C. and Mitchell, A.R. (1996). Relationship between blood pressure and stress-prone temperament in dogs. *Physiology and Behavior*, 60, 135–138.

Virányi, Zs., Topál, J., Gácsi, M., Miklósi, Á., and Csányi, V. (2004). Dogs can recognize the focus of attention in humans. *Behavioural Processes*, 66, 161–172.

Virányi, Zs., Topál, J., Miklósi, Á., and Csányi, V. (2006). A nonverbal test of knowledge attribution: a comparative study on dogs and children. *Animal Cognition, 9,* 13–26.

Virányi, Zs., Gácsi, M., Kubinyi, E. *et al.* (2008). Comprehension of human pointing gestures in young human-reared wolves and dogs. *Animal Cognition, 11,* 373–387.

Vogel, H.Hs. Scott, J.P., and Marston, M.V. (1950). Social facilitation and allelomimetic behaviour in dogs. *Behaviour, 2,* 121–134.

Voith, V.L. Wright, J.C. and Danneman, P.J. (1992). Is there a relationship between canine behaviour problems and spoiling activities, anthropomorphism, and obedience training? *Applied Animal Behaviour Science, 34,* 263–272.

Vollmer, P.J. (1977). Do mischievous dogs reveal their 'guilt'? *Veterinary Medicine Small Animal Clinician, 72,* 1002–1005.

Vucetich, J.A., Peterson, R.O., and Waite, T.A. (2004). Raven scavenging favours group foraging in wolves. *Animal Behaviour, 67,* 1117–1126.

Wabakken, P., Sand, H., Liberg, O., and Bjarvall, A. (2001). The recovery, distribution and population dynamics of wolves on the Scandinavian peninsula, 1978–1998. *Canadian Journal of Zoology, 79,* 710–725.

Walker, D.B., Walker, J.C., Cavnar, P.J. *et al.* (2006). Naturalistic quantification of canine olfactory sensitivity. *Applied Animal Behaviour Science, 97,* 242–254.

Wang, X.R. Tedford, H., Valkenburgh, B.V., and Wayne, R.K. (2004). Ancestry: Evolutionary history, molecular systematics, and evolutionary ecology of Canidae. In: MacDonald, D.W. and Sillero-Zubiri, C., eds. *The biology and conservation of wild canids,* pp. 39–54. Oxford University Press, Oxford.

Ward, C. and Smuts, B. (2007). Quantity-based judgments in the domestic dog (*Canis lupus familiaris*). *Animal Cognition, 10,* 71–80.

Warden, C.J. and Warner, L.H. (1928). The sensory capacities and intelligence of dogs, with a report on the ability of the noted dog 'Fellow' to respond to verbal stimuli. *Quarterly Review of Biology, 3,* 1–28.

Watson, J.S., Gergely, G., Topál, J., Gácsi, M., Sárközi, Zs., and Csányi, V. (2001). Distinguishing logic versus association in the solution of an invisiable displacement task by children and dogs: using negation of disjunction. *Journal of Comparative Psychology, 115,* 219–226.

Wayne, R.K. (1986a). Limb morphology of domestic and wild canids: the influence of development on morphologic change. *Journal of Morphology, 187,* 301–319.

Wayne, R.K. (1986b). Cranial morphology of domestic and wild canids: the influence of development on morphological change. *Evolution, 40,* 243–261.

Wayne, R.K. (1993). Molecular evolution of the dog family. *Trends in Genetics, 9,* 218–224.

Wayne, R.K. and Vila, C. (2001). Phylogeny and origin of the domestic dog In: Ruvinsky, A. and Sampson, J., eds. *The genetics of the dog,* pp. 1–14. CABI Publishing, Wallingford.

Wayne, R.K., Geffen, E., Girman, D.J., Koepfli, K.P., Lau, L.M., and Marshall, C. (1997). Molecular systematics of the Canidae. *Systematic Biology, 4,* 622–653.

Wells, D.L. (2004). The facilitation of social interactions by domestic dogs. *Anthrozoös, 17,* 340–352.

Wells, D.L. and Hepper, P.G. (1992). The behaviour of dogs in a rescue shelter. *Animal Welfare, 1,* 171–186.

Wells, D.L. and Hepper, P.G. (1998). A note on the influence of visual conspecific contact on the behaviour of sheltered dogs. *Applied Animal Behaviour Science, 60,* 83–88.

Wells, D.L. and Hepper, P.G. (2000). The influence of environmental change on the behaviour of sheltered dogs. *Applied Animal Behaviour Science, 68,* 151–162.

Wells, D.L. and Hepper, P.G. (2003). Directional tracking in the domestic dog, *Canis familiaris. Applied Animal Behaviour Science, 84,* 297–305.

Wells, D.L. and Hepper, P.G. (2006). Prenatal olfactory learning in the domestic dog. *Animal Behaviour, 72,* 681–686.

Wells, D.L. Graham, L., and Hepper, P.G. (2002). The influence of length of time in a rescue shelter on the behaviour of kennelled dogs. *Animal Welfare, 11,* 317–325.

West, R.E. and Young, R.J. (2002). Do domestic dogs show any evidence of being able to count? *Animal Cognition, 5,* 183–186.

West-Eberhard, M.J. (2003). *Developmental plasticity and evolution.* Oxford University Press, Oxford.

West-Eberhard, M.J. (2005). Developmental plasticity and the origin of species differences. *Proceedings of the National Academy of Sciences of the USA, 102,* 6543–6549.

Whiten, A. (2000). Chimpanzee cognition and the question of mental re-representation. In: Sperber , D., ed. *Metarepresentation: a multidisciplinary perspective,* pp. 139–167. Oxford University Press, Oxford.

Whiten, A. and Byrne, R.W. (1988). Tactical deception in primates. *Behavioral and Brain Sciences, 11,* 233–273.

Whiten, A. and Ham, R. (1992). On the nature and evolution of imitation in the animal kingdom: reappraisal of a century of research. In: Slater, P.J.B. *et al.*, eds. *Advances in the study of behaviour,* pp. 239–283. Academic Press, New York.

Wickler, W. (1976). The ethological analysis of attachment. Sociometric, motivational and sociophysiological aspects. *Zeitschrift für Tierpsychologie, 42,* 12–28.

Williams, M. and Johnston, J.M. (2002). Training and maintaining the performance of dogs (*Canis familiaris*) on an increasing number of odor discriminations in a controlled setting. *Applied Animal Behaviour Science*, 78, 55–65.

Wilson, C.C. (1991). The pet as an anxiolytic intervention. *Journal of Nervous and Mental Disease*, 179, 482–489.

Wilson, C.C. and Turner, D.C. (1998). *Companion animals in human health.* Sage, London.

Wilson, P.J. Grewal, S., Lawford, I.D. *et al.* (2000). DNA profiles of the eastern Canadian wolf and the red wolf provide evidence for a common evolutionary history independent of the gray wolf. *Canadian Journal of Zoology*, 78, 2156–2166.

Wilsson, E. and Sundgren, P.E. (1998). Behaviour test for eight-week old pups—heritabilities of tested behaviour traits and its correspondence to later behaviour. *Applied Animal Behaviour Science*, 58, 151–162.

Woodbury, C.B. (1943). The learning of stimulus patterns by dogs. *Journal of Comparative Psychology*, 35, 29–40.

Wright, J.C. (1980). The development of social structure during the primary socialization period in German shepherds. *Developmental Psychobiology*, 13, 17–24.

Wright, C.J. and Nesselrote, M.S. (1987). Classification of behavioural problems in dogs: distribution of age, breed, sex and reproductive status. *Applied Animal Behaviour Science*, 19, 169–178.

Wyrwicka, W. (1958). Studies on detour behaviour. *Behaviour*, 14, 240–264.

Yin, S. (2002). A new perspective on barking in dogs (*Canis familiaris*). *Journal of Comparative Psychology*, 116, 189–193.

Yohe, R.M. and Pavesic, M.G. (2000). Early domestic dogs from western Idaho, USA. In: Crockford, S.J., ed. *Dogs through time: an archaeological perspective*, pp. 93–104. British Archaeological International Reports Series 889, Oxford.

Young, A. and Bannasch, D. (2006). Morphological variation in the dog. In: Ostrander, E.A., Giger, U., and Lindbladh, K., eds. *The dog and its genome*, pp. 47–67. Cold Spring Harbor Laboratory Press, New York.

Young, C.A. (1991) Verbal commands as discriminative stimuli in domestic dogs (*Canis familiaris*). *Applied Animal Behaviour Science*, 32, 75–89.

Zentall, T.R. (2001). Imitation in animals: evidence, function, and mechanisms. *Cybernetics and Systems*, 32, 53–96

Zimen, E. (1982). A wolf pack sociogram. In: Harrington, F.H. and Paquet, P.C., eds. *Wolves of the world: perspectives of behavior, ecology and conservation*, pp. 282–322. Noyes Publications, Park Ridge, NJ.

Zimen, E. (1987). Ontogeny of approach and flight behavior towards humans in wolves, poodles and wolf-poodle hybrids. In: Frank, H., ed. *Man and wolf*, pp. 275–292. Junk Publishers, Dordrecht.

Zimen, E. (2000). Der Wolf: Verhalten. Ökologie und Mythos. Knesebeck, München.

Index